This book presents the newly developed theory of nonharmonic Fourier series and its applications to the control of distributed parameter systems. The authors extend the theory to include vector exponential series.

The first part of the book presents the modern theory of nonharmonic Fourier series based on the geometry of Hilbert spaces. This approach permits the successful development of the theory of scalar exponential families and vector exponential families. The development of this mathematical apparatus paves the way for the second part of the book, which extends and upgrades the method of moments – one of the most powerful tools in the flourishing theory of the control of distributed parameter systems. The authors present the Fourier method for evolutionary equations in Hilbert space and investigate the deep connections between controllability and properties of exponentials. They go on to discuss the controllability of systems described by parabolic and hyperbolic PDEs for internal, boundary, initial, and pointwise control. Finally, they consider a number of applications to control problems of connected strings systems.

FAMILIES OF EXPONENTIALS

FAMILIES OF EXPONENTIALS

The Method of Moments in Controllability Problems for Distributed Parameter Systems

SERGEI A. AVDONIN
St. Petersburg State University

SERGEI A. IVANOV
St. Petersburg State University

CAMBRIDGE
UNIVERSITY PRESS

CAMBRIDGE UNIVERSITY PRESS
Cambridge, New York, Melbourne, Madrid, Cape Town, Singapore,
São Paulo, Delhi, Dubai, Tokyo, Mexico City

Cambridge University Press
The Edinburgh Building, Cambridge CB2 8RU, UK

Published in the United States of America by Cambridge University Press, New York

www.cambridge.org
Information on this title: www.cambridge.org/9780521144957

© Cambridge University Press 1995

First published 1995
First paperback printing 2010

A catalogue record for this publication is available from the British Library

Library of Congress Cataloguing in Publication data
Avdonin, Sergei A.
Families of exponentials : the method of moments in
controllability problems for distributed parameter systems / Sergei
A. Avdonin, Sergei A. Ivanov.
p.   cm.
Includes bibliographical references.
ISBN 0-521-45243-0
1. Fourier analysis.   2. Moments method (Mathematics)
3. Exponential functions   4. Control theory.   5. Distributed
parameter systems.   I. Ivanov, Sergei A.   II. Title.
QA403.5.A93   1995
515'.2433 – dc20   94-40490
                                    CIP

ISBN 978-0-521-45243-4 Hardback
ISBN 978-0-521-14495-7 Paperback

Contents

Acknowledgments	*page*	*xi*
Notation		*xii*
Introduction		1
I	**Elements of Hilbert space theory**	**17**
	1 Families of vectors and families of subspaces	17
	1.1 Orthoprojectors and angles between subspaces	17
	1.2 Skew projectors	20
	1.3 Families of Hilbert space vectors	23
	1.4 Families of subspaces of Hilbert space	30
	1.5 Invariance of family properties under mappings	31
	1.6 Minimality of families of elements and their projections	32
	2 Abstract problem of moments	34
II	**Families of vector-valued exponentials**	**39**
	1 Hardy spaces and operator functions	39
	1.1 Hardy spaces	39
	1.1.1 Definitions and Hilbert structure	39
	1.1.2 Relationship between $H^2(\mathfrak{R})$ and $H^2_+(\mathfrak{R})$ spaces	40
	1.1.3 Calculation of Cauchy integrals by means of residues	41
	1.1.4 Simple fractions	41
	1.1.5 Paley–Wiener theorem and connections between simple fractions and exponentials	41
	1.1.6 Decomposition of the space $L^2(\mathbb{R})$ and the Riesz projector	42

1.2	Inner and outer operator functions	42
	1.2.1 Definition of inner functions	42
	1.2.2 Classification of scalar inner functions	43
	1.2.3 Factorization of scalar inner functions	43
	1.2.4 Classification of inner operator functions	44
	1.2.5 Factorization of inner operator functions	44
	1.2.6 Scalar outer functions	45
	1.2.7 Factorization of scalar functions	46
	1.2.8 Outer operator functions	46
	1.2.9 Factorization of strong functions	47
	1.2.10 Semisimple zeros	47
	1.2.11 Existence of the Blaschke–Potapov product	47
1.3	Nagy–Foias model operator	48
	1.3.1 Subspaces K_S	48
	1.3.2 Definitions of model operators in $\mathbb{D}$ and $\mathbb{C}_+$	49
	1.3.3 The spectrum and eigenfunctions of model operators in $\mathbb{C}_+$	49
	1.3.4 The spectrum and eigenfunctions of model operators in $\mathbb{D}$	50
	1.3.5 Discrete spectrum subspaces of model operators in $\mathbb{C}_+$	51
	1.3.6 Invariant subspaces of a model operator in $\mathbb{D}$	51
	1.3.7 Countable sets in the upper half-plane	52
	1.3.8 Indications for a set to be Carlesonian	52
	1.3.9 Separability of simple fraction families	54
	1.3.10 Minimality and the Blaschke condition	55
	1.3.11 $\mathscr{L}$-basis property of simple scalar fraction families	56
	1.3.12 $\mathscr{L}$-basis property of simple fraction families and Carleson constant	56
	1.3.13 Carleson–Newman condition and simple fractions	56
	1.3.14 The inequalities for exponential sums on an interval	57
	1.3.15 Property of functions from the subpsace K_a	58
	1.3.16 Isomorphisms transferring exponentials into exponentials	58
1.4	Entire functions	59
	1.4.1 Entire functions of the exponential type	59
	1.4.2 Growth indicator and indicator diagram	59
	1.4.3 Functions of the Cartwright class	60

	1.4.4 Zeros of Cartwright-class functions	60
	1.4.5 Summability and growth of functions on the real axis	61
	1.4.6 Functions of the sine type	61
2	Vector exponential families on the semiaxis	61
	2.1 Minimality	64
	2.2 The basis property	67
3	Families of vector exponentials on an interval	74
	3.1 Geometrical indications of minimal and basis properties	74
	3.2 Minimal vector families and the generating function	80
	3.3 Hilbert operator and exponential families on an interval	90
	3.4 Indications for Hilbert operator to be bounded	94
4	Minimality and the basis property of scalar exponential families in $L^2(0, T)$	99
	4.1 Minimality	99
	4.2 The basis property of family $\{e^{i\lambda_n t}\}$	101
	4.3 Basis subfamilies of $\{e^{i\lambda_n t}\}$	107
	4.4 Algorithm of a basis subfamily extraction	110
	4.5 Riesz bases of elements of the form $t^s e^{i\lambda_n t}$	113
	4.6 Complementation of a basis on subinterval up to a basis on interval	114
	4.7 On the families of exponentials with the imaginary spectrum	115
5	Additional information about vector exponential theory	117
	5.1 Relationship of minimality of vector and scalar exponential families on an interval	117
	5.2 Perturbation of basis families	120
	5.3 Exponential bases in Sobolev spaces	129
	5.4 Relationship between minimality of exponential families of parabolic and hyperbolic types	133
6	W-linear independence of exponential families	136
	6.1 Dirichlet series	137
	6.2 Parabolic family on an interval	138
	6.3 Hyperbolic family on the semiaxis	139
	6.4 Hyperbolic family on an interval	140
	6.5 ω-linear independence in $L^2(0, T)$	141
	6.6 Excessiveness of the exponential family arising in the problem of controlling rectangular membrane oscillation	142

III **Fourier method in operator equations and controllability** **146**
types
   1   Evolution equations of the first order in time   146
   2   Evolution equations of the second order in time   154
   3   Controllability types and their relationship with
      exponential families   161

IV **Controllability of parabolic-type systems**   **173**
   1   Control with spatial support in the domain   175
      1.1   Infinite dimensional control   175
      1.2   Finite dimensional control   178
      1.3   Pointwise control   180
   2   Boundary control   181
      2.1   Infinite dimensional control   181
      2.2   Finite dimensional control   186
      2.3   Pointwise control   187

V **Controllability of hyperbolic-type systems**   **190**
   1   Control with spatial support in the domain   190
      1.1   Infinite dimensional control   190
      1.2   Finite dimensional control   194
      1.3   Pointwise control   197
   2   Boundary control   205
      2.1   Infinite dimensional control   205
      2.2   Finite dimensional control   211
      2.3   Pointwise control   214

VI **Control of rectangular membrane vibrations**   **216**
   1   Boundary control   216
      1.1   Regularity of the solution   216
      1.2   Lack of controllability   221
      1.3   Estimate of the Carleson constant   223
      1.4   Pointwise boundary control   226
   2   Initial and pointwise control   228
      2.1   Principal results   228
      2.2   Initial controllability   229
      2.3   Lack of pointwise controllability   232
      2.4   Construction of basis subfamily   233

VII **Boundary control of string systems**   **238**
   1   System of connected homogeneous strings controlled
      at the ends   238

Contents

2 System of strings connected elastically at one point   248
3 Control of multichannel acoustic system   264
4 Controllability of a nonhomogeneous string controlled
   at the ends   283

References   290
Index   301

Acknowledgments

We are very grateful to B. S. Pavlov for the many years of attention and support he has given to our scientific work.

Various aspects of this study were discussed with C. Bardos, N. Burq, H. O. Fattorini, E. M. Il'in, I. Joó, V. A. Kozlov, J. E. Lagnese, G. Leugering, S. N. Naboko, V. G. Osmolovskiĭ, D. L. Russell, G. Schmidt, T. I. Seidman, and V. I. Vasyunin. Their comments were greatly appreciated.

We would also like to thank A. S. Silbergleit for his great help in the translation of the book.

At the final stage of the preparation of this manuscript, we obtained valuable support from the Commission of the European Communities in the framework of the EC–Russia collaboration (contract ESPRIT P9282 ACTCS). Our research was also supported in part by INTAS (grant 93 1424), the International Sciences Foundation (grants NSI000 and NSI300), and the Russian Foundation of Fundamental Research (grant 95-01-00360 a). We express our deep gratitude to these foundations.

Notation

M, card M – number of elements of M if M is finite and ∞ if M is infinite

$\varphi_{\mathfrak{H}}(\mathfrak{M}, \mathfrak{N})$ – angle between subspaces   18

$\rightharpoonup$ – weak convergence of elements of a Hilbert space

codim Ξ – dimension of orthogonal complement of a linear set Ξ or dimension of the space orthogonal to all elements of family Ξ

$\langle \cdot, \cdot \rangle$ – scalar product in auxiliary Hilbert space $\mathfrak{N}$

$\langle\!\langle \cdot \rangle\!\rangle$ – norm in an auxiliary Hilbert space $\mathfrak{N}$

$\delta(\sigma)$ – Carleson constant   52

(A_2) – Muckenhoupt condition   95

K_s – a subspace

$$H^2_+(\mathfrak{N}) \ominus SH^2_+(\mathfrak{N}) = \{ f \in H^2_+(\mathfrak{N}) \mid f \perp SH^2_+(\mathfrak{N}) \}$$ 48

K_a – a subspace K_s with $S = \exp(ika)$ 48

l^2_r – Hilbert space of sequences with the norm

$$\|c\|_r = \left[\sum_{n=1}^{\infty} |c_n|^2 (\lambda_n + \alpha)^r \right]^{1/2}$$ 147

W_r – Hilbert space of functions $\sum\limits_{n=1}^{\infty} c_n \varphi_n$ with the norm $\|f\|_{W_r} := \|\{c_n\}\|_r$ 147

$\mathcal{W}_{r+1}$ – Hilbert space $W_{r+1} \oplus W_r$ 154

$W(\rho_n)$ – Hilbert space of series $\sum\limits_{n=1}^{\infty} c_n \varphi_n$ with the norm $(\sum |c_n|^2 \rho_n^2)^{1/2}$ 163

φ_n – eigenfunction of operator A

λ_n – eigenvalue of operator A or a point of spectrum of exponential family

ω_n	– eigenfrequency of operator A	
κ_n	– multiplicity of eigenvalue λ_n of operator A	
$\mathscr{E}$	– exponential family $\{e_n\}$, with elements $e_n = \exp(i\lambda_n)\eta_n$, where η_n belongs to an auxiliary Hilbert space	
$\tilde{\mathscr{E}}$	– exponential family $\{r_n e_n\}$	163, 170
$\mathscr{E}_0$	– exponential family $\{\rho_n e_n\}$	163, 170
x_λ	– simple fraction $x_\lambda(k) = \sqrt{\dfrac{\operatorname{Im}\lambda}{\pi}}\dfrac{1}{k - \bar{\lambda}}$	41
$\mathscr{X}_\Pi$	– family of vector simple fractions $\{x_\lambda \eta_\lambda\}_{\lambda \in \sigma}$	47
$\mathscr{X}'_\Pi$	– family biorthogonal to the family $\mathscr{X}_\Pi$	
$\mathscr{X}'_{\Theta,\Pi}$	– family biorthogonal to the family $P_\Theta \mathscr{X}_\Pi$	

Operations

Cl, Cl$_H$	– closure of a set in the norm of space H
Lin Ξ	– linear span of family Ξ
$\dotplus$	– direct sum of linear sets
$\bigvee\Xi$	– closure of the linear span of family Ξ

Sets

$\mathbb{R}$	– set of real numbers	
$\mathbb{R}_+$	– set of positive numbers	
$\mathbb{Q}$	– set of rational numbers	
$\mathbb{Z}$	– set of integers	
$\mathbb{K}$	– set of nonzero integers	
$\mathbb{N}$	– set of positive integers	
$\mathbb{C}$	– set of complex numbers (complex plane)	
$\mathbb{C}^n$	– Cartesian product of n complex planes	
$\mathbb{C}_+$ ($\mathbb{C}_-$)	– open upper (lower) half-plane, i.e., set of complex numbers with positive (negative) imaginary part	
$\mathbb{D}$	– open unit disk in $\mathbb{C}$ with the center in origin	
$\mathbb{T}$	– unit circle in $\mathbb{C}$ with the center in origin	
$\mathscr{L}(X, Y)$	– set of linear bounded operators acting from space X into space Y	
$D(A)$	– domain of operator A	
D_Ξ	– domain of operator of the problem of moments with respect to family Ξ	34

| R_Ξ | – image of operator of the problem of moments with respect to family Ξ | 34 |
| Ker A | – null space of operator A | |

Operators

$P_{\mathfrak{M}}$	– orthoprojector on subspace $\mathfrak{N}$	
$\mathscr{P}_{\mathfrak{M}}^{\|\mathfrak{N}}$	– skew projector on subspace $\mathfrak{M}$ parallel to subspace $\mathfrak{N}$	20
I	– identical operator	
$\mathscr{I}_\Xi$	– operator of the problem of moments with respect to family Ξ	
P_s	– projector onto subspace K_s	48
$P_\pm$	– projectors onto subspace $H_\pm^2$	42

Conditions on countable sets of the upper half-plane

(B)	– Blaschke condition	43
(R)	– rareness condition	52
(C)	– Carleson condition	52
(CN)	– Carleson–Newman condition	52

Property of families of elements or subspaces of a Hilbert space

$\Xi \in (LB)$	– family Ξ forms a Riesz basis in closure of its linear span	26, 30
$\Xi \in (UM)$	– family Ξ of elements is *-uniformly minimal or family Ξ of subspaces is uniformly minimal	26, 30
$\Xi \in (M)$	– family Ξ is minimal	25, 30
$\Xi \in (W)$	– family Ξ is W-linear independent	24

Types of controllability

B controllability	162, 169
E controllability	162, 169
UM controllability	162, 169
M controllability	162, 169
W controllability	162, 169

Cancellations

$\mathscr{L}$-basis	– Riesz basis in closure of linear span	26, 30
BP	– Blaschke product	43
BPP	– Blaschke–Potapov product	44
DPS	– distributed parameter system	
ESF	– entire singular (operator) function	44
STF	– sine-type function	61
GF	– generating function	80, 101, 113

Signs

$A := B$ or $B =: A$ – object A is equal to object B by definition

$f(x) \asymp g(x), x \in X$ – this relation means that there are positive constants c and C such that for all $x \in X$ the inequalities $cg(x) \le f(x) \le Cg(x)$ are valid

$f(x) \prec g(x)$ or $g(x) \succ f(x), x \in X$ – this means that there exists a positive constant C such that for all $x \in X$ the inequality $f(x) \le Cg(x)$ is valid

Introduction

This book deals with the controllability problem in distributed parameter systems (DPS). As the following chapters explain, one way to handle this problem is to reduce it to the problem of moments and to study the resulting exponential families $\mathscr{E}_{par} = \{\eta_n\,e^{-\lambda_n t}\}_{n=1}^{\infty}$ (the "parabolic" case) or $\mathscr{E}_{hyp} = \{\eta_n\,e^{\pm i\sqrt{\lambda_n}t}\}_{n=1}^{\infty}$ (the "hyperbolic" case). Here, $\{\lambda_n\}$ is the spectrum of a system, and vectors η_n belong to an auxiliary Hilbert space $\mathfrak{N}$, dim $\mathfrak{N} \le \infty$, determined by the space of control actions.

Over the past twenty-five years, DPS control theory has been developing rapidly, both because of its important technological applications and its usefulness in resolving a variety of mathematical problems. Indeed, the theory has been described at length in various monographs (Butkovskiĭ 1965, 1975; Lions 1968, 1983, 1988a; Lurie 1975; Curtain and Pritchard 1978; Egorov A. I. 1978; Litvinov 1987; Lagnese and Lions 1989; Krabs 1992).[1]

The controllability question occupies a prominent place in DPS control theory for a number of reasons. First, many practical problems in various fields of engineering, physics, and chemistry are formulated as controllability problems, that is, as questions about how to describe reachability sets. Second, it is essential to have some insight into controllability in order to resolve DPS optimal control problems. Furthermore, controllability plays a vital role in the stabilization and identification of DPS. Recent studies have demonstrated its profound connection to the classical inverse problems of mathematical physics, for example (Belishev 1989;

The first version of this book was published in Russian in 1989 (Avdonin and Ivanov 1989b). For the present edition, the book has been thoroughly revised and new results were added.
[1] Since the number of publications pertinent to the subject of this book is enormous, we are unable to present an exhaustive list of references; an extensive bibliography may be found in Fleming (1988).

Avdonin, Belishev, and Ivanov, 1991a). Note, too, that DPS controllability studies have brought to light a host of interesting and complex questions in several branches of mathematics, such as PDE theory, operator theory, the theory of functions, and the theory of numbers.

The various techniques used to investigate DPS controllability can be divided into three fundamental approaches. The first one is based on theoretical operator methods (see, e.g., Fattorini 1966, 1967; Tsujioka 1970; Fuhrmann 1972; Weiss 1973; Triggiani 1978; Nefedov and Sholokhovich 1985; Sholokhovich 1987). These are rather general methods that make it possible to treat a broad range of systems described by equations in Hilbert and Banach spaces; however, they are not always effective in addressing concrete problems.

The second approach employs a specific technique of the theory of partial differential equations. This technique has been the subject of considerable research and has such broad applications that we can mention only a few of the works that have focused on it.

D. L. Russell (1971a, 1971b, 1972, 1973) and J. E. Lagnese (1983), for example, apply the method of characteristics for hyperbolic equations and the Holmgreen uniqueness theorem (see also Littman 1986).

J.-L. Lions (1986) suggested the Hilbert Uniqueness Method, which is based on the duality between controllability and observability and on a priori estimates of solutions of nonhomogeneous boundary value problems. This method was developed by L. F. Ho (1986), P. Grisvard (1987), E. Zuazua (1987), A. Haraux (1988), and I. Lasiecka and R. Triggiani (1989). A number of searchers (Chen et al. 1987; Leugering and Schmidt 1989; Schmidt 1992; Lagnese, Leugering, and Schmidt 1993) have applied the method to networks of strings and beams.

Bardos, Lebeau, and Rauch (1988a, 1988b, 1992) have developed an approach to the controllability problems for hyperbolic equations using microlocal analysis and propagations of singularities. This approach made it possible to solve the problem of exact controllability in cases where controls act on a part of the boundary or on a subdomain. See also Emanuilov (1990).

The third approach, which reduces the control problem to the problem of moments relative to a family of exponentials, is known as the moment method. It is a powerful tool of control theory in that it provides solutions to many kinds of problems. N. N. Krasovskiĭ (1968) applied the method to the systems described by ordinary differential equations. It has also been used in DPS control theory to investigate optimum time control problems (Egorov Yu. V. 1963a, 1963b; Butkovskiĭ 1965, 1975; Gal'chuk

1968; Korobov and Sklyar 1987) and to solve optimal problems with the quadratic quality criterion (Plotnikov 1968; Egorov A. I. 1978; Vasil'ev, Ishmukhametov, and Potanov 1989). In addition, this method has been used in combination with the Pontryagin maximum principle to address .the optimal boundary control problem to parabolic vector equations (Kuzenkov and Plotnikov 1989), and to study some bilinear control problems (Egorov A. I. and Shakirov 1983), as well as observation problems in parabolic-type equations (Mizel and Seidman 1969, 1972; Seidman 1976, 1977).

In investigations of DPS controllability, the moment method has most often been used for systems with one spatial variable and for a scalar control function (see, e.g., Russell 1967, 1978; Fattorini and Russell 1971; Butkovskiĭ 1975; Reid and Russell 1985). Work has also been done on controllability problems associated with several control actions (Fattorini 1968; Sakawa 1974).

The moment method has also been used to analyze controllability in systems permitting separation of spatial variables. Here, the method has made it possible to reduce the controllability problem to a series of scalar problems (see Graham and Russell 1975; Fattorini 1975, 1979; Krabs, Leugering, and Seidman 1985). For the exponential family $\mathscr{E}$ arising in the transition from a control problem for a moment one, the role of auxiliary space $\mathfrak{N}$ is filled by the space to which the values of control actions belong. In the case of a single control action dim $\mathfrak{N} = 1$, the usual "scalar" families of exponentials appear. If there is a finite number N of scalar control actions, then dim $\mathfrak{N} = N$, and a family of vector exponentials with the values in a finite dimensional space arises. A string vibration equation with the control actions at both boundary points serves as an example (where $N = 2$). If, for instance, a control acts on the boundary Γ of a multidimensional spatial domain Ω, then it is natural to choose $L^2(\Gamma)$ as $\mathfrak{N}$, whereupon dim $\mathfrak{N} = \infty$.

The solvability of the resulting problem of moments, and hence of the primary control problem, depends on the properties of the corresponding exponential family. The study of scalar exponential families (nonharmonic Fourier series) dates back to the 1930s (Paley and Wiener 1934) and since then has become a well-known branch of the mathematical analysis. Thus, questions concerning the completeness, minimality, and basis property of such families in space $L^2(0, T)$ have been investigated in some depth (Ingham 1934; Levinson 1940; Duffin and Eachus 1942; Duffin and Schaffer 1952; Levin 1956, 1961; Kadets 1964; Redheffer 1968; Katsnelson 1971; Young 1980). In addition, B. S. Pavlov (1973, 1979) has suggested a geometric (in a Hilbert space sense) approach that has provided a basis property criterion. (For a detailed exposition of this approach and its relation to other problems connected with the theory of functions, see Hrushchev, Nikol'skiĭ, and Pavlov 1981.) We use Pavlov's geometrical approach in this book to develop a theory of exponentials in a space of vector functions. This work has also enabled us to shed new light on scalar families and thus is proving to be useful in DPS control problems.

In fact, DPS control problems were the very reason that we decided to investigate exponential families. Without the results that we have obtained on vector exponential families, it would be very difficult to apply the moment method to problems that cannot be treated in the terms of a scalar exponential family or a series of such families of simple enough structure. In other words, the extension of the moment method to a wider class of DPS is one of the principal objectives of this book.

The book consists of seven chapters, each of which is divided into numbered sections, which in turn contain enumerated assertions (remarks, theorems, corollaries, and so on). When referring to a statement or a formula within a chapter section, we omit the number of that chapter section (e.g., we refer to Proposition 17(a) rather than Proposition I.1.17(a)). When referring to a formula of another section, we add the number of that section. Sections are divided into subsections. Although this arrangement may seem unwieldy, it is difficult to treat this complex subject in any other way.

Chapter I presents the basic information needed to understand projectors in Hilbert spaces, families of elements, and families of subspaces, as well as the problem of moments. Although we cannot claim to be presenting original results (except, perhaps, for some assertions on the problem of moments solvability) or to have made any methodological discoveries, we have brought together for the first time all basic information concerning this subject.

The discussion opens in Section I.1 with the geometry of Hilbert spaces. For two subspaces $\mathfrak{M}$ and $\mathfrak{N}$ of a Hilbert space $\mathfrak{H}$, we introduce the concept of an angle $\varphi(\mathfrak{M}, \mathfrak{N})$ between them,

$$\varphi(\mathfrak{M}, \mathfrak{N}) = \arccos \sup_{m \in \mathfrak{M}, n \in \mathfrak{N}} \frac{|(m, n)|}{\|m\| \, \|n\|}.$$

In terms of the angles, we elucidate the properties of operators $P_{\mathfrak{M}}|_{\mathfrak{N}}$ (orthoprojectors on $\mathfrak{M}$ restricted to $\mathfrak{N}$). In particular,

$$\|[P_{\mathfrak{M}}|_{\mathfrak{N}}]^{-1}\| = 1/\sin \varphi(\mathfrak{H} \ominus \mathfrak{M}, \mathfrak{N}).$$

Skew projector $\mathscr{P}_{\mathfrak{M}}^{\|\mathfrak{N}}$ from the direct sum $\mathfrak{M} \dotplus \mathfrak{N}$ to $\mathfrak{M}$ parallel to $\mathfrak{N}$ is studied further:

$$\mathscr{P}_{\mathfrak{M}}^{\|\mathfrak{N}}(m + n) = m \qquad \forall m \in \mathfrak{M}, n \in \mathfrak{N}.$$

This projector is bounded if and only if $\varphi(\mathfrak{M}, \mathfrak{N}) > 0$. Such projectors play a central role in the investigation of exponential families in $L^2(0, T)$.

Next, we study families $\Xi = \{\xi_n\}$ of element (and families of subspaces) from the perspective of the "degree" of their linear independence. The linear independence of any finite subfamilies of Ξ is the weakest one. We let it be denoted by $\Xi \in (L)$. The next step is to introduce W-linear independence (notation $\Xi \in (W)$). This property, in somewhat simplified terms, means that the weak convergence of series $\sum c_n \xi_n$ to zero implies that all coefficients c_n are zeros. A stronger property is minimality ($\Xi \in (M)$). It means that for any n, element ξ_n does not lie in the closure of linear span of the remaining elements. To put it another way: there exists a family $\Xi' = \{\xi'_n\}$, called the biorthogonal family, such that

$$(\xi_n, \xi'_n) = \delta_n^m.$$

If, along with this, $\|\xi'_n\|$ are jointly bounded, then the family is said to be *-uniformly minimal (notation $\Xi \in (UM)$). For the almost normed families ($\|\xi_n\| \asymp 1$), the (UM) property is equivalent to

$$\varphi\left(\xi_n, \bigvee_{m \neq n} \xi_m\right) \geq \delta > 0.$$

A family for which the latter relation holds is said to be uniformly minimal.

The strongest property, which is the $\mathscr{L}$-basis or Riesz basis property in the closure of the linear span of the family, means that family Ξ is an image of an orthonormal one under the action of some isomorphism; we write $\Xi \in (LB)$ in this case.

Families of subspaces are classified by similar definitions. Thus the hierarchy of the "independence"

$$(LB) \Rightarrow (UM) \Rightarrow (M) \Rightarrow (W) \Rightarrow (L)$$

is established.

In Section I.2, we turn to the problem of moments. For a given family Ξ and some element $c \in \ell^2$, one has to find $f \in \mathfrak{H}$ such that $\{(f, \xi_n)\} = c$. Operator

$$\mathscr{J}_\Xi : f \mapsto \{(f, \xi_n)\} \tag{1}$$

is called the operator of the problem of moments. We focus our attention on the "quality" of the solvability of the moment problem or, more precisely, on the image R_Ξ of operator $\mathcal{J}_\Xi$. The solvability of the problem of moments is directly associated with "the degree of linear independence" of family Ξ. In particular,

$$\Xi \in (LB) \Rightarrow R_\Xi = \mathfrak{H}, \qquad R_\Xi = \mathfrak{H} \Rightarrow \Xi \in (UM),$$

$$Cl_\mathfrak{H} R_\Xi = \mathfrak{H} \Leftrightarrow \Xi \in (W).$$

We also prove that R_Ξ is closed if a Riesz basis can be found in Ξ.

Chapter II examines the properties of family $\mathscr{E}_T$ of vector exponentials in $L^2(0, T; \mathfrak{N})$,

$$\mathscr{E}_T = \{e_n\}_{n\in\mathbb{Z}}, \qquad e_n = e^{-i\bar{\lambda}_n t}\eta_n, \qquad \text{Im } \lambda_n > 0,$$

in detail. Here, $\eta_n \in \mathfrak{N}$, dim $\mathfrak{N} < \infty$, $T \leq \infty$. By the Paley–Wiener theorem, the inverse Fourier transform turns $L^2(0, \infty)$ into Hardy space H_+^2, which consists of analytic functions in the upper half-plane $\mathbb{C}_+$ whose traces are squarely summable over the real axis. Here, the exponentials turn into simple fractions $x_n(k) = (k - \bar{\lambda}_n)^{-1}$ belonging to H_+^2 for Im $\lambda_n > 0$. This makes it possible to invoke the powerful theory of Hardy spaces in the study of exponentials. Section II.1 explains these spaces and simple fraction families. It should be pointed out that the reader will require some knowledge of the basics of the Hardy space theory in order to understand the theory developed in this book.

To begin with, one needs to be familiar with the concepts associated with inner–outer factorization. Consider, for simplicity, the case of functions bounded in $\mathbb{C}_+$. If such functions have a unit absolute value almost everywhere on the real axis (and are analytic in $\mathbb{C}_+$), then they are said to be inner functions. Among them, Blaschke products (BP), $B(k)$, are recognized,

$$B(k) = \prod_{n\in\mathbb{Z}} \varepsilon_n \frac{k - \lambda_n}{k - \bar{\lambda}_n}, \qquad \text{Im } \lambda_n > 0,$$

where ε_n are the phase factors, $|\varepsilon_n| = 1$, and numbers λ_n – the zeros of the BP – satisfy the Blaschke condition

$$\sum_{n\in\mathbb{Z}} \frac{\text{Im } \lambda_n}{1 + |\lambda_n|^2} < \infty. \tag{B}$$

Functions exp(ika), $a \geq 0$, are obviously inner functions as well; in contrast to BP's, they have no zeros at all. Such functions are called entire singular inner functions (they have the essentially singular point at infinity). Functions bounded and analytic in $\mathbb{C}_+$ possess factorization

$f = f_i f_e$ in which f_i is an inner function and f_e is an outer function of the form

$$f_e(k) = \exp\left(\frac{i}{\pi} \int_{\mathbb{R}} \frac{h(t)}{k - t} \, dt\right)$$

with $h(t) = \log|f(t)|$. Outer functions have no zeros in $\mathbb{C}_+$. In contrast to entire singular inner functions, they cannot decrease exponentially at $\mathrm{Im}\, k + \infty$.

Section II.1 introduces the concepts of inner and outer functions for analytic operator functions with the values in finite-dimensional space $\mathfrak{N}$, along with the Blaschke–Potapov (BPP) product, which is the analog of BP, and the entire singular inner operator function (ESF). Matrix exponential $\exp(ikQ)$ with nonnegative matrix (operator) Q is an ESF. Without going into detailed definitions here, suffice it to say that an operator function belongs to the corresponding class if its determinant is a function from a similar scalar class. For analytic operator functions F bounded in $\mathbb{C}_+$ there also exists factorization

$$F = \Pi \Theta F_e^+,$$

where Π is a BPP, Θ is an ESF, and F_e^+ is an outer operator function.

Consider now the exposition of the known results on the properties of families $\mathscr{X}$ of simple fractions $x_n(k)$, $n \in \mathbb{Z}$. It appears that for the minimality of $\mathscr{X}$ (on H_+^2) the validity of Blaschke condition (B) is necessary and sufficient.

The criterion of uniform minimality of $\mathscr{X}$ is the Carleson condition

$$\inf_{m \in \mathbb{Z}} \prod_{n, n \neq m} \left| \frac{\lambda_n - \lambda_m}{\lambda_n - \bar{\lambda}_m} \right| > 0. \tag{C}$$

This condition is well known in the theory of interpolation of bounded analytic functions. Normalized families of simple fractions exhibit a surprising equivalence between the uniform minimality and the $\mathscr{L}$-basis property: that is, the Carleson condition proves to be the $\mathscr{L}$-basis criterion. In the strip $0 < c \leq \mathrm{Im}\, k \leq C$, condition (C) transforms into the separability condition $(\inf_{m \neq n}|\lambda_m - \lambda_n| > 0)$. Recall that these properties of family $\mathscr{X}$ are equivalent to similar properties of an exponential family in $L^2(0, \infty)$.

In Section II.2, minimality and $\mathscr{L}$-basis criteria are given for family $\mathscr{E}_\infty$ of vector exponentials. For family $\mathscr{E}_\infty$ to be minimal in $L^2(0, \infty; \mathfrak{N})$, it is necessary and sufficient that λ_n satisfy the Blaschke condition, so the finite dimensional case has no specific character of its own, in comparison with the scalar one.

The situation is more complicated when it comes to the $\mathscr{L}$-basis property. If one takes two Carlesonian sets σ_1 and σ_2, then their unification may be not Carlesonian (in contrast to the Blaschke condition, the Carleson one is not "additive"). Consequently, scalar family $\{e^{i\lambda t}\}_{\lambda \in \sigma_1 \cup \sigma_2}$ will not yet be an $\mathscr{L}$-basis. At the same time, vector exponential family

$$\{e^{i\lambda t}\eta_1\}_{\lambda \in \sigma_1} \cup \{e^{i\lambda t}\eta_2\}_{\lambda \in \sigma_2}$$

evidently constitutes an $\mathscr{L}$-basis if η_1 and η_2 are linearly independent. It is known that $\mathscr{L}$-basis family $\mathscr{E}_\infty$ allows a splitting into dim $\mathfrak{N}$ subfamilies, each of which has a Carlesonian spectrum. This single condition is not enough for the $\mathscr{L}$-basis property of $\mathscr{E}_\infty$. To obtain the $\mathscr{L}$-basis criterion it is necessary to demand in addition that every group of "close points" λ has vectors η_λ, which are "linear independent" uniformly in groups. The exact formulation of the criterion is presented in Subsection II.2.2.

A criterion for vector exponential family $\mathscr{E}_T$ to form a basis in space $L^2(0, T; \mathfrak{N})$ is established in Section II.3 in terms of the generating function (GF).

The GF concept was formulated some time ago (Paley and Wiener 1934) and since then has been widely used (see, e.g., Levin 1956, 1961) for the investigation of scalar exponential families $\{e^{i\lambda_n t}\}$ in $L^2(0, T)$. The GF is constructed by its zeros λ_n with the help of the following formula:

$$f(k) = e^{ikT/2}f_0, \qquad f_0 = \text{p.v.} \prod_n (1 - k/\lambda_n), \tag{2}$$

under the assumption that f_0 has the same exponential type $T/2$ both in the upper and the lower half-planes. Later, we assume that the spectrum $\sigma = \{\lambda_n\}$ lies in the strip $0 < c \le \text{Im } \lambda_n \le C$; note that the shift $\lambda_n \mapsto \lambda_n + i\delta$ does not change the minimality and $\mathscr{L}$-basis properties of the family. By employing the GF, B. S. Pavlov (1979) managed to obtain the basis criterion: family $\{e^{i\lambda_n t}\}$ constitutes a Riesz basis in $L^2(0, T)$ if and only if

(i) $\{\lambda_n\}$ is separable and
(ii) $|f(x)|^2$ satisfies the so-called Muckenhoupt condition

$$\sup_{I \in \mathscr{T}} \frac{1}{|I|} \int_I |f(x)|^2 \, dx \, \frac{1}{|I|} \int_I |f(x)|^{-2} \, dx < \infty,$$

where $\mathscr{T}$ is the set of intervals of the real axis.

Condition (i) is equivalent to the $\mathscr{L}$-basis property of family $\{e^{i\lambda_n t}\}$ in $L^2(0, \infty)$. Condition (ii) leads to several equivalent statements. One of them, which has just appeared in the Pavlov approach, requires the Hilbert operator

$$(Hu)(x) = \frac{1}{\pi} \text{p.v.} \int_{\mathbb{R}} \frac{u(t)}{x - t} dt \qquad (3)$$

to be bounded in the space of functions squarely integrable on the line with the weight $|f(x)|^2$.

Extending this approach, we obtain the necessary and sufficient condition that the vector exponential family forms a basis (see Section II.3). We find that GF f from (ii) may be defined as a function fulfilling factorization conditions

$$f(k) = B(k)f_e^+(k), \qquad k \in \mathbb{C}_+,$$

$$f(k) = e^{ikT}f_e^-(k), \qquad k \in \mathbb{C}_-.$$

Here B is the BP constructed by $\{\lambda_n\}$ while $f_e^\pm$ are outer functions in $\mathbb{C}_+$ and $\mathbb{C}_-$, respectively. These are precisely the relations that provide the grounds for the definition of the GF in the vector case. Entire operator function F with a factorization

$$F = \Pi F_e^+ = e^{ikT}F_e^- \qquad (4)$$

is said to be a GF for family $\mathscr{E}_T = \{e^{i\lambda_n t}\eta_n\}$ in $L^2(0, T; \mathfrak{N})$. Here, $F_e^\pm$ are outer operator functions in $\mathbb{C}_\pm$, and $\Pi(k)$ is the BPP constructed by λ_n and η_n (i.e., the determinant of Π is the BP with zeros λ_n and $\eta_n \in \text{Ker } \Pi^*(\lambda_n)$). Family $\mathscr{E}_T$ is shown to form a Riesz basis in $L^2(0, T; \mathfrak{N})$ if and only if (i) family $\mathscr{E}_\infty$ forms an $\mathscr{L}$-basis on the semiaxis, and (ii) the Hilbert operator (3) is bounded in the space of vector functions squarely summable on the line with the matrix weight $F^*(x)F(x)$.

Properties of scalar exponential families are examined in Section II.4. The known results concerning the minimality and basis property are discussed first, and then some new findings presented. For example, it is now thought that sine-type functions play a significant role in exponential families. Entire function f of the exponential type is called a sine-type function if its zeros lie in the strip $|\text{Im } k| \leq C$ and if both f and $1/f$ are bounded on some line parallel to the real axis. The proximity of numbers λ_n to the zeros of some sine-type function is the known sufficient condition for family $\{e^{i\lambda_n t}\}$ to be a Riesz basis. The converse statement is also proved to be true, and this fact is used to demonstrate the following result.

If family $\mathscr{E}_T = \{e^{i\lambda_n t}\}$ forms a Riesz basis in $L^2(0, T)$, then for any $T' < T$ there exists a subfamily $\mathscr{E}' \subset \mathscr{E}_T$ constituting a Riesz basis in $L^2(0, T')$. We also elaborate an algorithm for the construction of such a subfamily.

In the same section, we prove that statements similar to the one formulated above are valid for families of a more general form:

$$\{t^m e^{i\lambda_n t}\}, \qquad m = 0, 1, \ldots, r_n, \ n \in \mathbb{Z}.$$

In Section II.5, we look at vector exponential families further. We show, for example, that when family $\mathscr{E}_T = \{e^{i\lambda_n t}\eta_n\}$ is minimal in $L^2(0, T; \mathbb{C}^N)$, then scalar family $\{e^{i\lambda_n t}\}$ generally is not minimal in $L^2(0, NT)$, but becomes minimal after N arbitrary elements are removed from it.

Another assertion we make there deals with the stability of the basis property. If $\mathscr{E}_T$ forms a Riesz basis in $L^2(0, T; \mathfrak{N})$, then $\varepsilon > 0$ may be found such that any family $\tilde{\mathscr{E}}$ of the form $\{e^{i\tilde{\lambda}_n t}\tilde{\eta}_n\}$ is also a Riesz basis in $L^2(0, T; \mathfrak{N})$ as soon as

$$|\lambda_n - \tilde{\lambda}_n| + \|\eta_n - \tilde{\eta}_n\| < \varepsilon.$$

Chapter II closes with a discussion of the conditions that provide the weak convergence in $L^2(0, T)$ to zero of series $\sum a_n e^{-\mu_n t}$ (the "parabolic" case) or $\sum a_n e^{i\lambda_n t}$ (the "hyperbolic" case), which implies all the coefficients to be zeros. In the parabolic case, it takes place under very weak limitations on $\{\mu_n\}$. However, to make this implication hold in the hyperbolic case, stringent restrictions on $\{a_n\}$ have to be imposed. As becomes clear later in the book, differences in the behavior of the exponential family lead to a qualitative distinction in the controllability of parabolic and hyperbolic systems.

Evolution equations of the first and second order in time

$$\dot{x}(t) + Ax(t) = f(t), \tag{5}$$

$$\ddot{x}(t) + Ax(t) = f(t) \tag{6}$$

are treated in Chapter III. Here, A is a self-adjoint, semibounded-from-below operator in Hilbert space H; operator $A_\alpha := A + \alpha I$ is positive definite. We assume A to have a set of eigenvalues $\{\lambda_n\}_{n=1}^\infty$ with corresponding eigenfunctions φ_n forming an orthonormal basis in H.

We introduce a scale of Hilbert spaces W_s, $s \in \mathbb{R}$. For $s > 0$, W_s is the domain of operator $A_\alpha^{s/2}$; for $s < 0$, $W_s = W'_{-s}$ is the space dual to W_{-s} with respect to inner product in H, $W_0 = H = H'$.

We use the Fourier method to construct generalized solutions to equations (5) and (6), which are continuous functions in time with the values in one of the spaces from the above scale. That is to say, we demonstrate in Section III.1, that for $f \in L^2(0, T; W_{r-1})$ and initial condition $x(0) = x_0 \in W_r$, equation (5) has the unique solution of class $C([0, T]; W_r)$. In Section III.2, we arrive at a similar result for equation (6): with f from the class under initial conditions $x(0) = x_0 \in W_r$, $\dot{x}(0) = x_1 \in W_{r-1}$, equation (6) has a unique solution x such that $x \in C([0, T]; W_r)$, $\dot{x} \in C([0, T]; W_{r-1})$. It is convenient to write the latter inclusions in the form $(x, \dot{x}) \in C([0, T]; \mathscr{W}_r)$, $\mathscr{W}_r = W_r \oplus W_{r-1}$. Note that these results are sharp ones.

Next, we consider controls systems

$$\dot{x}(t) + Ax(t) = Bu(t), \tag{7}$$

$$\ddot{x}(t) + Ax(t) = Bu(t). \tag{8}$$

Here, control u belongs to space $\mathscr{U} := L^2(0, T; U)$, where U is a Hilbert space, and B is a bounded operator from U to W_{r-1}. With the help of the Fourier method, control problems for systems (7) and (8) are reduced to the problem of moments. Namely, reachability set $R(T)$ of system (7) in time T (for $x(0) = 0$) is shown to be isomorphic to the image of operator (1) of the problem of moments for family

$$\tilde{\mathscr{E}}_{\mathrm{par}} = \{(\lambda_n + \alpha)^{r/2} e^{-\lambda_n t} B^* \varphi_n\} \subset \mathscr{U}.$$

For system (8) with initial conditions $x(0) = \dot{x}(0) = 0$, reachability set $\mathscr{R}(T)$,

$$\mathscr{R}(T) = \{(x(T), \dot{x}(T)) \mid u \in \mathscr{U}\},$$

is isomorphic to the image of the operator of the problem of moments for family

$$\tilde{\mathscr{E}}_{\mathrm{hyp}} = \{(\lambda_n + \alpha)^{\frac{r-1}{2}} e^{\pm i\sqrt{\lambda_n} t}\}_{n=1}^{\infty}.$$

(For the sake of brevity, we assume here that numbers λ_n are separated from zero.)

Section III.3 then relates the questions of controllability for systems (7) and (8) to the properties of family $\mathscr{E}$ in space $\mathscr{U}$. Previous studies have focused mainly on two types of DPS controllability: the exact one, when the reachability set includes some explicitly described space, and the approximate one, when the reachability set is dense in the phase space of the system. From a practical point of view, the distinction between the possibility of getting exactly into some state and the possibility of finding the system in its ε-vicinity of this state is not too significant. The behavior of the control norm when $\varepsilon \to 0$ is important, however, as is the stability of the controllability type under small perturbations of the system parameters. We suggest a new classification of controllability types that refines the common one by taking into account both of these points and relates the "quality" of controllability directly to the properties of exponential families.

Suppose that in system (7) H_0 is a Hilbert space densely and continuously embedded into W_r, which comprises all the eigenfunctions φ_n (in particular, H_0 may coincide with W_r). The conventional definition of the exact controllability (we call this property E controllability relative to H_0 in time T) means that $R(T) \supset H_0$. It is natural to extract the case of the equality, $R(T) = H_0$, which we call B controllability (relative to H_0 in time T). In this case, the final state $x(T)$ may be achieved with the help of control u, whose norm is equivalent to that of $x(T)$:

$$\|x(T)\|_{H_0} \asymp \|u\|_{\mathcal{U}}.$$

The approximate controllability (we called it W controllability in time T) means that $\mathrm{Cl}_{W_r} R(T) = W_r$. It makes physical sense to distinguish the situation when reachability set $R(T)$ contains all (finite) linear combinations of eigenfunctions φ_n. This type is called M controllability.

Similar definitions of controllability types are also introduced for a system of the second order in time. For instance, system (8) is said to be M controllable in time T if $\mathcal{R}(T)$ contains all linear combinations of states of the form $(\varphi_n, 0)$ and $(0, \varphi_n)$.

Many studies dealing with the proof of the approximate controllability for DPS of this or that kind, have demonstrated their M controllability. D. L. Russell (1978: 699) has recognized this case and called it the eigenfunction controllability.

Next, concentrate on the spaces H_0 of a more special type, such as W_s for system (7) or spaces $\mathcal{W}_s$ for system (8). From the isomorphicity of reachability sets to the image of the moment problem operator, one is able to relate controllability types to the properties (types of linear independence) of vector exponential families.

Suppose that in system (7) $H_0 = W_r$. Then it follows that

(a) the system is B-controllable relative to H_0 in time T if and only if $\tilde{\mathscr{E}}_{\mathrm{par}}$ forms an $\mathscr{L}$-basis in $\mathcal{U}$;

(b) the system is M-controllable in time T if and only if $\tilde{\mathscr{E}}_{\mathrm{par}}$ is minimal in $\mathcal{U}$;

(c) the system is W-controllable in time T if and only if $\tilde{\mathscr{E}}_{\text{par}}$ is W-linear independent in $\mathscr{U}$;

(d) the unification of sets $R(T)$ in $T > 0$ is dense in W_r if and only if family $e^{-\alpha t}\tilde{\mathscr{E}}_{\text{par}}$ is W-linearly independent in $\mathscr{U}_\infty := L^2(0, \infty; U)$.

Quite similar assertions are correct for system (8) when $\tilde{\mathscr{E}}_{\text{par}}$ is replaced by $\tilde{\mathscr{E}}_{\text{hyp}}$ and W_r by $\mathscr{W}_r$.

The discussion then turns to an abstract version of the Hilbert Uniqueness Method suggested by J.-L. Lions (1986) for systems (7) and (8).

Chapters IV and V are denoted to controllability in systems described by differential equations of parabolic and hyperbolic types. These problems are discussed in the framework of Chapter III, in which A is an elliptic differential operator of the second order in a bounded domain $\Omega \subset \mathbb{R}^N$. In this case, spaces W_r are related to the Sobolev spaces by the following inclusions:

$$H_0^r(\Omega) \subset W_r \subset H^r(\Omega), \qquad r \in [0, r_0],$$

where r_0 depends on the smoothness of operator coefficients and the boundary Γ of domain Ω.

Operator B has various forms, depending on the kinds of control actions. We first discuss controls entering the right-hand side of the differential equation (controls with a spatial support in domain Ω). If, for instance, a control is of the form $u(x, t)$ and its x support lies in a subdomain Ω', then B is the operator of multiplication by the characteristic function of Ω' and $U = L^2(\Omega')$. If the control is finite-dimensional, that is, the right-hand side is $\sum_{p=1}^{m} b_p(x)u_p(t)$ (functions b_p are specified), then $U = \mathbb{C}^m$ and

$$B\eta = \sum_{p=1}^{m} \eta_p b_p(\cdot): \mathbb{C}^m \to W_{r-1}.$$

In particular, functions b_p may be generalized functions, for example, $b_p(x) = \delta(x - x_p)$, $x_p \in \Omega$. In the latter case, we speak about pointwise control.

Within the framework of the same scheme, we also consider boundary controls that enter the boundary conditions of the corresponding initial boundary-value problem. So, for a boundary condition acting on a part Γ' of the boundary in the Dirichlet problem $U = L^2(\Gamma')$ and the value of functional $B\eta$, $\eta \in L^2(\Gamma')$ on element φ is given by the formula

$$\langle B\eta, \varphi \rangle = \int_{\Gamma'} \eta(s) \frac{\partial \varphi(s)}{\partial \nu_A} \, ds$$

(a derivative along the conormal stands under the integral). For classical solutions of initial boundary-value problems, the formula is obtained by means of integration by parts. For the generalized solutions concerned, just the specification of operator B allows one to actually assign a rigorous meaning to the initial boundary-value problems in question. For controls with both spatial and boundary support, one is able to distinguish the case of finite-dimensional, particularly pointwise, control.

In Chapters IV and V the unified approach (see Chapter III), is used to investigate a variety of control actions, and a number of new findings are reported. In addition, further information is provided on the properties of exponential families. Note that in some cases the clarification of the relationship between DPS controllability and exponential families permits us to use results on the controllability provided by other methods (not associated with the moment approach) to establish new properties of vector exponential families.

To mention some of the more interesting results concerning parabolic system controllability (Chapter IV), consider a system controlled over a piece Γ' of the boundary (Γ' is relatively open and nonempty). For $U = L^2(\Gamma')$ a parabolic system happens to be W controllable in any time. Moreover, the same fact takes place for a narrower class of controls:

$$u(x, t)|_{x \in \Gamma'} = f(x)g(t),$$

where f runs through $L^2(\Gamma')$, while g – through $L^2(0, T)$.

For any kind of finite dimensional controls for dim $\Omega > 1$, the system proves to be not M-controllable in any time. Lack of M controllability implies a lack of E controllability relative to Sobolev space with any exponent. Simultaneously, the W controllability is equivalent to the so-called rank criterion; when the latter is broken, the exponential family becomes linearly dependent.

Hyperbolic system controllability (Chapter V) has, on the one hand, a great deal in common with the controllability of parabolic systems under the same types of control actions. So, for any kind of finite-dimensional controls, the system is not M-controllable. On the other hand, hyperbolic systems are boundary controllable only for large enough T (it is associated with the finite propagation velocity of perturbations). Furthermore, in many cases when the corresponding parabolic system allows for only M controllability, the hyperbolic system tends to be B-controllable relative to certain phase space $\mathscr{W}_r$. According to our approach, the distinctions are due to essentially differing features of exponential families with real (in the parabolic case) and imaginary (in the hyperbolic case) exponents.

Chapter VI turns to boundary, initial, and pointwise control problems of the vibrations of a homogeneous rectangular membrane. These problems demonstrate clearly the advantages of the moment method. In particular, in Section VI.1, we arrive at a sharp result pertaining to the smoothness of solution to the wave equation in rectangle $\Omega = (0, a) \times (0, b)$ with a nonhomogeneous Neumann boundary condition of class $L^2(0, T; L^2(\Gamma))$ and a zero initial condition. We prove that relation

$$(z, \dot{z}) \in C([0, T]; H^{3/4}(\Omega) \oplus H^{-1/4}(\Omega))$$

holds for solution z of the problem.

The chapter also presents interesting negative results that prove the lack of approximate controllability. In Section VI.1 we consider the case of boundary condition u, $(\partial y/\partial v) = b(x)u(t)$, with a fixed function b. In Section VI.2 we deal with the problems of initial and pointwise control. We find any number of membrane points can be made to have arbitrary trajectories, if the initial conditions are appropriate. This problem is dual to the problem of pointwise control. We show that for any $T > 0$ and any number of pointwise controls, reachability sets $\mathcal{R}(T)$ are not dense in the phase space of the system.

The following conjecture therefore seems reasonable. For a hyperbolic equation in an arbitrary domain $\Omega \subset \mathbb{R}^N$, $N > 1$, the system reachability set under a finite-dimensional control of any kind is not dense in the phase space of the system for any $T > 0$.

Chapter VII considers boundary control problems connected with hyperbolic systems for vector functions with one spatial variable. With the help of the moment method, we reduce the problems to the study of vector exponential family with the values in a finite-dimensional space. The theory developed in Chapter II is particularly relevant to this case, and the reduction to the moment problem allows us to investigate system controllability properties in a comprehensive fashion.

The kinds of problems found in this chapter can be illustrated first by a network of connected homogeneous strings controlled at the nodes. We prove the system to be B-controllable (in a large enough time) if the graph representing the network is a tree. In the opposite case, the system may be M-controllable if there is no cycle of strings with commensurable optical lengths. If such a cycle can be found, however, the system is not W-controllable.

We also treat a control problem for a multichannel acoustic system

$$\begin{cases} Q(x)\dfrac{\partial^2 y}{\partial t^2} = \dfrac{\partial^2 y}{\partial x^2}, & 0 < x < 1, 0 < t < T, \\[2mm] y(0, t) = u(t), & u \in L^2(0, T; \mathbb{C}^N), \\[2mm] y(l, t) = 0. \end{cases}$$

Here, Q is a positive definite matrix function. The system B controllability in space $L^2(0, l; \mathbb{C}^N) \oplus H^{-1}(0, l; \mathbb{C}^N)$ in time $T \geq T_0$, $T_0 := 2 \int_0^l \|Q(x)\|^{1/2} \, dx$, is proved. In this problem, as well as in other problems of this kind that have a clear physical meaning, one manages to construct a generating matrix function (4) for the family of vector exponentials, as expressed via solutions of the corresponding Helmholtz equation, and then to describe the properties of the vector exponential family that appears.

I

Elements of Hilbert space theory

To understand the theory put forward in this book, it is necessary to be familiar with the following concepts:

(a) Skew projectors and angles between subspaces.
(b) Families of elements of Hilbert spaces: W-linear independence, minimality, and Riesz-basis property.
(c) Solvability of moment problem in Hilbert space.

Throughout this chapter, $\mathfrak{H}$ denotes separable Hilbert space, and $\mathfrak{M}$ and $\mathfrak{N}$ are subspaces of $\mathfrak{H}$ (a subspace is any closed linear subset of $\mathfrak{H}$).

For a linear span of the elements $\xi_j \in \mathfrak{H}$ or subspaces Ξ_j, we write $\mathrm{Lin}\{\xi_j\}(\mathrm{Lin}\{\Xi_j\})$. A closure of a linear span is denoted by

$$\mathrm{Cl}_{\mathfrak{H}} \mathrm{Lin}\{\xi_j\}(\mathrm{Cl}_{\mathfrak{H}} \mathrm{Lin}\{\Xi_j\}),$$

or simply by $\bigvee\{\xi_j\}(\bigvee\{\Xi_j\})$ if it is clear in which metrics the closure is taken. If $\mathfrak{M}$ and $\mathfrak{N}$ are two subspaces of $\mathfrak{H}$, then $\mathfrak{M}^\perp$ and $\mathfrak{N}^\perp$ refer to their orthogonal complements in

$$\bigvee\{\mathfrak{M}, \mathfrak{N}\} =: \mathfrak{M} \bigvee \mathfrak{N}.$$

The notation $P_{\mathfrak{M}}$ is reserved for the orthoprojector on a subspace $\mathfrak{M}$ in $\mathfrak{H}$. In what follows we denote by $A|_B$ the restriction of (acting in Hilbert spaces $\mathfrak{H}$ operator A on set $B \subset \mathfrak{H}$). Notice that we distinguish operators $[P_{\mathfrak{M}}|_{\mathfrak{N}}: \mathfrak{N} \mapsto \mathfrak{H}]$ and $[P_{\mathfrak{M}}|_{\mathfrak{N}}: \mathfrak{N} \mapsto \mathfrak{M}]$.

1. Families of vectors and families of subspaces

1.1. Orthoprojectors and angles between subspaces

Let $\mathfrak{M}$ and $\mathfrak{N}$ be two subspaces of $\mathfrak{H}$.

Definition I.1.1. A number $\varphi = \varphi_{\mathfrak{H}}(\mathfrak{M}, \mathfrak{N})$, $\varphi \in [0, \pi/2]$, is said to be an angle between the subspaces $\mathfrak{M}$ and $\mathfrak{N}$ if

$$\cos \varphi = \sup_{m \in \mathfrak{M},\, n \in \mathfrak{N}} \frac{|(m, n)|}{\|m\|\,\|n\|}.$$

If $\mathfrak{M}$ is a one-dimensional subspace, we usually write $\varphi(m, \mathfrak{N})$ instead of $\varphi(\mathfrak{M}, \mathfrak{N})$, with $m \in \mathfrak{M}$, $m \neq 0$.

For two one-dimensional subspaces, the definition coincides with the usual one for an angle between the straight lines. On the other hand, this definition differs from that of an angle between planes in $\mathbb{R}^3$ (in our sense, the angle between two planes equals zero, since they contain a nonzero common element).

It is evident that if $\mathfrak{M}$ and $\mathfrak{N}$ have nonzero common elements, the angle between them is zero. However, it is possible that $\varphi(\mathfrak{M}, \mathfrak{N}) = 0$ though $\mathfrak{M} \cap \mathfrak{N} = \{0\}$.

Example I.1.2. Let $\{\xi_j^0\}_{j \in \mathbb{N}}$ be an orthonormal basis in $\mathfrak{H}$,

$$\mathfrak{M} := \bigvee_{j \in \mathbb{N}} \{\xi_{2j-1}^0\}, \qquad \mathfrak{N}^0 := \mathrm{Lin}\{\xi_{2j-1}^0 + \varepsilon_j \xi_{2j}^0\}, \qquad \mathfrak{N} = \mathrm{Cl}\,\mathfrak{N}^0,$$

where $\varepsilon_j > 0$, $j \in \mathbb{N}$, and $\varepsilon_j \xrightarrow[j \to \infty]{} 0$. Then $\mathfrak{M} \cap \mathfrak{N} = \{0\}$, but $\varphi(\mathfrak{M}, \mathfrak{N}) = 0$.

Let us check whether $\mathfrak{M}$ and $\mathfrak{N}$ do not have nonzero common elements. In fact, let a sequence $\{f_n\}_{n \in \mathbb{N}}$ be found, with the elements from $\mathfrak{N}^0$, which converges to $f \in \mathfrak{M}$. If

$$f_n = \sum_j c_j^{(n)}(\xi_{2j-1}^0 + \varepsilon_j \xi_{2j}^0),$$

then, obviously, sequence $\{c_j^{(n)}\}$ converges to some c_j, that is,

$$f = \sum_j c_j \xi_{2j-1}^0 + \sum_j \varepsilon_j c_j \xi_{2j}^0.$$

Clearly, $f \in \mathfrak{M}$ if and only if all $c_j = 0$, that is, when $f = 0$. On the other hand,

$$\cos^2 \varphi(\mathfrak{M}, \mathfrak{N}) \geq \cos^2 \varphi(\xi_{2j-1}^0, \xi_{2j-1}^0 + \varepsilon_j \xi_{2j}^0) = \frac{|(\xi_{2j-1}^0, \xi_{2j-1}^0 + \varepsilon_j \xi_{2j}^0)|^2}{\|\xi_{2j-1}^0\|^2 \|\xi_{2j-1}^0 + \varepsilon_j \xi_{2j}^0\|^2}$$

$$= (1 + \varepsilon_j^2)^{-1} \xrightarrow[j \to \infty]{} 1.$$

Lemma I.1.3. Let dim $\mathfrak{M} < \infty$ and $\varphi(\mathfrak{M}, \mathfrak{N}) = 0$. Then $\mathfrak{M}$ and $\mathfrak{N}$ have a nonzero common element.

PROOF. Since the angle between the subspaces is zero, sequences $\{m_j\}_{j\in\mathbb{N}} \subset \mathfrak{M}$ and $\{n_j\}_{j\in\mathbb{N}} \subset \mathfrak{N}$ of elements with unit norms may be found such that

$$(m_j, n_j) \xrightarrow[j \to \infty]{} 1.$$

Then $\|m_j - n_j\| \to 0$. Since the dimension of $\mathfrak{M}$ is finite, one can choose a subsequence $\{m_{j_r}\}$ converging to some $m \in \mathfrak{M}$. One has $\|m\| = 1$, and the subsequence $\{n_{j_r}\}$ also tends to m. The lemma is proved.

In the next lemma, the angles between subspaces are related with the orthogonal projectors on them.

Lemma I.1.4. The following formulas are valid:

(a) $\cos \varphi(\mathfrak{M}, \mathfrak{N}) = \|P_{\mathfrak{M}}|_{\mathfrak{N}}\| = \|P_{\mathfrak{M}} P_{\mathfrak{N}}\|$,
(b) $\sin \varphi(\mathfrak{M}^{\perp}, \mathfrak{N}) = \|[P_{\mathfrak{M}}|_{\mathfrak{N}}]^{-1}\|^{-1}$,
(c) $\sin \varphi(\mathfrak{M}, \mathfrak{N}) = \inf\limits_{m \in \mathfrak{M}, \, \|m\| = 1; \, n \in \mathfrak{N}} \|m - n\|$.

PROOF.

(a) Directly from the definition of an angle the relation follows

$$\cos \varphi(\mathfrak{M}, \mathfrak{N}) = \sup_{m \in \mathfrak{M}, n \in \mathfrak{N}, m \neq 0, n \neq 0} \frac{|(m, P_{\mathfrak{M}} n + P_{\mathfrak{M}^{\perp}} n)|}{\|m\| \, \|n\|}$$

$$= \sup_{n \in \mathfrak{N}, n \neq 0} \sup_{m \in \mathfrak{M}, m \neq 0} \frac{|(m, P_{\mathfrak{M}} n)|}{\|m\| \, \|n\|} = \sup_{n \in \mathfrak{N}, n \neq 0} \frac{\|P_{\mathfrak{M}} n\|}{\|n\|}.$$

(b) $\|[P_{\mathfrak{M}}|_{\mathfrak{N}}]^{-1}\|^{-2} = \inf\limits_{n \in \mathfrak{N}, \|n\| = 1} \|P_{\mathfrak{M}} n\|^2 = \inf\limits_{n \in \mathfrak{N}, \|n\| = 1} (1 - \|P_{\mathfrak{M}^{\perp}} n\|^2)$

$$= 1 - \sup_{\|n\| = 1} \|P_{\mathfrak{M}^{\perp}} n\|^2 = 1 - \|P_{\mathfrak{M}^{\perp}}|_{\mathfrak{N}}\|^2.$$

Now (a) implies (b).

(c) $\inf\limits_{m \in \mathfrak{M}, \|m\| = 1; \, n \in \mathfrak{N}} \|m - n\|^2 = \inf\limits_{m \in \mathfrak{M}, \|m\| = 1; \, n \in \mathfrak{N}} \|P_{\mathfrak{N}} m + P_{\mathfrak{N}^{\perp}} m - n\|^2$

$$= \inf\limits_{m \in \mathfrak{M}, \|m\| = 1} \|P_{\mathfrak{N}^{\perp}} m\|^2$$

$$= 1 - \sup_{m \in \mathfrak{M}, \|m\| = 1} \|P_{\mathfrak{N}} m\|^2.$$

Formula (a) provides (c). The lemma is proved.

1.2. Skew projectors

An algebraic sum of two linear sets $\mathcal{M}$ and $\mathcal{N}$ is said to be their direct sum $\mathcal{M} \dotplus \mathcal{N}$ if $\mathcal{M} \cap \mathcal{N} = \{0\}$ holds.

Definition I.1.5. Operator $\mathscr{P} = \mathscr{P}_{\mathcal{M}}^{\|\mathcal{N}}$ acting from the direct sum $\mathcal{M} \dotplus \mathcal{N}$ according to the rule

$$(m + n) \mapsto \mathscr{P}(m + n) = m, \qquad m \in \mathcal{M}, n \in \mathcal{N},$$

is a skew projector on $\mathcal{M}$ parallel to $\mathcal{N}$.

Obviously, the definition is reasonable and $\mathscr{P}$ is a projector, that means, $\mathscr{P}^2 = \mathscr{P}$. It so happens that any projector has the form of $\mathscr{P}_{\mathcal{M}}^{\|\mathcal{N}}$.

Lemma I.1.6. Let $\mathscr{P}$ be a projector in $\mathfrak{H}$ (i.e., $\mathscr{P}^2 = \mathscr{P}$). Then $\mathscr{P} = \mathscr{P}_{\mathcal{M}}^{\|\mathcal{N}}$, where $\mathcal{M} = \operatorname{Im} \mathscr{P}$, $\mathcal{N} = \operatorname{Ker} \mathscr{P}$.

PROOF.

(i) Let us first establish that $\mathcal{M} \cap \mathcal{N} = \{0\}$, that is, operator $\mathscr{P}_{\mathcal{M}}^{\|\mathcal{N}}$ is defined properly.

   If $l \in \mathcal{M} \cap N$, then $l = \mathscr{P}w$ for some $w \in \mathscr{D}(\mathscr{P})$ and $Pl = 0$. Then, $l = \mathscr{P}w = \mathscr{P}^2 w = \mathscr{P}(\mathscr{P}w) = \mathscr{P}l = 0$.

(ii) Since $\mathscr{P}^2 = \mathscr{P}$, $\mathcal{M} \subset \mathscr{D}(\mathscr{P})$.

(iii) Let us verify that

$$\mathscr{P}|_{\mathcal{M} \dotplus \mathcal{N}} = \mathscr{P}_{\mathcal{M}}^{\|\mathcal{N}}, \qquad \text{i.e., } \mathscr{P} \supset \mathscr{P}_{\mathcal{M}}^{\|\mathcal{N}}.$$

Actually, for $m \in \mathcal{M}$, $n \in \mathcal{N}$, we have

$$\mathscr{P}(m + n) = \mathscr{P}m + \mathscr{P}n = \mathscr{P}m.$$

Further, $\mathscr{P}(m - \mathscr{P}m) = 0$, so that $m - \mathscr{P}m \in \mathcal{N}$. It is clear that $m - \mathscr{P}m \in \mathcal{M}$; hence, according to (i), $m - \mathscr{P}m = 0$, that is, $\mathscr{P}(m + n) = m$.

Let us now take some arbitrary $l \in \mathscr{D}(\mathscr{P})$ and $m := \mathscr{P}l \in \mathcal{M}$. Then $\mathscr{P}(l - m) = \mathscr{P}l - \mathscr{P}^2 l = 0$ holds, which means $n := l - m \in \mathcal{N}$. In this way, the opposite inclusion, $\mathcal{M} \dotplus \mathcal{N} \supset \mathscr{D}(\mathscr{P})$, is obtained which completes the lemma.

By means of the representation, we establish the form of an operator adjoint to a projector. Recall that $\mathfrak{M}$ and $\mathfrak{N}$ are subspaces.

Lemma I.1.7. *The operator* $\mathscr{P}^* = [\mathscr{P}_{\mathfrak{M}}^{\|\mathfrak{N}}]^*$ *coincides with the operator*

$$\hat{\mathscr{P}} := \mathscr{P}_{\mathfrak{N}^\perp}^{\|\mathfrak{M}^\perp}.$$

PROOF. We have

$$\mathfrak{M}^\perp \cap \mathfrak{N}^\perp = \{m \in \mathfrak{M} \bigvee \mathfrak{N} \mid m \perp \mathfrak{M}, m \perp \mathfrak{N}\} = \{0\}.$$

Therefore operator $\hat{\mathscr{P}}$ is defined correctly.

(i) Let us check that

$$\hat{\mathscr{P}} \subset \mathscr{P}^*,$$

that is, for any $l = m + n \in \mathscr{D}(\mathscr{P})$ and arbitrary $w = m^\perp + n^\perp$ $(m \in \mathfrak{M}, n \in \mathfrak{N}, m^\perp \in \mathfrak{M}^\perp, n^\perp \in \mathfrak{N}^\perp)$ an equality

$$(\mathscr{P}l, w) = (l, \hat{\mathscr{P}}w)$$

holds.

Actually

$$(\mathscr{P}(m + n), w) = (m, m^\perp + n^\perp) = (m, n^\perp) = (m + n, \hat{\mathscr{P}}w).$$

(ii) To show that $\mathscr{P}^*$ is a projector, let $w \in \mathscr{D}(\mathscr{P}^*)$, that is,

$$(\mathscr{P}l, w) = (l, \mathscr{P}^*w) \qquad \forall l \in \mathscr{D}(\mathscr{P}).$$

Hence, for every $l \in \mathscr{D}(\mathscr{P})$, the equality is true:

$$(\mathscr{P}(\mathscr{P}l), w) = (\mathscr{P}l, \mathscr{P}^*w).$$

As $\mathscr{P}^2 = \mathscr{P}$,

$$(\mathscr{P}(\mathscr{P}l), w) = (l, \mathscr{P}^*w).$$

Comparing the right-hand sides of the two latter equalities, one finds that $\mathscr{P}^*w \in \mathscr{D}(\mathscr{P}^*)$ and $\mathscr{P}^*(\mathscr{P}^*w) = \mathscr{P}^*w$.

By the force of Lemma 6 for $\mathscr{M} = \operatorname{Im} \mathscr{P}^*$, $\mathscr{N} = \operatorname{Ker} \mathscr{P}^*$ one has

$$\mathscr{P}^* = \mathscr{P}_{\mathscr{M}}^{\|\mathscr{N}}.$$

If one demonstrates that $\mathscr{M} \subset \mathfrak{N}^\perp$, $\mathscr{N} \subset \mathfrak{M}^\perp$, then the assertion of our lemma will follow from (i). Let $l \in \mathscr{D}(\mathscr{P}^*)$. Since $0 = (\mathscr{P}n, l) = (n, \mathscr{P}^*l)$, for $n \in \mathfrak{N}$ one finds $\mathscr{P}^*l \perp \mathfrak{N}$, so that $\mathscr{M} = \operatorname{Im} \mathscr{P}^* \subset \mathfrak{N}^\perp$. If $\mathscr{P}^*l = 0$, then for $m \in \mathfrak{M}$ $0 = (m, \mathscr{P}^*l) = (\mathscr{P}m, l) = (m, l)$ and $\mathscr{N} = \operatorname{Ker} \mathscr{P}^*$ is orthogonal to $\mathfrak{M}$.

The lemma is proved.

We now establish a relationship between skew and orthogonal projectors.

Lemma I.1.8. *Operators* $\mathscr{P} = \mathscr{P}_{\mathfrak{M}}^{\|\mathfrak{N}}$ *and* $\tilde{\mathscr{P}} := [P_{\mathfrak{N}^{\perp}}|_{\mathfrak{M}}]^{-1}P_{\mathfrak{N}^{\perp}}$ *exist simultaneously and are equal.*

PROOF. $\tilde{\mathscr{P}}$ is defined correctly if and only if $P_{\mathfrak{N}^{\perp}}|_{\mathfrak{M}}$ is nondegenerate, which is equivalent to the equality $\mathfrak{M} \cap \mathfrak{N} = \{0\}$, since for $P_{\mathfrak{N}^{\perp}}n = 0$, $n \in (\mathfrak{N}^{\perp})^{\perp} = \mathfrak{N}$. In turn, the equality is equivalent to the existence of the skew projector $\mathscr{P}$.

We check an inclusion

$$\mathscr{P} \subset \tilde{\mathscr{P}}$$

(i.e., $\mathscr{D}(\mathscr{P}) \subset \mathscr{D}(\tilde{\mathscr{P}})$ and $\tilde{\mathscr{P}}|_{\mathscr{D}(\mathscr{P})} = \mathscr{P}$).

If $l = m + n$, $m \in \mathfrak{M}$, $n \in \mathfrak{N}$, then $P_{\mathfrak{N}^{\perp}}l = P_{\mathfrak{N}^{\perp}}m$ and $\tilde{\mathscr{P}}l = m = \mathscr{P}l$.

Let $l \subset \mathscr{D}(\tilde{\mathscr{P}})$. Since $\mathscr{D}([P_{\mathfrak{N}^{\perp}}|_{\mathfrak{M}}]^{-1}) = \mathrm{Im}(P_{\mathfrak{N}^{\perp}}|_{\mathfrak{M}}) = P_{\mathfrak{N}^{\perp}}\mathfrak{M}$, $P_{\mathfrak{N}^{\perp}}l = P_{\mathfrak{N}^{\perp}}m$ for some $m \in \mathfrak{M}$. Therefore, $P_{\mathfrak{N}^{\perp}}(l - m) = 0$ and, consequently, $l - m = n \in \mathfrak{N}$. Thus, $\mathscr{D}(\tilde{\mathscr{P}}) = \mathscr{D}(\mathscr{P})$, which completes the proof.

Lemma I.1.9. *Assume that* $\mathfrak{M} \cap \mathfrak{N} = \{0\}$. *Then:*

(a) *operator* $\mathscr{P}_{\mathfrak{M}}^{\|\mathfrak{N}}$ *is bounded if and only if* $\varphi(\mathfrak{M}, \mathfrak{N}) > 0$ *and in this case*

$$\|\mathscr{P}_{\mathfrak{M}}^{\|\mathfrak{N}}\|^{-1} = \sin \varphi(\mathfrak{M}, \mathfrak{N}); \tag{1}$$

(b) *linear set* $\mathfrak{M} \dotplus \mathfrak{N}$ *is closed if and only if* $\varphi(\mathfrak{M}, \mathfrak{N}) > 0$;

(c) $\varphi(\mathfrak{M}, \mathfrak{N}) = \varphi(\mathfrak{M}^{\perp}, \mathfrak{N}^{\perp})$.

PROOF OF THE LEMMA.

(a)

$$\|\mathscr{P}_{\mathfrak{M}}^{\|\mathfrak{N}}\|^{-1} = \inf_{m \in \mathfrak{M}, n \in \mathfrak{N}} \frac{\|m - n\|}{\|\mathscr{P}_{\mathfrak{M}}^{\|\mathfrak{N}}(m - n)\|} = \inf_{m, n} \frac{\|m - n\|}{\|m\|}.$$

From Lemma 4(c) one obtains assertion (a).

(b) If $\mathfrak{M} \dotplus \mathfrak{N}$ is a subspace, operator $\mathscr{P}_{\mathfrak{M}}^{\|\mathfrak{N}}$ is defined on a closed set. It is not difficult to show that the operator is closed; thus it is bounded according to the closed graph theorem. From (a) it follows that $\varphi(\mathfrak{M}, \mathfrak{N}) > 0$.

Conversely, if angle $\varphi(\mathfrak{M}, \mathfrak{N}) > 0$, then, by the force of (a), operator $\mathscr{P}_{\mathfrak{M}}^{\|\mathfrak{N}}$ is bounded. Therefore, the convergence of sequence $\{m_j + n_j\}$, $m_j \in \mathfrak{M}$, $n_j \in \mathfrak{N}$, implies the convergence of $\mathscr{P}_{\mathfrak{M}}^{\|\mathfrak{N}}(m_j + n_j) = m_j$. Since $\mathfrak{M}$ is a subspace, then m_j converges to some element of $\mathfrak{M}$. Then sequence $\{n_j\}$ converges to an element of $\mathfrak{N}$ as well; that is, $\mathfrak{M} \dotplus \mathfrak{N}$ is a closed set.

Assertion (c) follows from (a), Lemma 7, and the fact that the norms of an operator and its adjoint coincide. The lemma is proved.

In Chapter II, exponentials on an interval are considered projection (restrictions) of exponentials on the positive semiaxis. The properties of such an operation are studied in the next lemma, which applies to the general situation. Notice that along with operator $P_{\mathfrak{M}}|_{\mathfrak{N}}$ acting from $\mathfrak{N}$ into $\mathfrak{H}$ we consider operator $[P_{\mathfrak{M}}|_{\mathfrak{N}}: \mathfrak{N} \mapsto \mathfrak{M}]$ acting into $\mathfrak{M}$. The operator adjoint to the last one is expressed in a simpler form.

Lemma I.1.10.

(a) *Projection operator $P_{\mathfrak{M}}$ contracted to $\mathfrak{N}$ is an isomorphism onto its image if and only if*

$$\varphi(\mathfrak{M}^{\perp}, \mathfrak{N}) > 0. \qquad (2)$$

(b) *A relation is true*

$$[P_{\mathfrak{M}}|_{\mathfrak{N}}: \mathfrak{N} \mapsto \mathfrak{M}]^* = P_{\mathfrak{N}}|_{\mathfrak{M}}: \mathfrak{M} \mapsto \mathfrak{N}. \qquad (3)$$

(c) *Operator $P_{\mathfrak{M}}|_{\mathfrak{N}}: \mathfrak{N} \mapsto \mathfrak{M}$ acts as an isomorphism of subspaces $\mathfrak{N}$ and $\mathfrak{M}$ if and only if (2) holds and*

$$\varphi(\mathfrak{M}, \mathfrak{N}^{\perp}) > 0. \qquad (4)$$

PROOF. Part (a) of the lemma follows immediately from Lemma 4(b).

Assertion (b) is a direct consequence of an identity

$$(P_{\mathfrak{M}} n, m) = (n, P_{\mathfrak{N}} m), \qquad m \in \mathfrak{M}, n \in \mathfrak{N}.$$

With regard to (c), operator $P_{\mathfrak{M}}: \mathfrak{N} \mapsto \mathfrak{M}$ acts as an isomorphism of subspaces $\mathfrak{N}$ and $\mathfrak{M}$ if and only if it is an isomorphism onto its image, and the image coincides with $\mathfrak{M}$. The image of this operator is dense in $\mathfrak{M}$ only in the case where the adjoint operator is invertible. From (3) one finds that the invertibility of $[P_{\mathfrak{M}}: \mathfrak{N} \mapsto \mathfrak{M}]^*$ is equal to $\mathfrak{N}^{\perp} \cap \mathfrak{M} = \{0\}$. Hence, operator $P_{\mathfrak{M}}: \mathfrak{N} \mapsto \mathfrak{M}$ is isomorphic only when (2) and $\mathfrak{N}^{\perp} \cap \mathfrak{M} = \{0\}$ hold simultaneously. Under condition (2), subspace $\mathfrak{N}^{\perp} \bigvee \mathfrak{M}$ coincides with $\mathfrak{M} \bigvee \mathfrak{N}$ (otherwise a nonzero element of $\mathfrak{M}^{\perp} \cap \mathfrak{N}$ would be found). Therefore, from Lemma 9(c) it follows that (2) and $\mathfrak{N}^{\perp} \cap \mathfrak{M} = \{0\}$ are simultaneously correct if and only if both (2) and (4) hold. The lemma is proved.

1.3. Families of Hilbert space vectors

Let $\Xi = \{\xi_j\}_{j \in \mathbb{N}}$ be an arbitrary family of elements (vectors) ξ_j of Hilbert space $\mathfrak{H}$. In this section we define various types of "independence" for the elements of Ξ.

Definition I.1.11. Family Ξ is *W*-linearly independent if there exists no nonzero sequence $\{a_j\} \in \ell^2$ such that for any element $f \in \mathfrak{H}$ satisfying $\sum_{j=1}^{\infty} |(f, \xi_j)|^2 < \infty$ series $\sum_{j=1}^{\infty} a_j(f, \xi_j)$ converges to zero. The property of Ξ to be *W*-linearly independent is denoted as $\Xi \in (W)$.

We consider primarily the situation with $\sum_{j=1}^{\infty} |(f, \xi_j)|^2 < \infty$ for any $f \in \mathfrak{H}$. In this case, the *W*-linear independence means the uniqueness of the weak sum $\sum a_j \xi_j$:

$$\left\{ \sum_{j=1}^{\infty} |a_j|^2 < \infty, \ \sum_{j=1}^{\infty} a_j \xi_j = 0 \text{ weakly in } \mathfrak{H} \right\} \Rightarrow \{a_j = 0 \ \forall j\}.$$

In the literature (see Gokhberg and Krein 1965), one may find the following concept of ω-linear independence:

$$\left\{ \sum_{j=1}^{\infty} |a_j|^2 < \infty, \ \sum_{j=1}^{\infty} a_j \xi_j = 0 \right\} \Rightarrow \{a_j = 0 \ \forall j\}.$$

It is easily seen that the property of *W*-linear independence that has been introduced is not weaker than ω-linear independence. The example shows that *W*-linear independence is, in fact, stronger.

Example I.1.12. Assume that family $\Xi^0 = \{\xi_j^0\}_{j \in \mathbb{N}}$ forms an orthonormal basis in $\mathfrak{H}$. Let us settle

$$\xi_1 := \xi_1^0; \qquad \xi_j := j(\xi_j^0 - \xi_{j-1}^0), \quad j = 2, 3, \dots$$

and demonstrate that the family $\Xi = \{\xi_j\}_{j \in \mathbb{N}}$ is ω-linearly independent but not *W*-linearly independent.

Suppose that series $\sum_{j=1}^{\infty} c_j \xi_j$ converges to zero. Then multiplying it by ξ_j^0, one arrives at

$$(j+1)c_{j+1} = jc_j = \text{const.}$$

Hence, the partial sums of the series may be written as

$$\sum_{j=1}^{R} c_j \xi_j = c_1[\xi_1^0 + (\xi_2^0 - \xi_1^0) + \cdots + (\xi_R^0 - \xi_{R-1}^0)] = c_1 \xi_R^0. \tag{5}$$

So, series $\sum_1^{\infty} c_j \xi_j$ converges to zero if and only if $c_1 = c_2 = \cdots = 0$, that is, Ξ is ω-linearly independent.

Series $\sum_1^{\infty} (\xi_j/j)$ converges weakly to zero, since $\sum_{j=1}^{R} (\xi_j/j) = \xi_R^0$. Thus $\Xi \notin (W)$.

Definition I.1.13. Family Ξ is minimal (notation $\Xi \in (M)$), if for any j element ξ_j does not belong to the closure of all the remaining elements:

$$\xi_j \notin \bigvee_{i \neq j} \xi_i =: \Xi^{(j)}. \tag{6}$$

If $\Xi \in (M)$, there exists family $\Xi' = \{\xi_j'\}$, which is said to be biorthogonal to Ξ, such that

$$(\xi_j, \xi_i') = \delta_j^i = \begin{cases} 1 & i = j \\ 0 & i \neq j. \end{cases} \tag{7}$$

The biorthogonal family is specified by formula (7) up to any elements from the orthogonal complement to $\bigvee \Xi$. Condition $\xi_j' \in \bigvee \Xi$ extracts biorthogonal elements with minimum norm.

According to Lemma 3, condition (6) is equivalent to the property of angles $\varphi(\xi_j, \Xi^{(j)})$, $j = 1, 2, \ldots$, to be positive. That is why in the subspace $\bigvee \Xi$ there exist bounded operators

$$\mathscr{P}_j := \mathscr{P}_{\xi_j}^{\|\Xi^{(j)}}$$

and, as is easily verified,

$$\xi_j' = \frac{\mathscr{P}_j^* \xi_j}{\|\xi_j\|^2} \tag{8}$$

and

$$\mathscr{P}_j = (\cdot, \xi_j')\xi_j. \tag{9}$$

It follows that

$$\|\mathscr{P}_j\| = \frac{\|\mathscr{P}_j \xi_j'\|}{\|\xi_j'\|} = \|\xi_j'\|\,\|\xi_j\|.$$

Hence for the biorthogonal family elements from Lemma 9(a) an equality follows

$$\|\xi_j'\|\,\|\xi_j\| = [\sin \varphi(\xi_j, \Xi^{(j)})]^{-1} \geq 1. \tag{10}$$

Example I.1.14. Let $\Xi = \{t^j\}_{j=0}^{\infty}$, $\mathfrak{H} = L^2(0, 1)$. Then $\Xi \in (W)$, but $\Xi \notin (M)$.

The family of polynomials $P_n(t)$, $n \geq 0$, is dense in $L^2(0, 1)$, and the same is true for polynomial family $t^R\{P_n(t)\}$, where $R \in \mathbb{N}$ is fixed. Therefore, the family $\{t^R, t^{R+1}, \ldots\}$ is complete and $t \in \bigvee_{j \neq k} t^j$; that is, $\Xi \notin (M)$ (details may be found in Sadovnichiĭ 1979). Let us prove the inclusion

$\Xi \in (W)$. Let f_s be the characteristic function of segment $[0, s]$. Then

$$(f_s, t^j)_{L^2(0,1)} = \int_0^s t^j \, dt = \frac{s^{j+1}}{j+1}. \tag{11}$$

and

$$\sum_{j=0}^{\infty} |(f_s, t^j)|^2 < \infty.$$

Suppose that $\Xi \notin (W)$. Then function $F(s) = (f_s, \sum_{j=0}^{\infty} c_j t^j)$ turns identically into zero for some nontrivial sequence $\{c_j\} \in \ell^2$. By the force of (11), $F(s)$ takes the form

$$F(s) = \sum_{j=0}^{\infty} c_j \frac{s^{j+1}}{j+1}$$

and therefore is analytical inside the unit disc. So $c_j = 0$ for every j.

Definition I.1.15. Family Ξ is *-uniform minimal (notation $\Xi \in (UM)$), if there exists a biorthogonal family Ξ' with uniformly bounded norms of the elements.

Recall that under a uniformly minimal family (without *), such a minimal one is meant, for which the numbers $\|\xi_j\| \cdot \|\xi_j'\|$ are uniformly bounded, or for which $\varphi(\xi_j, \Xi^{(j)}) \geq c > 0$. The above definition is more convenient for our purposes. For almost normed families (i.e., under the condition $\|\xi_j\| \asymp 1$), the definitions coincide.

Definition I.1.16. Family Ξ is said to be an $\mathscr{L}$-basis or a Riesz basis in the closure of its linear span (notation $\Xi \in (LB)$), if Ξ is an image of an isomorphic mapping V of some orthonormal family.
   Ξ is said to be a Riesz basis, if $\Xi \in (LB)$ and Ξ is a complete family: $\bigvee \Xi = \mathfrak{H}$.

Operator V^{-1} is said to be an orthogonalizer of the family Ξ. In what follows, we always use the term "basis" to mean Riesz basis.
   The introduced properties of the families are interrelated in such a way that every property is implied by the next one:

$$(LB) \Rightarrow (UM) \Rightarrow (M) \Rightarrow (W) \Rightarrow (\omega\text{-linear independence})$$

$$\Rightarrow (\text{linear independence}).$$

Proposition I.1.17 (Bari theorem, Bari 1951; Gokhberg and Krein 1965; Nikol'skiĭ 1980).

(a) *If* $\Xi = \{\xi_j\}_{j=1}^{\infty} \in (LB)$, *then the biorthogonal to* Ξ *family* $\Xi' = \{\xi_j'\}_{j=1}^{\infty}$ *such that* $\Xi' \subset \bigvee \Xi$ *is also an* $\mathscr{L}$*-basis. In addition,* Ξ' *is expressed via the orthogonalizer* V^{-1} *of family* Ξ *and the orthonormal basis* $\Xi^0 = \{\xi_j^0\}_{j=1}^{\infty}$, $\xi_j^0 = V^{-1}\xi_j$, *of the space* $V^{-1}(\bigvee\Xi)$ *by the formula* $\xi_j' = (V^{-1})^* V^{-1}\xi_j$.

(b) $\Xi \in (LB)$ *if and only if for any finite sequence* $\{c_j\}$ *an estimate holds*

$$\left\| \sum_{j=1}^{\infty} c_j \xi_j \right\|^2 \asymp \sum_{j=1}^{\infty} |c_j|^2.$$

In this case, each element $f \in \bigvee\Xi$ has a series expansion

$$f = \sum_{j=1}^{\infty} (f, \xi_j')\xi_j.$$

Example I.1.18. Let $\Xi^0 = \{\xi_j^0\}_{j=0}^{\infty}$ be an orthonormal basis in $\mathfrak{H}$ and $\Xi = \{\xi_j\}_{j=1}^{\infty}$, $\xi_j := \xi_0^0 + \xi_j^0$. Then

(a) Ξ is a complete family.
(b) $\Xi \in (UM)$.
(c) $\Xi \notin (LB)$, but its biorthogonal family Ξ' is an $\mathscr{L}$-basis.
(d) Projection of element ξ_0^0 on any element ξ_j of the (complete) family parallel to the closure of the linear span of all the remaining elements of the family equals zero.

Let us check these assertions.

(a) If $f \perp \bigvee\Xi$ and $f = \sum_{j=0}^{\infty} c_j \xi_j^0$, then by means of scalar multiplication by ξ_j with $j > 0$, one finds $c_0 + c_j = 0$. So $c_j = -c_0$, $j \in \mathbb{N}$, and for $c_0 \neq 0$, $\{c_j\} \notin \ell^2$.
(b), (c) It is easy to see that $\Xi' = \{\xi_j^0\}_{j=1}^{\infty}$. If one takes a sequence $\{c_j\} \in \ell^2 \backslash \ell^1$, then the series $\sum_1^{\infty} c_j \xi_j$ diverges in $\mathfrak{H}$ since

$$\sum_1^N c_j \xi_j = \sum_1^N c_j \xi_j^0 + \left(\sum_1^N c_j \right) \xi_0^0$$

and the Bari theorem (Proposition 17(b)) provides $\Xi \notin (LB)$.
(d) From (9) it follows that the skew projector has the form

$$\mathscr{P}_j = (\cdot, \xi_j^0)\xi_j, \qquad j > 1,$$

which produces an equality $\mathscr{P}_j \xi_0^0 = 0$ for all $j \in \mathbb{N}$.

We now introduce a subject convenient for the description of properties of elements ξ_j belonging to some family $\Xi = \{\xi_j\}_{j \in \mathcal{N}}$, where $\mathcal{N}$ is a subset of $\mathbb{N}$.

Definition I.1.19. A matrix with elements

$$\Gamma_{ij} = (\xi_i, \xi_j); \qquad i, j \in \mathcal{N} \subset \mathbb{N}$$

is called the Gram matrix of the family $\Xi = \{\xi_j\}_{j \in \mathcal{N}}$.

For a countable family Ξ (i.e., for $\mathcal{N} = \mathbb{N}$) the Gram matrix is infinite, whereas for a finite one ($\mathcal{N} = \{1, 2, \ldots, N\}$) it has N^2 entries. A family of elements is determined by the Gram matrix up to unitary equivalence. The bilinear form of the Gram matrix reads

$$(c, \Gamma d) = (\Gamma c, d) = \left(\sum_{j=1}^{\infty} c_j \xi_j, \sum_{j=1}^{\infty} d_j \xi_j \right)_{\mathfrak{H}} \tag{12}$$

(for the case $\mathcal{N} = \mathbb{N}$). This form is defined over the elements $c, d \in \ell^2$ such that the series $\sum c_j \xi_j$ and $\sum d_j \xi_j$ converge.

Theorem I.1.20. The following assertions are valid.

(a) *The square form of the Gram matrix is positive if and only if the family Ξ is ω-linearly independent.*
(b) *The bilinear form of the Gram matrix generates a bounded and boundedly invertible operator Γ if and only if $\Xi \in (LB)$.*
(c) *If $\Xi \in (LB)$, then the Gram matrix for family Ξ' is the inverse matrix to the Gram matrix for Ξ:*

$$(\Gamma^{-1})_{ij} = (\xi_i', \xi_j'). \tag{13}$$

PROOF. Assertion (a) follows from the equality

$$(\Gamma c, c)_{\ell^2} = \left\| \sum_j c_j \xi_j \right\|^2$$

and the definition of ω-linear independence.

Assertion (b) follows from (12) and the Bari theorem (Proposition 17(b)). For (c), set $\Gamma^{-1} \zeta_j =: x^j = \{x_n^j\}_{n=1}^{\infty}$, where $\{\zeta_j\}_{j=1}^{\infty}$ is the standard basis in ℓ^2. Then

$$(\Gamma x^j, \zeta_i)_{\ell^2} = (\zeta_j, \zeta_i)_{\ell^2} = \delta_j^i.$$

On the other hand, (12) implies

$$(\Gamma x^j, \zeta_i)_{\ell^2} = \left(\sum_n x_n^j \xi_n, \xi_i \right)_{\mathfrak{H}},$$

and we obtain

$$\xi_j' = \sum_n x_n^j \xi_n.$$

Moreover,

$$(\xi_j', \xi_i')_{\mathfrak{H}} = \left(\sum_n x_n^j \xi_n, \sum_n x_n^i \xi_n \right)_{\mathfrak{H}} = (\Gamma x^j, x^i)_{\ell^2} = (\zeta_j, x^i)$$

$$= (\zeta_j, \Gamma^{-1}\zeta_i) = (\overline{\Gamma^{-1}})_{ji} = (\Gamma^{-1})_{ij}.$$

The theorem is proved.

Relation (13) is valid for a minimal family Ξ if $\{\zeta_j\}$ belongs to the image of operator Γ. However, it is possible for the family Ξ to be minimal while the basis vectors ζ_j do not lie in the domain of the operator Γ^{-1}. As can be seen from (13), it happens when

$$\sum_i |(\xi_i', \xi_j')|^2 = \infty.$$

We now prove the lemma on the conservation of the W-linear independence property for the families whose elements differ by scalar factors.

Lemma I.1.21. Suppose that the families $\Xi = \{\xi_j\}_{j=1}^\infty$ and $\tilde{\Xi} = \{\tilde{\xi}_j\}_{j=1}^\infty$ are related by the equalities $\tilde{\xi}_j = \alpha_j \xi_j$, where $|\alpha_j| < 1$ and $\alpha_j \neq 0$ for each j. If $\Xi \in (W)$, then also $\tilde{\Xi} \in (W)$.

PROOF. Assume $\{a_j\} \in \ell^2$ and $\sum_{j=1}^\infty a_j(f, \tilde{\xi}_j) = 0$ for any element f satisfying condition

$$\sum_{j=1} |(f, \tilde{\xi}_j)|^2 < \infty.$$

For family $\{\xi_j\}$ it means that

$$\sum_{j=1}^\infty a_j \bar{\alpha}_j (f, \xi_j) = 0 \tag{14}$$

for any f such that

$$\sum_{j=1}^\infty |\alpha_j|^2 |(f, \xi_j)|^2 < \infty.$$

Particularly (14) is true for all f such that $\sum_j |(f, \xi_j)|^2 < \infty$. Therefore, using property $\Xi \in (W)$, we obtain $a_j \bar{\alpha}_j = 0$ and, consequently, $a_j = 0$ for all j. The lemma is proved.

1.4. Families of subspaces of Hilbert space

Let Ξ_j, $j \in \mathbb{N}$ be subspaces of Hilbert space $\mathfrak{H}$ and $\Xi = \{\Xi_j\}_{j \in \mathbb{N}}$. Define for the family Ξ properties similar to those of families of elements.

Definition I.1.22. Family Σ is minimal (notation $\Xi \in (M)$), if for any j there exists a bounded skew projector $\mathscr{P}_j$ on the subspace Ξ_j parallel to $\Xi^{(j)} := \bigvee_{i \neq j} \Xi_i$. Family Ξ' of subspaces $\Xi'_j := \mathscr{P}_j^* \Xi_j$ is said to be biorthogonal to Ξ.

Family Ξ is uniformly minimal (notation $\Xi \in (UM)$) if $\|\mathscr{P}_j\| \prec 1$.

Family Ξ is said to be a Riesz basis in the closure of its linear span (an $\mathscr{L}$-basis, notation $\Xi \in (LB)$), if there exists an isomorphism V such that the family of subspaces $\{V^{-1}\Xi_j\}$ is orthogonal: $V^{-1}\Xi_j \perp V^{-1}\Xi_i$, $i \neq j$. Operator V^{-1} is called an orthogonalizer of the family Ξ.

Family Ξ is a basis, if $\Xi \in (LB)$ and Ξ is complete: $\bigvee \Xi = \mathfrak{H}$.

Remark I.1.23. Sometimes it is convenient to abandon the restriction that the biorthogonal to Ξ family must lie in $\bigvee \Xi$. Therefore, we call any family of subspaces $\{\Xi'_j\}$ such that

$$P_{\bigvee \Xi} \Xi'_j = \mathscr{P}_j^* \Xi_j, \qquad \mathscr{P}_j = \mathscr{P}_{\Xi_j}^{\|\Xi^{(j)}} : \bigvee \Xi \mapsto \Xi_j,$$

biorthogonal to Ξ.

Definition 22 is consistent with the corresponding definitions for families of elements in the sense that when choosing in every subspace Ξ_j an orthonormal basis $\{\xi_{j,n}\}_{n=1}^{\dim \Xi_j}$, one gets a family of *elements* $\{\xi_{j,n}\}_{j,n}$ with the similar property.

We now give an analogy of the Bari theorem (Proposition 17) for a family of subspaces.

Proposition I.1.24 (Nikol'skiĭ 1980: lecture VI, sec. 4). The following conditions are equivalent:

(a) $\Xi \in (LB)$,

(b) *in the subspace $\bigvee \Xi$ the norms $\| \cdot \|_{\mathfrak{H}}$ and $[\sum_{j=1}^{\infty} \|\mathscr{P}_j \cdot \|^2]^{1/2}$ are equivalent:*

$$\|m\|^2 \asymp \sum_{j=1}^{\infty} \|\mathscr{P}_j m\|^2, \qquad m \in \bigvee \Xi,$$

(c) $\varepsilon > 0$ *may be found, such that for any partition of the family* Ξ *in two subfamilies* Ξ_I *and* Ξ_{II} $(\Xi_I \cup \Xi_{II} = \Xi, \Sigma_I \cap \Xi_{II} = \varnothing)$ *we have*

$$\varphi(\bigvee\Xi_I, \bigvee\Xi_{II}) > \varepsilon.$$

Note that its turn condition (c) is equivalent to the uniform boundedness of all the projectors

$$\mathscr{P}^{\|\bigvee\Xi_{II}}_{\bigvee\Xi_I},$$

and in such a case one easily finds that

$$\mathscr{P}^{\|\bigvee\Xi_{II}}_{\bigvee\Xi_I} = \sum_{j:\, \Xi_j \in \Xi_I} \mathscr{P}_j.$$

1.5. Invariance of family properties under mappings

Theorem I.1.25. Let $\mathfrak{U}$ *be an isomorphism in* $\mathfrak{H}$ *and* $\Xi = \{\Xi_j\}$ *be a family of subspaces. Then* Ξ *has any of the following properties: (a) minimality, (b) uniform minimality, (c) an* $\mathscr{L}$-*basis, and (d) a basis, if and only if a family* $\mathfrak{U}\Xi = \{\mathfrak{U}\Xi_j\}$ *has the same property.*

PROOF. For properties (d) and (c) the results are obvious. Indeed, if V^{-1} is an orthogonalizer for the family Ξ, then operator $\tilde{V} = V^{-1}\mathfrak{U}^{-1}$ is an orthogonalizer for $\mathfrak{U}\Xi$, and vice versa.

To prove conservation of minimality and uniform minimality properties, we need the following proposition.

Proposition I.1.26 (Pavlov 1971). *If* $\mathfrak{M}$ *and* $\mathfrak{N}$ *are subspaces in* $\mathfrak{H}$ *and* $\mathfrak{U}$ *is an isomorphism of the spaces* $\mathfrak{H}$, *then*

$$\sin \varphi(\mathfrak{U}\mathfrak{M}, \mathfrak{U}\mathfrak{N}) \geq \|\mathfrak{U}^{-1}\|^{-1}\|\mathfrak{U}\|^{-1} \sin \varphi(\mathfrak{M}, \mathfrak{N}).$$

PROOF. The formula is implied by inequalities

$$\left\|\frac{\mathfrak{U}m}{\|\mathfrak{U}m\|} - \mathfrak{U}n\right\| \geq \frac{1}{\|\mathfrak{U}^{-1}\|}\left\|\frac{m}{\|\mathfrak{U}m\|} - n\right\| \geq \frac{\|\mathfrak{U}\|^{-1}}{\|\mathfrak{U}^{-1}\|}\left\|\frac{m}{\|m\|} - \frac{\|\mathfrak{U}m\|}{\|m\|}n\right\|$$

and Lemma 4(c).

Assertions (a) and (b) of our theorem now follow from the definitions and Proposition 26.

Remark I.1.27. A similar statement is true not only for families of subspaces, but also for families of elements. For the latter, it is not difficult to demonstrate that the property of W-linear independence also survives under an isomorphism.

1.6. Minimality of families of elements and their projections

Let $\Xi = \{\xi_j\}_{j=1}^{\infty}$ be a family of elements in Hilbert space $\mathfrak{H}$ and $\mathfrak{M} := \bigvee \Xi$. Assume $\mathfrak{N}$ is a subspace of $\mathfrak{H}$ and $P_{\mathfrak{N}}\Xi$ is a family of projections on $\mathfrak{N}$ of the elements from Ξ. Let $\Xi' = \{\xi_j'\}_{j=1}^{\infty}$ be a family biorthogonal to Ξ, and $\Xi'_{\mathfrak{N}} = \{\zeta_{\mathfrak{N},j}'\}_{j=1}^{\infty}$ be a family biorthogonal to $P_{\mathfrak{N}}\Xi$ (we suppose that families Ξ and $P_{\mathfrak{N}}\Xi$ are both minimal). Let us assume that biorthogonal families Ξ' and $\Xi'_{\mathfrak{N}}$ lie in the subspace $\mathfrak{M}$ and $\mathfrak{N}$, respectively. In this section, we present a number of results on the properties of $P_{\mathfrak{N}}\Xi$, which are needed to study exponential families in the space L^2 over an interval: the exponentials over an interval will be considered projections of exponentials over a semiaxis.

Lemma I.1.28. Let a family of elements $P_{\mathfrak{N}}\Xi$ be minimal ($P_{\mathfrak{N}}\Xi$ is *-uniformly minimal, $P_{\mathfrak{N}}\Xi \in (UM)$). Then Ξ is also minimal (Ξ is *-uniformly minimal, $\Xi \in (UM)$), and for the elements of biorthogonal families Ξ' and $\Xi'_{\mathfrak{N}}$ a relation holds:

$$P_{\mathfrak{M}}\zeta_{\mathfrak{N},j}' = \xi_j'. \tag{15}$$

PROOF. The first assertion of the lemma follows immediately from formula (15). For the latter we have

$$\delta_j^i = (\zeta_{\mathfrak{N},j}', P_{\mathfrak{N}}\xi_i) = (\zeta_{\mathfrak{N},j}', \xi_i) = (P_{\mathfrak{M}}\zeta_{\mathfrak{N},j}', \xi_i)$$

where δ_j^i is Kronecker's delta. The lemma is proved.

Remark I.1.29. If $P_{\mathfrak{N}}\Xi$ is a uniformly minimal family of *subspaces* or $P_{\mathfrak{N}}\Xi$ is a uniformly minimal family of elements (*without* *), then the family Ξ may not be a uniformly minimal one, as an example shows. Assume $\{\xi_j^0\}_{j=0}^{\infty}$ to be an orthonormal basis in $\mathfrak{H}$, and $\mathfrak{N}$ to be a subspace of all the elements orthogonal to ξ_0^0. Set $\xi_j := \alpha_j \xi_0^0 + \xi_j^0$, $j > 0$, where $\alpha_j \to \infty$ when $j \to \infty$. Introduce a family Ξ of one-dimensional subspaces spanned over the elements ξ_j. Since $P_{\mathfrak{N}}\Xi$ is an orthogonal family of subspaces spanned on ξ_j^0, $P_{\mathfrak{N}}\Xi \in (UM)$. But the angle between the subspaces Ξ_j and Ξ_{j+1} tends to zero with $j \to \infty$. Such a difference in properties of the families of elements (Lemma 28) and the families of subspaces is associated with the fact that our definition of *-uniform minimality depends on the normalization of the elements.

A biorthogonal family Ξ' may be incomplete in $\mathfrak{M} = \bigvee \Xi$ (see Example 18). Let us give a criterion for a biorthogonal family to be complete in $\mathfrak{M}$.

In order to do this, consider in Hilbert space $\mathfrak{H}$ an operator V defined on the span of an orthonormalized basis $\Xi^0 = \{\xi_j^0\}_{j=1}^\infty$ and set $\Xi = \{\xi_j\}_{j=1}^\infty$, $\xi_j := V\xi_j^0$.

Lemma I.1.30. If a family Ξ is both complete and minimal, then the biorthogonal family $\Xi' = \{\xi_j'\}_{j=1}^\infty$ is complete in the space $\mathfrak{H}$ if and only if operator V allows a closure.

PROOF. The condition for Ξ to be minimal implies $\xi_j' \in \mathscr{D}(V^*)$ and $V^*\xi_n' = \xi_j^0$. Indeed, for $x = \sum_{j=1}^M c_j\xi_j^0$ one has $(Vx, \xi_j') = c_j = (x, \xi_j^0)$. The assertion of the lemma now follows from the fact that operator V^* is densely defined if and only if V allows a closure (Birman and Solomyak 1980; chap. 3, sec. 3, th. 7).

Theorem I.1.31. Suppose that families Ξ and $P_{\mathfrak{N}}\Xi$ are minimal, family $P_{\mathfrak{N}}\Xi$ is complete in $\mathfrak{N}$, and family Ξ' is complete in $\mathfrak{M}$:

$$\bigvee \Xi' = \bigvee \Xi =: \mathfrak{M}. \tag{16}$$

Then

(a) operator $P_{\mathfrak{M}}|_{\mathfrak{N}}$ is invertible and has a dense image;

(b) if a family Ξ is a unification of a finite number of $\mathscr{L}$-basis subspaces, $\Xi = \bigcup_{j=1}^N \Xi_j$, $\Xi_j \in (LB)$, then the family biorthogonal to $P_{\mathfrak{N}}\Xi$ is complete in the subspace $\mathfrak{N}$.

PROOF.

(a) Operator $P_{\mathfrak{N}}|_{\mathfrak{M}}: \mathfrak{M} \mapsto \mathfrak{N}$ adjoint to $P_{\mathfrak{M}}|_{\mathfrak{N}}: \mathfrak{N} \mapsto \mathfrak{M}$ has a dense image since subspace $\mathfrak{N}$, according to the condition, is a closure of the linear span of the family $P_{\mathfrak{N}}\Xi$. Therefore operator $P_{\mathfrak{M}}|_{\mathfrak{N}}$ is invertible. From (15) and condition (16) it follows that operator $P_{\mathfrak{M}}|_{\mathfrak{N}}$ has a dense image

$$\bigvee P_{\mathfrak{M}}\xi_{\mathfrak{N},j}' = \bigvee \Xi' = \mathfrak{M}.$$

(b) Denote $\{\xi_j^0\}$ an orthonormal basis in the subspace $\mathfrak{M}$ and, just as in Lemma 30, set $V\xi_j^0 = \xi_j$. Let us check that under the condition (b) of the Theorem 31 operator V is bounded.

Represent family Ξ_j, $j = 1, \ldots, N$, in the form

$$\Xi_j = \{\xi_n\}_n \in \mathcal{N}_j, \qquad \mathcal{N}_j \subset \mathbb{N}.$$

Let V_j^{-1} be an orthogonalizer of the family Ξ_j:

$$V_j^{-1}\xi_n = \xi_n^0, \qquad n \in \mathcal{N}_j.$$

Since $\Xi_j \in (LB)$, V_j is an isomorphism of the spaces $\bigvee\{\xi_n^0\}_{n \in \mathcal{N}_j}$ and $\bigvee\{\xi_n\}_{n \in \mathcal{N}_j}$. Hence operator $V = \sum_{j=1}^{N} V_j$ is bounded. The boundedness of V implies the boundedness of the operator $P_{\mathfrak{N}} V$ transferring orthonormal basis $\{\xi_j^0\}$ into family $P_{\mathfrak{N}} \Xi$.

Assertion (b) now follows from Lemma 30.

2. Abstract problem of moments

We now move to the investigation of the abstract problem of moments, that is, the problem of moments in relation to an arbitrary family of elements of Hilbert space. In this investigation we explain which properties of the family influence the solvability of the problem of moments, and in which way.

Let $\Xi = \{\xi_j\}_{j=1}^{\infty}$ be an arbitrary family of elements of separable Hilbert space $\mathfrak{H}$. We introduce the operator of the problem of moments $\mathscr{J}_{\Xi} : \mathfrak{H} \to \ell^2$ according to the rule

$$\mathscr{J}_{\Xi} f = \{(f, \xi_j)_{\mathfrak{H}}\}_{j=1}^{\infty}.$$

The domain Dom $\mathscr{J}_{\Xi}$ of the operator $\mathscr{J}_{\Xi}$, that is, the set $\{f \in \mathfrak{H} \mid \mathscr{J}_{\Xi} f \in \ell^2\}$, we denote by D_{Ξ}. and its image Im $\mathscr{J}_{\Xi}$ by R_{Ξ}. We recall

$$c_j = (f, \xi_j), \qquad j \in \mathbb{N},$$

the moment equalities, and we refer to the problem of determination of f by $\mathscr{J}_{\Xi} f$ as the problem of moments.

One can easily see that operator $\mathscr{J}_{\Xi}$ is closed. Indeed, if $f_n \to 0$ and $\mathscr{J}_{\Xi} f_n \to \ell \in \ell^2$ for $n \to \infty$, then $l = 0$, since any component (f_n, ξ_j) of the element $\mathscr{J}_{\Xi} f_n$ tends to zero.

If family Ξ is not complete in $\mathfrak{H}$, which means that the span closure $\bigvee \Xi$ does not coincide with $\mathfrak{H}$, then operator $\mathscr{J}_{\Xi}$ has a nontrivial null-space $\mathfrak{H} \ominus \bigvee \Xi$. Let $\mathscr{J}_{\Xi}^0$ be a restriction of $\mathscr{J}_{\Xi}$ on $\bigvee \Xi$: $\mathscr{J}_{\Xi}^0 = \mathscr{J}_{\Xi}|_{\bigvee \Xi}$. Operator $\mathscr{J}_{\Xi}^0$ is nondegenerate and, evidently, the images of $\mathscr{J}_{\Xi}$ and $\mathscr{J}_{\Xi}^0$ coincide. Therefore, in order to study the solvability of the problem of moments it is sufficient to consider the invertible operator $\mathscr{I}_\Xi^0$. The operator $(\mathscr{I}_\Xi^0)^{-1}$ inverse to the closed one is also known to be closed.

Theorem I.2.1. The following assertions are valid:

(a) *Operator $\mathscr{I}_\Xi^0$ is an isomorphism between $\bigvee \Xi$ and ℓ^2 if and only if $\Xi \in (LB)$. In particular, $\Xi \in (LB) \Rightarrow R_\Xi = \ell^2$.*

(b) *$R_\Xi = \ell^2 \Rightarrow \Xi \in (UM)$.*

(c) *$\Xi \in (M) \Leftrightarrow \ell^2$ basis vectors belong to R_Ξ.*

(d) *Cl $R_\Xi = \ell^2 \Leftrightarrow \Xi \in (W)$.*

(e) *Assume family Ξ to be represented as a unification of two families Ξ_0 and Ξ_1 such that $\Xi_0 \in (LB)$ and $\Xi_1 \subset \bigvee \Xi_0$. Then the set R_Ξ is a (closed) subspace of ℓ^2 and codimension of Ξ is equal to the number of elements of Ξ_1 (the dimension of the space orthogonal to all the elements of family Ξ is called the codimension of Ξ).*

PROOF. Let $\{\zeta_j\}$ be the standard basis in ℓ^2.

(a) If $\mathscr{I}_\Xi^0$ is an isomorphism, then family $\{(\mathscr{I}_\Xi^0)^{-1}\zeta_j\}_{j \in \mathbb{N}}$ is a Riesz basis in $\bigvee \Xi$ by definition. Let us show that the family is the biorthogonal to Ξ family:

$$\{((\mathscr{I}_\Xi^0)^{-1}\zeta_j, \xi_i)_{\mathfrak{H}}\}_{i \in \mathbb{N}} = \mathscr{I}_\Xi(\mathscr{I}_\Xi^0)^{-1}\zeta_j = \mathscr{I}_\Xi^0(\mathscr{I}_\Xi^0)^{-1}\zeta_j = \zeta_j.$$

So by the force of Proposition 1.17(a), Ξ is a Riesz basis (in $\bigvee \Xi$).

If Ξ is an $\mathscr{L}$-basis, then $\mathscr{I}_\Xi^0$ is an isomorphism according to Proposition 1.17(b).

(b) Since $R_\Xi = \ell^2$, the elements $\xi'_j := (\mathscr{I}_\Xi^0)^{-1}\zeta_j$ exist and in such a case, as demonstrated earlier, form a biorthogonal to the Ξ family. By the closed graph theorem, operator $(\mathscr{I}_\Xi^0)^{-1}$ is bounded. Therefore,

$$\sup_j \|\xi'_j\|_{\mathfrak{H}} \leq \|(\mathscr{I}_\Xi^0)^{-1}\| \sup_j \|\zeta_j\|_{\ell^2} = \|(\mathscr{I}_\Xi^0)^{-1}\| < \infty.$$

(c) The assertion follows from equality (7), Section I.1

$$(\xi'_i, \xi_j) = \delta^i_j.$$

(d) Let $\Xi \notin (W)$; i.e., a nonzero vector $a = \{a_n\} \in \ell^2$ may be chosen such that for any $f \in D_\Xi$

$$\left(f, \sum_{j=1}^N a_n \xi_n \right) \xrightarrow[N \to \infty]{} 0.$$

Then $a \perp R_\Xi$: indeed, for every $f \in D_\Xi$

$$(\mathscr{J}_\Xi f, a) = \lim_{N \to \infty} \sum_{j=1}^{N} (\mathscr{J}_\Xi f)_j \bar{a}_j = \lim_{N \to \infty} \sum_{j=1}^{N} (f, \xi_j) \bar{a}_j$$

$$= \lim_{N \to \infty} \left(f, \sum_{j=1}^{N} a_j \xi_j \right) = 0.$$

These arguments may be easily inverted.

(e) Set $\Xi_0 = \{\xi_{j,0}\}_{j=1}^{\infty} \in (LB)$ and $\Xi_1 = \{\xi_{j,1}\}_{j=1}^{\omega}$, $\omega \leq \infty$. Denote $\{\xi_j'\}_{j=1}^{\infty}$ the family biorthogonal to Ξ_0, which belongs to $\bigvee \Xi_0$. Expand the elements of family Ξ_1 over the basis Ξ_0,

$$\xi_{j,1} = \sum_{j=1}^{\infty} (\xi_{j,1}, \xi_j') \xi_{n,0},$$

and denote by b_j sequence $\{(\xi_{j,1}, \xi_j')\}_{n=1}^{\infty}$. Since $\{\xi_j'\}$ is an $\mathscr{L}$-basis simultaneously with $\{\xi_{n,0}\}$, $b_j \in \ell^2$.

For an arbitrary element $f \in D_\Xi$, its image $\mathscr{J}_\Xi f$ lies in $\ell^2 \oplus \ell$, where $\ell = \ell^2$ for $\omega = \infty$ and $\ell = \mathbb{C}^N$ for $\omega = N < \infty$ (the first component corresponds to family Ξ_0 and the second – to Ξ_1). At the same time, if f has the form $f = \sum c_j \xi_j'$ then $\mathscr{J}_\Xi f$ has the form of the ordered pair $\{c, Bc\}$ with $c = \{c_j\} \in \ell^2$, while operator $B \colon \ell^2 \to \ell$ acts according to the formula

$$Bc = \{(c, b_j)_{\ell^2}\}_{j=1}^{\omega}.$$

From this it follows that set R_Ξ consists of all the ordered pairs $\{c, Bc\}$, $c \in \ell^2$. That is why the closure of R_Ξ is equivalent to the fact that B is closed. But operator B coincides with the operator of the problem of moments for the family $\{b_j\}_{j=1}^{\omega}$ and, hence, is closed.

It remains only to prove formula $\operatorname{codim} R_\Xi = \operatorname{card} \Xi_1$. Let us establish that every pair $\kappa_j = (b_j, -\zeta_j)$ is orthogonal to set R_Ξ. And in reality, the scalar product of the pair κ_j and any element (c, Bc) of set R_Ξ is

$$(b_j, c)_{\ell^2} - \overline{(Bc)}_j = (b_j, c) - \overline{(c, b_j)} = 0.$$

Since the set of pairs $\{\kappa_j\}_{j=1}^{\omega}$ is linearly independent, assertion 5, and with it our theorem, are proved.

Example I.2.2. For the operator of the problem of moments concerning the family Ξ involved in Example 1.18, $D_\Xi = \{f \mid f \perp \xi_0\}$. Thus, we have an example of a non–$\mathscr{L}$-basis family with an always solvable problem of moments.

Another example of this kind gives any orthogonal family with element norms approaching infinity.

An example of the nonminimal family for which the problem of moments is solvable over a dense set, serves, by the force of assertions (d) and (e) of the theorem, any nonminimal but W-linearly independent family, say, the one from example 1.14.

It is not difficult to see that for a minimal family Ξ a formal solution to the problem of moments $\mathcal{J}_\Xi f = \{a_j\}$ reads $f = \sum a_j \xi_j'$. In the following theorem we ascertain when an informal meaning may be attached to this series.

Theorem I.2.3.

(a) *Suppose* $\Xi \in (M)$ *and* $a = \{a_j\} \in \ell^2$. *Then inclusion* $a \in R_\Xi$ *is equivalent to the following: there exists an element* $f \in \mathfrak{H}$ *such that*

$$\left(\sum_{j=1}^{N} a_j \xi_j', g \right)_{\mathfrak{H}} \xrightarrow[N \to \infty]{} (f, g) \tag{1}$$

for each (finite) linear combination $g = \sum c_j \xi_j$ *of elements from* Ξ.

(b) *If* $R_\Xi = \ell^2$, *then series* $\sum a_j \xi_j'$ *converges for any* $a = \{a_j\} \in \ell^2$ *to some* $f \in \mathfrak{H}$ *and* $\mathcal{J}_\Xi f = a$.

PROOF. (a) Let $a \in R_\Xi$, which means that there exists $f \in \mathfrak{H}$ such that $a = \mathcal{J}_\Xi f$. Set $g = \sum c_j \xi_j$, were c is an arbitrary finite vector. Then for large enough N,

$$\left(\sum_{j=1}^{N} a_j \xi_j', g \right) = \sum_{j=1}^{N} a_j \bar{c}_j = \sum_{j=1}^{N} (f, \xi_j) \bar{c}_j = \left(f, \sum_{j=1}^{N} c_j \xi_j \right) = (f, g).$$

Conversely, let relation (1) be valid for some $f \in \mathfrak{H}$ and any finite linear combination $g = \sum c_j \xi_j$. Taking element ξ_n for g and observing that for $N \geq n \, (\sum_{j=1}^{N} a_j \xi_j', \xi_n)_{\mathfrak{H}} = a_n$, one gets $(f, \xi_n) = a_n$. Hence $a \in R_\Xi$.

Assertion (b) of the theorem follows immediately from that proved in Theorem 1(b) boundedness of operator $(\mathcal{J}_\Xi^0)^{-1}$, since

$$(\mathcal{J}_\Xi^0)^{-1} \{a_j\}_{j=1}^{N} = \sum_{j=1}^{N} a_j \xi_j'.$$

The theorem is proved.

Remark I.2.4. As the theorem shows, if series $\sum_{j=1}^{\infty} a_j \xi_j'$ converges weakly, its sum serves a solution f of the problem of moments $\mathcal{J}_\Xi f = \{a_n\}$.

Theorem 3 allows us to produce some rough tests for a problem of moments to be solvable based on the norm estimates of the elements of a biorthogonal family. Let us bring in two statements of the kind.

Corollary I.2.5. Let $\Xi \in (M)$. Then the assertions are true:

(a) if $a \in \ell^2$ and $\sum_{j=1}^{\infty} |a_j| \, \|\xi_j'\| < \infty$, then $a \in R_\Xi$.

(b) if there exists $a \in \ell^2$ such that $\sum_{j=1}^{\infty} |a_j|^2 \, \|\xi_j\|^2 = \infty$, then $R_\Xi \neq \ell^2$.

PROOF.

(a) Series $\sum_{j=1}^{\infty} a_j \xi_j'$ converges in the norm and the more so, weakly. By Theorem 3(a), $a \in R_\Xi$.

(b) By the Orlichz lemma (Nikol'skiǐ 1980: lecture VI), there may be found for any N vectors $\{a_j^{(N)}\}_{j=1}^{N}$ such that $|a_j^{(N)}| = |a_j|$ and

$$\left\| \sum_{j=1}^{N} a_j^{(N)} \xi_j' \right\|^2 \geq \sum_{j=1}^{N} |a_j|^2 \, \|\xi_j'\|^2.$$

If $R_\Xi = \ell^2$, then operator $(\mathscr{I}_\Xi^0)^{-1}$ is bounded. Therefore

$$\left\| \sum_{j=1}^{N} a_j^{(N)} \xi_j' \right\|^2 = \|(\mathscr{I}_\Xi^0)^{-1}\{a_j^{(N)}\}\|^2 < \sum_{j=1}^{N} |a_j^{(N)}|^2 = \sum_{j=1}^{N} |a_j|^2 \leq \|a\|^2.$$

The obtained contradiction just proves the second assertion.

II

Families of vector-valued exponentials

1. Hardy spaces and operator functions

In this section we set forth the theory of vector and scalar Hardy spaces without the proofs. We then describe the properties of both the inner and outer operator functions and the shift operator. Other sources should be consulted for the scalar theory of functions in Hardy spaces (Privalov 1950; Hoffman 1962; Duren 1970; Koosis 1980; Nikol'skiĭ 1980; Garnett 1981) and for the theory of vector spaces and shift operators (Helson 1964; Sz.-Nagy and Foias 1970; Nikol'skiĭ 1980). Needless to say, any pertinent details that have not yet found their way into the literature are explained wherever necessary.

The symbols $\mathbb{C}_+$ and $\mathbb{C}_-$ denote, respectively, the upper and the lower open half-plane. $\mathbb{D}$ is the unit disc: $\mathbb{D} = \{z \mid |z| < 1\}$, and $\mathbb{T} = \{z \mid |z| = 1\}$. $\mathfrak{N}$ means an auxiliary Hilbert space, dim $\mathfrak{N} \leq \infty$. In the space $\mathfrak{N}$ we specify an orthonormal basis $\{\xi_p^0\}$, $p = 1, 2, \ldots$, dim $\mathfrak{N}$. The scalar product and the norm in $\mathfrak{N}$ are $\langle \cdot, \cdot \rangle$ and $\langle\langle \cdot \rangle\rangle$, respectively.

1.1. Hardy spaces

1.1.1. Definitions and Hilbert structure

Definition II.1.1. Hardy space $H^2(\mathfrak{N})$ of $\mathfrak{N}$-valued functions is a Hilbert space of analytical in $\mathbb{D}$ functions of the form

$$F(z) = \sum_{n=0}^{\infty} \hat{F}_n z^n,$$

where $\hat{F}_n$ (Fourier coefficients) belong to $\mathfrak{N}$. The scalar product in $H^2(\mathfrak{N})$ is given by

$$(F, G)_{H^2(\mathfrak{N})} = \sum_{n=0}^{\infty} \langle \hat{F}_n, \hat{G}_n \rangle.$$

It follows from the definition that

$$\|F\|_{H^2(\mathfrak{N})}^2 = \sum_{n=0}^{\infty} \langle\langle F_n \rangle\rangle^2 < \infty$$

and that the family $\{\xi_p^0 z^n\}$, $p = 1, 2, \ldots, \dim \mathfrak{N}$; $n = 0, 1, 2, \ldots$ forms an orthonormal basis in $H^2(\mathfrak{N})$.

It is known that $F(z)$ may be identified with its boundary values $F(e^{i\theta})$ by means of the limits along a nontangential to the unit circle paths (those limits exist almost everywhere by the Fatou theorem). The corresponding scalar function $\langle\langle F(e^{i\theta}) \rangle\rangle$ is then squarely summable on the unit circle $\mathbb{T}$.

Definition II.1.2. Hardy space $H_+^2(\mathfrak{N})$ $[H_-^2(\mathfrak{N})]$ of $\mathfrak{N}$-valued functions in $\mathbb{C}_+$ $[\mathbb{C}_-]$ is the set of functions f analytical in $\mathbb{C}_+$ $[\mathbb{C}_-]$ and such that

$$\sup \int_{\mathbb{R}} \langle\langle f(x + iy) \rangle\rangle^2 \, dx < \infty,$$

where the sup is taken over all $y > 0$ for $H_+^2(\mathfrak{N})$ and all $y < 0$ for $H_-^2(\mathfrak{N})$. Along with it,

$$\|f\|_{H_+^2(\mathfrak{N})}^2 = \int_{\mathbb{R}} \langle\langle f(x) \rangle\rangle^2 \, dx.$$

In the scalar case ($\mathfrak{N} = \mathbb{C}$), we write $H_\pm^2$ instead of $H_\pm^2(\mathbb{C})$. Note also that for any $\varepsilon > 0$

$$|f(x + iy)| \le \sqrt{2/\pi y} \, \|f\|_{H_+^2(\mathfrak{N})}, \qquad f \in H_+^2, \, y \ge \varepsilon \tag{1}$$

(see Nikol'skiĭ 1980: lecture XI, sec. 1, formula (4)).

As in the case of Hardy space in the unit disc, functions $f \in H_+^2(\mathfrak{N})$ ($f \in H_-^2(\mathfrak{N})$) are identified with their boundary values belonging to the space $L^2(\mathbb{R}; \mathfrak{N})$, and the scalar product in $H_\pm^2(\mathfrak{N})$ is determined by the formula

$$(f, g)_{H_\pm^2(\mathfrak{N})} = \int_{\mathbb{R}} \langle f(x), g(x) \rangle \, dx.$$

It is evident that for $f \in H_+^2(\mathfrak{N})$, functions $\overline{f(k)}$ and $f(-k)$ belong to the space $H_-^2(\mathfrak{N})$.

1.1.2. Relationship between $H^2(\mathfrak{N})$ and $H_\pm^2(\mathfrak{N})$ spaces

Hardy spaces in $\mathbb{D}$ and $\mathbb{C}_+$ are related by a unitary mapping U:

$$H^2(\mathfrak{N}) \ni f(x) \overset{U}{\mapsto} \frac{1}{\sqrt{\pi}} \frac{1}{x + i} f\left(\frac{x - i}{x + i}\right) \in H_+^2(\mathfrak{N}).$$

Here, $k \xrightarrow{\omega} (k - i)/(k + i)$ is a conformal mapping of $\mathbb{C}_+$ on $\mathbb{D}$, and a factor $1/[\sqrt{\pi}(x + i)]$ corresponds to a transformation of Lebesgue measures on the unit circle and the real axis.

1.1.3. Calculation of Cauchy integrals by means of residues

For $f \in H_+^2(\mathfrak{N})$ and $\lambda \in \mathbb{C}_+$, the relations

$$\int_{\mathbb{R}} \frac{f(k)}{(k - \lambda)} \, dk = 2\pi i f(\lambda), \qquad \int_{\mathbb{R}} \frac{f(k)}{(k - \bar{\lambda})} \, dk = 0$$

hold.

1.1.4. Simple fractions

Simple scalar (x_λ and $\hat{x}_\mu$) and vector ($x_\lambda \eta$ and $\hat{x}_\mu \eta$; $\eta \in \mathfrak{N}$) fractions play an important role in what follows:

$$x_\lambda(k) = \sqrt{\frac{\operatorname{Im} \lambda}{\pi}} \frac{1}{k - \bar{\lambda}} \in H_+^2, \qquad \lambda \in \mathbb{C}_+, \ \|x_\lambda\|_{H_+^2} = 1,$$

$$\hat{x}_\mu(z) = \sqrt{1 - |\mu|^2} \frac{1}{1 - \bar{\mu} z} \in H^2, \qquad \mu \in \mathbb{D}, \ \|\hat{x}_\mu\|_{H^2} = 1.$$

These fractions are interrelated by a transformation U:

$$U\hat{x}_\mu = c_\mu x_\lambda, \qquad \mu = \omega(\lambda) = \frac{\lambda - i}{\lambda + i}, \qquad c_\mu = \frac{|\mu + i|}{\mu + i}.$$

Simple fractions constitute a complete family in Hardy space:

$$\bigvee_{\lambda \in \mathbb{C}_+} x_\lambda = H_+^2$$

(see also Subsection 1.3.10).

1.1.5. Paley–Wiener theorem and connections between simple fractions and exponentials

Simple fractions are generated by exponentials by means of the Fourier transform.

Let $\mathscr{F}$ denote the inverse Fourier transform of functions f from $L^1(\mathbb{R}_+) \cap L^2(\mathbb{R}_+)$, $\mathbb{R}_+ := \{x \mid x > 0\}$, by a formula

$$(\mathscr{F}f)(k) = \frac{1}{\sqrt{2\pi}} \int_0^\infty f(t) \, e^{ikt} \, dt.$$

According to the Paley–Wiener theorem (Paley and Wiener 1934), mapping $\mathcal{F}$ is extended to unitary mapping of $L^2(\mathbb{R}_+)$ onto $H^2_+(\mathfrak{N})$. By the mapping, exponentials transform into simple fractions:

$$\mathcal{F}(-i\sqrt{2\operatorname{Im}\lambda}\,\bar{e}^{i\bar{\lambda}t}] = x_\lambda.$$

1.1.6. Decomposition of $L^2(\mathbb{R})$ and the Riesz projector

If, as already mentioned, one identifies functions from $H^2_\pm(\mathfrak{N})$ with their boundary values from $L^2(\mathbb{R}, \mathfrak{N})$, then

$$L^2(\mathbb{R}, \mathfrak{N}) = H^2_+(\mathfrak{N}) \oplus H^2_-(\mathfrak{N}).$$

Orthoprojectors $P_\pm$ from $L^2(\mathbb{R}, \mathfrak{N})$ on $H^2_\pm(\mathfrak{N})$ are expressed via the Hilbert operator $\mathcal{H}$,

$$(\mathcal{H}u)(x) := \frac{1}{\pi}\operatorname{p.v.}\int_\mathbb{R}\frac{u(t)}{x-t}\,dt,$$

according to the formula

$$P_\pm = \tfrac{1}{2}I_{L^2(\mathbb{R},\mathfrak{N})} \pm \frac{i}{2}\mathcal{H}. \tag{2}$$

These projectors are called the Riesz projectors.

As an exercise, one can check this formula first for rational functions from $L^2(\mathfrak{N})$ (representing them as a sum of two functions with poles in $\mathbb{C}_+$ and $\mathbb{C}_-$) and then for arbitrary functions from $L^2(\mathbb{R})$ approximated by rational ones.

1.2. Inner and outer operator functions

1.2.1. Definition of inner functions

Definition II.1.3. The analytical operator function S bounded in $\mathbb{C}_+$ [$\mathbb{C}_-$] is called an inner function in the upper [lower] half-plane if for almost every $k \in \mathbb{R}$, $S(k)$ is a unitary operator in $\mathfrak{N}$. Functions of the form $\hat{S}(z) = S(\omega^{-1}(z)) = S[i(1+z)/(1-z)]$, where S is an inner function in $\mathbb{C}_+$, are inner functions in $\mathbb{D}$.

Note that if S is an inner operator function in $\mathbb{C}_+$, then $S(-k)$, $S^*(\bar{k})$ are inner operator functions in $\mathbb{C}_-$. Inner operator functions satisfy a relation

$$S^*(\bar{k}) = S^{-1}(k), \tag{3}$$

which is obtained with the help of analytical continuation from the real axis into a complex plane.

1.2.2. Classification of scalar inner functions

Definition II.1.4.

(a) An inner function of the form

$$B(k) = e^{i\alpha} \prod_{j \in \mathbb{N}} \varepsilon_j \frac{k - \lambda_j}{k - \bar{\lambda}_j}$$

is said to be a Blaschke product (BP) constructed by the zeros $\lambda_j \in \mathbb{C}_+$, $j \in \mathbb{N}$. Here, $\alpha \in \mathbb{R}$, ε_j are phase factors with unit modulus and are such that each cofactor $\varepsilon_j(i - \lambda_j)/(i - \bar{\lambda}_j)$ in BP is positive at the point $k = i$.

(b) For $a \geq 0$ and $\alpha \in \mathbb{R}$, function $\exp(ika + i\alpha)$ is said to be an entire singular inner function.

(c) Function S_{sing} of the form

$$\exp\left\{\frac{i}{\pi} \int_{-\infty}^{\infty} \left(\frac{1}{k - t} + \frac{1}{1 + t^2}\right) d\rho(t) + i\alpha\right\}; \qquad \alpha \in \mathbb{R}$$

is called a singular inner function, where $d\rho$ is singular over the axis measure such that

$$\int_{-\infty}^{\infty} \frac{d\rho(t)}{1 + t^2} < \infty.$$

Note that singular inner functions have no zeros in $\mathbb{C}_+$ and that for the convergence of the Blaschke product it is necessary and sufficient for zeros λ_j to satisfy the Blaschke condition

$$\sum_{j=1}^{\infty} \frac{\text{Im } \lambda_j}{1 + |\lambda_j|^2} < \infty. \tag{B}$$

Under the mapping of H_+^2 to H^2, the difference between singular and entire singular functions can be seen in a (singular) measure with loads in $\mathbb{T}\backslash i$ corresponds to a singular inner in $\mathbb{D}$ function, while a measure with a single load at $z = i$ corresponds to an entire singular one.

1.2.3. Factorization of scalar inner functions

Any scalar inner function f in $\mathbb{C}_+$ may be represented as a product of three factors: BP $B(k)$, entire singular function e^{ika} (or unitary constant), and singular inner function S_{sing}:

$$f(k) = B(k) \, e^{ika} \, S_{\text{sing}}(k).$$

1.2.4. Classification of inner operator functions

Further, in Subsections 1.2.4–1.2.11, let be dim $\mathfrak{N} < \infty$.

Definition II.1.5. Inner operator function S in $\mathbb{C}_+$ is called a Blaschke–Potapov product (BPP), entire singular function (ESF), or singular inner function if $\det S$ is a Blaschke product, entire singular inner function, or singular inner function, respectively.

We denote BPP by Π, $\tilde{\Pi}$, and so on, and ESF by Θ, $\tilde{\Theta}$, and so on.

Following are some examples of entire singular inner functions. Of these, the simplest meaningful function is $\Theta(k) = \exp(ikA)$, with A being a nonnegative operator in $\mathfrak{N}$. Taking eigenvectors of A as a basis of $\mathfrak{N}$ we get $A = \mathrm{diag}[a_p]$, $a_p \geq 0$, and then

$$\Theta(k) = \mathrm{diag}[\exp(ika_p)],$$

$$\det \Theta(k) = \exp(ik \, \mathrm{Tr} \, A).$$

More complicated examples of ESF are finite products $\prod_{p=1}^{N} \exp(ikA_p)$, $A_p \geq 0$.

The general form of ESF may be described by the following proposition (Potapov 1955).

Any ESF Θ is the value at $t = 1$ of the solution of equation

$$\frac{d}{dt} Y(t, k) = ikA(t) Y(t, k),$$

with some nonnegative summable operator function $A(t)$. The solution is singled out by a condition $Y(0, k) = V$, where V is a constant unitary operator. Along with it,

$$\det \Theta(k) = c \exp\left(ik \int_0^1 \mathrm{Tr} \, A(t) \, dt \right), \qquad |c| = 1.$$

1.2.5. Factorization of inner operator functions

Every inner operator function S may be factorized by a BPP, an ESF, and a singular inner function, in any order.

It then follows that the zeros of an inner function (the zeros of $\det S$) satisfy the Blaschke condition.

Definition II.1.6. Inner operator function $\tilde{S}$ is a left divisor of inner operator function S if function $\hat{S} := [\tilde{S}]^{-1}S$ is an inner one. In other words,

S may be represented in the form

$$S = \tilde{S}\hat{S}.$$

So, for instance, if BPP $\Pi(k)$ has a zero at point λ and Ker $\Pi^*(\lambda) = \mathfrak{N}_\lambda$, then Π is divisible from the left by the (elementary) inner cofactor

$$\frac{k - \lambda}{k - \bar{\lambda}} \delta_\lambda + (I_{\mathfrak{N}} - \delta_\lambda),$$

where δ_λ is the orthoprojector in $\mathfrak{N}$ on $\mathfrak{N}_\lambda$.

1.2.6. Scalar outer functions

Definition II.1.7. Scalar function f_e is called an outer function in $\mathbb{C}_+$ if it can be represented in the form

$$f_e(k) = \exp\left[\frac{i}{\pi} \int_{\mathbb{R}} \left(\frac{1}{k - t} + \frac{t}{1 + t^2}\right) h(t)\, dt\right], \qquad k \in \mathbb{C}_+$$

with $h(t) \in \mathbb{R}$ and

$$\int_{\mathbb{R}} \frac{h(t)}{1 + t^2}\, dt < \infty.$$

Scalar function $f_e(k)$ is called an outer function in $\mathbb{C}_-$ if $f_e(-k)$ is an outer function in $\mathbb{C}_+$.

Outer (in $\mathbb{C}_+$) functions have no zeros in $\mathbb{C}_+$ and almost everywhere on $\mathbb{R}$

$$\log|f_e(t)| = h(t).$$

For Im $k > 0$, function $\log|f_e(k)|$ is a Poisson integral of function $h(t)$:

$$\log|f_e(x + iy)| = \frac{1}{\pi} \int_{\mathbb{R}} h(t) \, \text{Im}\, \frac{1}{t - x - iy}\, dt = \frac{y}{\pi} \int_{\mathbb{R}} h(t) \frac{dt}{(x - t)^2 + y^2}.$$

This formula implies that

$$\frac{1}{y} \log|f_e(x + iy)| \xrightarrow[y \to +\infty]{} 0. \tag{4}$$

Note that function $1/f_e$ is an outer one for any outer f_e.

1.2.7. Factorization of scalar functions

Let f be a scalar function analytical in $\mathbb{C}_+$ for which

$$\int_{\mathbb{R}} \frac{\log|f(x)|}{1 + x^2}\, dx < \infty.$$

Then f may be represented by a product of an outer function f_e and inner function f_i.

1.2.8. Outer operator functions

Definition II.1.8.

(a) Operator function F is said to be strong if

$$\frac{\langle\langle F \rangle\rangle}{1 + |k|} \in L^2(\mathbb{R}). \tag{5}$$

(b) Operator function F is said to be a strong $H_+^2 - [H_-^2 -]$ function if for any $\eta \in \mathfrak{N}$, $F(k)\eta/(k + i) \in H_+^2(\mathfrak{N})$ $[F(k)\eta/(k - i) \in H_-^2(\mathfrak{N})]$.

(c) A strong $H_+^2 - [H_-^2 -]$ (operator) function is said to be outer in $\mathbb{C}_+$ $[\mathbb{C}_-]$ if $\det F$ is an outer function in $\mathbb{C}_+$ [in $\mathbb{C}_-$].

It follows by definition that an additional condition (5), which is absent in the scalar case, is imposed on outer operator functions. As in the case of inner functions, functions outer in $\mathbb{D}$ are defined as images of the ones outer in $\mathbb{C}_+$ under conformal mapping $\omega(k)$. Strong H_+^2 functions are transformed by it in functions $\hat{F}$ satisfying the condition

$$\hat{F}(z)\eta \in H^2(\mathfrak{N}), \qquad \forall \eta \in \mathfrak{N}.$$

It is known (Nikol'skiĭ 1980: lecture I, sec. 6) that a strong H_+^2 function is an outer one if and only if

$$\bigvee_{\substack{n = 0,1,\dots \\ \eta \in \mathfrak{N}}} \left\{ F(k)\left(\frac{k - i}{k + i}\right)^n \frac{1}{k + i}\eta \right\} = H_+^2(\mathfrak{N}).$$

For strong operator functions in $\mathbb{D}$, this condition may be written in a more natural way:

$$\bigvee_{\substack{n = 0,1,\dots \\ \eta \in \mathfrak{N}}} \{\hat{F}(z)z^n\eta\} = H^2(\mathfrak{N}).$$

It means that the family of functions $\hat{F}(z)P_n(z)$, where $P_n(z)$ is an $\mathfrak{N}$-valued polynomial, is dense in $H^2(\mathfrak{N})$. We denote functions outer in $\mathbb{C}_\pm$ as $F_e^\pm$.

1.2.9. Factorization of strong functions

Let F be a strong H_+^2 function and det $F \not\equiv 0$. Then F may be represented as a product of an inner and an outer operator function taken in any order:

$$F = S\tilde{F}_e^+ = F_e^+ \tilde{S}.$$

(S and $\tilde{S}$ are inner and F_e^+ and $\tilde{F}_e^+$ are outer operator functions.) Here, $\tilde{F}_e^+$ and F_e^+ are bounded in $\mathbb{C}_+$ if and only if F is bounded.

1.2.10. Semisimple zeros

Definition II.1.9. Let operator function $F(k)$ be an analytical in the vicinity of $\lambda \in \mathbb{C}$ and det $F(\lambda) = 0$. We consider λ to be a semisimple zero of $F(k)$ if for any vectors $\eta_0 \in \text{Ker } F(\lambda)$, $\eta_0 \neq 0$, and $\eta \in \mathfrak{R}$,

$$F'(\lambda)\eta_0 + F(\lambda)\eta \neq 0.$$

In other words, for any analytical vector function $\varphi(k)$ such that $\varphi(\lambda) \neq 0$, vector function $F(k)\varphi(k)$ does not have a second-order zero at λ:

$$F(\lambda)\varphi(\lambda) = 0 \Rightarrow \frac{d}{dk}[F(k)\varphi(k)]|_{k=\lambda} \neq 0.$$

In a scalar situation, a semisimple zero is a simple one. In a finite-dimensional case, λ may be a semisimple but not a simple zero, as, for instance, for a function

$$F(k) = (k - \lambda)\delta + (I_{\mathfrak{R}} - \delta),$$

with δ being an orthoprojector in $\mathfrak{R}$, dim $\delta\mathfrak{R} > 1$.

1.2.11. Existence of the Blaschke–Potapov product

Proposition II.1.10 (Potapov 1955). Let σ be a countable subset of $\mathbb{C}_+$ satisfying the Blaschke condition (Subsection 1.2.2), and let $\{\mathfrak{R}_\lambda\}_{\lambda \in \sigma}$ be a family of subspaces in $\mathfrak{R}$. Then BPP $\Pi(k)$ may be found such that the set of zeros of function det $\Pi(k)$ coincides with σ, the zeros of Π are semisimple, and Ker $\Pi^*(\lambda) = \mathfrak{R}_\lambda$, $\lambda \in \sigma$.

Later we characterize a family of subspaces of simple vector fractions $\{x_\lambda \mathfrak{R}_\lambda\}_{\lambda \in \sigma}$, $x_\lambda \mathfrak{R}_\lambda \subset H_+^2(\mathfrak{R})$ under the condition $\sigma \in (B)$ by Blaschke–Potapov product Π and denote it as $\mathcal{X}_\Pi$. This operator function Π is called BPP, generated by the family $\{x_\lambda \mathfrak{R}_\lambda\}_{\lambda \in \sigma}$.

If $\sigma \notin (B)$ (does not satisfy the Blaschke condition), then BPP does not exist. However, we preserve the notation $\mathscr{X}_\Pi$ for families $(x_\lambda \mathfrak{N}_\lambda)_{\lambda \in \sigma}$ of simple fractions in such a situation as well, and also for the case of infinite dimensional space $\mathfrak{N}$.

1.3. Nagy–Foias model operator

1.3.1. Subspaces K_s

Definition II.1.11. Let S be an inner function in $\mathbb{C}_+$. Let K_S be a subspace

$$H^2_+(\mathfrak{N}) \ominus SH^2_+(\mathfrak{N}) = \{f \in H^2_+(\mathfrak{N}) \mid f \perp SH^2_+(\mathfrak{N})\}.$$

P_S is the orthoprojector from $H^2_+(\mathfrak{N})$ (or from $L^2(\mathbb{R}; \mathfrak{N})$ on K_S. If $S(k) = \exp(ika)I_\mathfrak{N}$, we write K_a and P_a instead of K_S and P_S, respectively.

For orthoprojector P_S a formula is valid

$$P_S = P_+ - SP_+S^*. \tag{6}$$

(Orthoprojector $P_+ : L^2(\mathbb{R}; \mathfrak{N}) \mapsto H^2_+(\mathfrak{N})$ is introduced in Subsection 1.1.6.)

Note that in the case $S(k) = \exp(ika)$ only functions that exponentially decrease with Im $k \to \infty$ enter subspace $SH^2_+(\mathfrak{N})$.

It is not difficult to establish that

$$P_S x_\lambda = \sqrt{\frac{\mathrm{Im}\ \lambda}{\pi}} \frac{1 - S(k)S^{-1}(\bar\lambda)}{k - \bar\lambda}.$$

For $S = \exp(ika)$, we get

$$P_a x_\lambda = \sqrt{\frac{\mathrm{Im}\ \lambda}{\pi}} \frac{1 - \exp(ia(k - \bar\lambda))}{k - \bar\lambda}. \tag{7}$$

The Paley–Wiener theorem (Paley and Wiener 1934) implies that

$$\mathscr{F}L^2(0, a; \mathfrak{N}) = K_a. \tag{8}$$

If Θ is an ESF, all the elements of subspace K_θ are entire functions. For $\Theta = \exp(ika)$, it follows immediately out of (8) that any element $f \in K_a$ is expressed via some function φ from $L^2(0, a)$ by a formula

$$f = \int_0^a \exp(ikt)\varphi(t)\,dt.$$

1.3.2. Definitions of model operators in $\mathbb{D}$ and $\mathbb{C}_+$

Definition II.1.12. The operator of multiplication by z is called a shift operator σ in $H^2(\mathfrak{N})$:

$$H^2(\mathfrak{N}) \ni f = \sum_{n=0}^{\infty} \hat{f}_n z^n \overset{\sigma}{\mapsto} \sum_{n=0}^{\infty} \hat{f}_n z^{n+1}.$$

As is easily proved, the adjoint operator is

$$(\mathfrak{S}^* f)(z) = \frac{f(z) - f(0)}{z}.$$

Definition II.1.13. Let $\hat{S}$ be an inner function in $\mathbb{D}$,

$$P_{\hat{S}} := P_{K_{\hat{S}}}, \quad K_{\hat{S}} := H^2(\mathfrak{N}) \ominus \hat{S} H^2(\mathfrak{N}).$$

Operators $P_{\hat{S}} \mathfrak{S}|_{K_{\hat{S}}}$ and $[P_{\hat{S}} \mathfrak{S}|_{K_{\hat{S}}}]^* = \mathfrak{S}^*|_{K_{\hat{S}}}$ are called Nagy–Foias model operators in $\mathbb{D}$ (in $H^2(\mathfrak{N})$).

Generator A_S of semigroup $P_S e^{ikt}|_{K_s}$ and operator $-A_S^*$ (generator of the adjoint semigroup) are called Nagy–Foias model operators in $\mathbb{C}_+$ (in $H^2_+(\mathfrak{N})$). Here, S is an inner function in $\mathbb{C}_+$.

Note that the Kelly transform of operator A_S is the image of the model operator $P_{\hat{S}} \mathfrak{S}|_{K_s}$ under the mapping U of Hardy space $H^2(\mathfrak{N})$ to $H^2_+(\mathfrak{N})$ (Subsection 1.1.2). Here, $S(k) = \hat{S}(z)$, $z = \omega(k)$.

1.3.3. The spectrum and eigenfunctions of model operators in $\mathbb{C}_+$

The eigensubspaces of A_S and A_S^* operators corresponding to the eigenvalues λ and $\bar{\lambda}$ are, respectively,

$$\frac{S(k)}{k - \lambda} \mathfrak{N}^{\lambda}, \quad \mathfrak{N}^{\lambda} := \operatorname{Ker} S(\lambda), \tag{9}$$

$$\frac{1}{k - \bar{\lambda}} \mathfrak{N}_{\lambda}, \quad \mathfrak{N}_{\lambda} := \operatorname{Ker} S^*(\lambda). \tag{10}$$

If $\dim \mathfrak{N} < \infty$, the discrete spectrum of operator A_S coincides with the set of zeros of function $\det S$.

Let $\dim \mathfrak{N} < \infty$ and an inner function S be a BPP, $S = \Pi$; further, let σ be a set of zeros of $\det \Pi$, where the zeros are simple and $\eta_\lambda \in \operatorname{Ker} \Pi^*(\lambda)$, $\langle\langle \eta_\lambda \rangle\rangle = 1$. Following formulas (9) and (10), let us find a family $\mathcal{X}'_\Pi := \{x'_{\mathfrak{N}, \lambda}\}_{\lambda \in \sigma}$ biorthogonal to the family of simple fractions $\mathcal{X}_\Pi := \{x_\lambda \eta_\lambda\}_{\lambda \in \sigma}$.

From (9) and (10) it follows that biorthogonal elements $x'_{\mathfrak{N},\lambda}$ are of the form

$$x'_{\mathfrak{N},\lambda}(k) = \alpha_\lambda \frac{\Pi(k)\eta^\lambda}{k - \lambda}, \qquad \eta^\lambda \in \operatorname{Ker} \Pi(\lambda), \langle\langle \eta^\lambda \rangle\rangle = 1. \tag{11}$$

Now calculate coefficient α_λ:

$$1 = (x'_{\mathfrak{N},\lambda}, x_\lambda \eta_\lambda)_{H^2_+(\mathfrak{N})} = \alpha_\lambda \sqrt{\frac{\operatorname{Im}\lambda}{\pi}} \int_{\mathbb{R}} \frac{\langle \Pi(k)\eta^\lambda, \eta_\lambda \rangle}{(k - \lambda)^2} dk$$

$$= \alpha_\lambda \sqrt{\frac{\operatorname{Im}\lambda}{\pi}} 2\pi i \operatorname*{res}_{\lambda} \frac{\langle \Pi(k)\eta^\lambda, \eta_\lambda \rangle}{(k - \lambda)^2}$$

$$= \alpha_\lambda 2i\sqrt{\pi \operatorname{Im}\lambda} \left\langle \frac{d}{dk}\Pi(k)|_{k=\lambda}\eta^\lambda, \eta_\lambda \right\rangle.$$

Thus

$$\alpha_\lambda = -\frac{i}{2\sqrt{\pi \operatorname{Im}\lambda}} \frac{1}{\langle (d/dk)\Pi(k)|_{k=\lambda}\eta^\lambda, \eta_\lambda \rangle}.$$

In the scalar situation, we can now calculate the norms of elements of the biorthogonal system:

$$\|x'_\lambda\|^2_{H^2_+} = |\alpha_\lambda|^2 \int_{\mathbb{R}} \frac{\Pi(k)\bar{\Pi}(k)}{(k - \lambda)(k - \bar{\lambda})} dk$$

$$= |\alpha_\lambda|^2 \int_{\mathbb{R}} \frac{1}{(k - \lambda)(k - \bar{\lambda})} dk = |\alpha_\lambda|^2 \frac{\pi}{\operatorname{Im}\lambda}.$$

As one easily checks,

$$\left| \frac{d}{dk}\Pi(k)|_{k=\lambda} \right| = \frac{1}{2\operatorname{Im}\lambda} \prod_{\mu\in\sigma, \mu\neq\lambda} \frac{|\mu - \lambda|}{|\mu - \bar{\lambda}|},$$

Hence,

$$\|x'_\lambda\|_{H^2_+} = \prod_{\mu\in\sigma, \mu\neq\lambda} \frac{|\mu - \bar{\lambda}|}{|\mu - \lambda|}. \tag{12}$$

1.3.4. The spectrum and eigenfunctions of model operators in $\mathbb{D}$

Let $\hat{S}$ be an inner function in $\mathbb{D}$. Then the eigensubspaces of model operators $P_{\hat{S}}\mathfrak{S}|_{K_{\hat{S}}}$ and $\mathfrak{S}^*|_{K_{\hat{S}}}$ corresponding to μ and $\bar{\mu}$ are, respectively,

$$\frac{\hat{S}(z)}{z - \mu}\hat{\mathfrak{N}}^\mu, \qquad \hat{\mathfrak{N}}^\mu := \operatorname{Ker}\hat{S}(\mu), \tag{13}$$

$$\frac{1}{1 - \bar{\mu}z}\hat{\mathfrak{N}}_\mu, \qquad \hat{\mathfrak{N}}_\mu := \operatorname{Ker}\hat{S}^*(\mu). \tag{14}$$

Remark II.1.14. Associate (root) functions of operator $\mathfrak{S}^*|_{K_S}$ are rational functions with multiple poles. (For the scalar case, root subspaces are described in Nikol'skiĭ 1980; lecture IV; for the vector one, Buslaeva 1974 and Ivanov 1987.) In transitting from Hardy space to the space $L^2(0, \infty; \mathfrak{N})$, one obtains families of the form $\{P_n(t) \exp(i\lambda_n t)\}$, where $P_n(t)$ is an $\mathfrak{N}$-valued polynomial. To make our exposition as clear as possible, we omit the details of root subspaces.

1.3.5. Discrete spectrum subspaces of model operators in $\mathbb{C}_+$

Let dim $\mathfrak{N} < \infty$ and S be an inner function with factorization

$$S = \Pi\tilde{\Theta} = \Theta\tilde{\Pi}.$$

Further, let $\sigma_d(A_S)$ be the discrete spectrum of operator A_S, $\mathfrak{H}_d$ and $\mathfrak{H}_d^*$ be subspaces of the discrete spectrum of model operators A_S and $-A_S^*$, respectively. Then

$$\mathfrak{H}_d = \bigvee_{\lambda \in \sigma_d(A_s)} \frac{S(k)}{k - \lambda} \mathfrak{N}^\lambda = \Theta K_{\tilde{\Pi}},$$

$$\mathfrak{H}_d^* = \bigvee_{\lambda \in \sigma_d(A_s)} \frac{1}{k - \bar{\lambda}} \mathfrak{N}_\lambda = K_{\Pi}.$$

From these formulas and (9) and (10), it follows for $S = \Pi$ that the family $\{x'_{\mathfrak{N}, \lambda}\}$ biorthogonal to the family $\{x_\lambda \mathfrak{N}_\lambda\}$ of simple fractions is complete in the subspace K_{Π} and $K_{\Pi} = \bigvee x_\lambda \mathfrak{N}_\lambda = \bigvee x'_{\mathfrak{N}, \lambda}$.

1.3.6. Invariant subspaces of a model operator in $\mathbb{D}$

Proposition II.1.15 (Nikol'skiĭ 1980: lecture I, sec. 6). *A subspace* $G \subset H^2(\mathfrak{N})$ *is invariant with respect to operator* $\mathfrak{S}^*$ *(i.e.,* $\mathfrak{S}^* G \subset G$*) if and only if* G *is of the form*

$$G = H^2(\mathfrak{N}) \ominus \mathscr{S} H^2(\mathfrak{N}_1),$$

where $\mathfrak{N}_1$ *is a subspace of* $\mathfrak{N}$ *and* $\mathscr{S}$ *is an analytical in* $\mathbb{D}$ *operator function,* $\mathscr{S}(z)$*:* $\mathfrak{N}_1 \mapsto \mathfrak{N}$*, which is an isometric operator for almost all* $z \in \mathbb{T}$.

For orthoprojector P_G: $H^2(\mathfrak{N}) \mapsto G$, a formula holds (Nikol'skiĭ 1980: lecture II, sec. 4) similar to formula (6) (see Subsection 1.3.1).

$$P_G = I|_{H^2(\mathfrak{N})} - \mathscr{S}\hat{P}_+\mathscr{S}^*. \tag{15}$$

Here, $\hat{P}_+ = P_{H^2(\mathfrak{N}_1)}$ is the orthoprojector from the space $L^2(\mathbb{T}, \mathfrak{N}_1)$ on subspace $H^2(\mathfrak{N}_1)$, while $\mathscr{S}$ and $\mathscr{S}^*$ are multiplication operators by $\mathscr{S}(\cdot)$ and $\mathscr{S}^*(\cdot)$, respectively.

For dim $\mathfrak{N} < \infty$, isometric operator $\mathscr{S}(z)$ is a unitary mapping if and only if $\mathfrak{M}_1 = \mathfrak{N}$. In this case $\mathscr{S}$ is an inner function and subspace G has the form K_S (the symbol K_S is used when S is an inner operator function (Definition 11)). Therefore, in the scalar case, any $\mathfrak{S}^*$-invariant subspace different from H^2 and $\{0\}$ has the form K_S.

We now present two assertions concerning $\mathfrak{S}^*$-invariant subspaces.

Proposition II.1.16 (Nikol'skiĭ 1980: lecture I, sec. 1).

(a) *If G is an invariant subspace of operator $\mathfrak{S}^*$ and for some inner operator function $\hat{S}$ there is the embedding $G \subset K_{\hat{S}}$, then subspace G is of the form $K_{\tilde{S}}$, where $\tilde{S}$ is also an inner operator function.*

(b) *If $K_{\hat{S}_1} \subset K_{\hat{S}_2}$, then inner operator function $\hat{S}_1$ is a left divisor of inner operator function $\hat{S}_2$.*

1.3.7. Countable sets in the upper half-plane

Let $\sigma = \{\lambda_j\}_{j \in \mathbb{N}}$ be a countable set in $\mathbb{C}_+$. Then introduce a number of classes of such sets. Their meaning for a family of eigensubspaces of the model operator will shortly become clear.

Definition II.1.17. A set σ is said to be separable if

$$\inf_{\lambda, \mu \in \sigma, \lambda \neq \mu} |\lambda - \mu| > 0.$$

A set σ is called a Carlesonian one if

$$\delta(\sigma) := \inf_{j} \prod_{i, i \neq j} \left| \frac{\lambda_i - \lambda_j}{\lambda_i - \bar{\lambda}_j} \right| > 0. \tag{C}$$

For such sets we use the notation $\sigma \in (C)$. The value $\delta(\sigma)$ is called a Carleson constant of set σ.

If σ is a finite unification of Carlesonian sets, we say that σ satisfies the Carleson–Newman condition (notation $\sigma \in (CN)$).

A set σ is said to be rare (notation $\sigma \in (R)$) if

$$\inf_{\lambda, \mu \in \sigma, \lambda \neq \mu} R(\lambda, \mu) > 0, \qquad R(\lambda, \mu) := \left| \frac{\lambda - \mu}{\lambda - \bar{\mu}} \right|. \tag{R}$$

1.3.8. Indications for a set to be Carlesonian

The following relationship exists between the introduced properties of sets:

$$(C) \Leftrightarrow (R) \,\&\, (CN). \tag{16}$$

For us the case is significant when σ lies in a strip parallel to the real axis

$$0 < \inf_{\lambda \in \sigma} \operatorname{Im} \lambda \le \sup_{\lambda \in \sigma} \operatorname{Im} \lambda < \infty.$$

Then the separability of σ is equivalent to its rareness and to its Carlesonianness:

$$\sigma \in (C) \Leftrightarrow \sigma \in (R) \Leftrightarrow \inf_{\lambda, \mu \in \sigma, \lambda \ne \mu} |\lambda - \mu| > 0.$$

Example II.1.18. The sets $\sigma = \{2^r i\}_{r \in \mathbb{Z}}$ and $\sigma = \{r + i\}_{r \in \mathbb{Z}}$ are Carlesonian.

In the following lemma we obtain an estimate of the Carleson constant for sets that often arise in applications.

Lemma II.1.19.

(a) *Let $\{\lambda_n\}_{n \in \mathbb{Z}}$ be a separable increasing sequence of real numbers:*

$$\inf(\lambda_{n+1} - \lambda_n) \ge d > 0.$$

Then the set $\sigma := \{v_n\}_{n \in \mathbb{Z}}$, $v_n := \lambda_n + i/2$, is a Carlesonian one and

$$\log \frac{1}{\delta(\sigma)} \le \frac{C}{d}, \qquad C := \int_0^\infty \log\left(1 + \frac{1}{y^2}\right) dy < \infty.$$

(b) *For the set $\sigma^0 := \{nd + i/2\}_{n \in \mathbb{Z}}$ and for $d \to 0$*

$$\log \frac{1}{\delta(\sigma)} = \frac{C}{d}(1 + o(1)).$$

PROOF. From the definition of Carleson constant we have

$$\log \frac{1}{\delta(\sigma)} = \sup_{m \in \mathbb{Z}} \log \prod_{m, n \ne n} \left|\frac{\lambda_m - \lambda_n}{\lambda_m - \lambda_n + i}\right| = \sup_{m \in \mathbb{Z}} \frac{1}{2} \sum_{m, m \ne n} \log\left(1 + \frac{1}{|\lambda_m - \lambda_n|^2}\right).$$

The separability condition for the set $\{\lambda_n\}$ implies

$$|\lambda_m - \lambda_n| \ge |n - m| d,$$

and we deduce

$$\log \frac{1}{\delta(\sigma)} \le \sup_{m \in \mathbb{Z}} \frac{1}{2} \sum_{m, m \ne n} \log\left(1 + \frac{1}{(n - m)^2 d^2}\right) = \sum_{p=1}^\infty \log\left(1 + \frac{1}{p^2 d^2}\right).$$

For the set σ^0, the inequality turns into equality

$$\log \frac{1}{\delta(\sigma^0)} = \sum_{p=1}^{\infty} \log\left(1 + \frac{1}{p^2 d^2}\right).$$

So we arrive at the relations

$$\log \frac{1}{\delta(\sigma)} \le \sum_{p=1}^{\infty} \log\left(1 + \frac{1}{p^2 d^2}\right) = \log \frac{1}{\delta(\sigma^0)}. \qquad (17)$$

Estimating sums by integral, we obtain

$$\sum_{p=1}^{\infty} \log\left(1 + \frac{1}{p^2 d^2}\right) = \sum_{p=1}^{\infty} \int_{p-1}^{p} \log\left(1 + \frac{1}{p^2 d^2}\right) dx$$

$$\le \sum_{p=1}^{\infty} \int_{p-1}^{p} \log\left(1 + \frac{1}{x^2 d^2}\right) dx = \int_{0}^{\infty} \log\left(1 + \frac{1}{x^2 d^2}\right) dx$$

$$= \frac{1}{d} \int_{0}^{\infty} \log\left(1 + \frac{1}{y^2}\right) dy = \frac{C}{d} \qquad (18)$$

and

$$\sum_{p=1}^{\infty} \log\left(1 + \frac{1}{p^2 d^2}\right) = \sum_{p=1}^{\infty} \int_{p}^{p+1} \log\left(1 + \frac{1}{p^2 d^2}\right) dx$$

$$\ge \sum_{p=1}^{\infty} \int_{p}^{p+1} \log\left(1 + \frac{1}{x^2 d^2}\right) dx = \int_{1}^{\infty} \log\left(1 + \frac{1}{x^2 d^2}\right) dx$$

$$= \frac{1}{d} \int_{d}^{\infty} \log\left(1 + \frac{1}{y^2}\right) dy. \qquad (19)$$

Formulas (17), (18), and (19) complete the proof.

1.3.9. Separability of simple fraction families

Later we will need some estimates demonstrating that separability in H_+^2 of a simple fraction family

$$\inf_{\lambda, \mu \in \sigma, \lambda \ne \mu} \|x_\lambda - x_\mu\|_{H_+^2} > 0$$

is equivalent to the rareness of set σ.

Lemma II.1.20. *For $\lambda, \mu \in \mathbb{C}_+$ inequalities hold*

$$\frac{1}{\sqrt{2}} R(\lambda, \mu) \le \|x_\lambda - x_\mu\| \le R(\lambda, \mu). \qquad (20)$$

PROOF. From Subsection 1.1.3,

$$(x_\lambda, x_\mu) = 2\pi i \frac{\sqrt{\operatorname{Im}\mu}}{\sqrt{\pi}} x_\lambda(\mu) = 2i \frac{\sqrt{\operatorname{Im}\lambda}\sqrt{\operatorname{Im}\mu}}{\mu - \bar{\lambda}}.$$

So, using the identity $\operatorname{Re}(i/z) = \operatorname{Im} z/|z|^2$, one finds

$$\|x_\lambda - x_\mu\|^2 = 2(1 - \operatorname{Re}(x_\lambda, x_\mu)) = 2\left[1 - \frac{2\sqrt{\operatorname{Im}\lambda \operatorname{Im}\mu}(\operatorname{Im}\lambda + \operatorname{Im}\mu)}{|\mu - \bar{\lambda}|^2}\right]$$

$$= \frac{1}{|\mu - \bar{\lambda}|^2}[(\operatorname{Re}\lambda - \operatorname{Re}\mu)^2 + (\operatorname{Im}\lambda + \operatorname{Im}\mu)(\sqrt{\operatorname{Im}\lambda} - \sqrt{\operatorname{Im}\mu})^2]$$

$$= \frac{1}{|\mu - \bar{\lambda}|^2}\left[(\operatorname{Re}\lambda - \operatorname{Re}\mu)^2 + \frac{(\operatorname{Im}\lambda + \operatorname{Im}\mu)(\operatorname{Im}\lambda - \operatorname{Im}\mu)^2}{(\sqrt{\operatorname{Im}\lambda} + \sqrt{\operatorname{Im}\mu})^2}\right].$$

Since

$$\frac{1}{2} \le \frac{(\operatorname{Im}\lambda + \operatorname{Im}\mu)(\operatorname{Im}\lambda - \operatorname{Im}\mu)^2}{(\sqrt{\operatorname{Im}\lambda} + \sqrt{\operatorname{Im}\mu})^2} \le 1,$$

(20) follows from the equality

$$\frac{1}{|\mu - \bar{\lambda}|^2}[(\operatorname{Re}\lambda - \operatorname{Re}\mu)^2 + (\operatorname{Im}\lambda - \operatorname{Im}\mu)^2] = R^2(\lambda, \mu).$$

The lemma is proved.

1.3.10. Minimality and the Blaschke condition

Scalar family $\mathscr{X}_\Pi = \{x_\lambda\}_{\lambda \in \sigma} \subset H^2_+$, $x_\lambda = \sqrt{\dfrac{\operatorname{Im}\lambda}{\pi}}(k - \bar{\lambda})^{-1}$,

is minimal if and only if $\sigma \in (B)$ (satisfies the Blaschke condition, see (B), Subsection 1.2.2). Along with it, $\bigvee_{\lambda \in \sigma} x_\lambda = K_\Pi$, where

$$\Pi(k) = \prod_{j \in \mathbb{N}} \varepsilon_j \frac{k - \lambda_j}{k - \bar{\lambda}_j}.$$

If $\sigma \notin (B)$, then $\bigvee_{\lambda \in \sigma} x_\lambda = H^2_+$.

In Section 2 we prove a criterion of minimality for a vector family with $\dim \mathfrak{N} < \infty$:

$$\mathscr{X}_\Pi \in (M) \Leftrightarrow \sigma \in (B).$$

1.3.11. $\mathscr{L}$-basis property of simple scalar fraction families

As one sees from (12) (see Subsection 1.3.3), the Carleson condition is the condition of *-uniform minimality (Definition I.1.15) of a scalar family $\mathscr{X}_\Pi = \{x_\lambda\}_{\lambda \in \sigma}$.

A remarkable fact is that *-uniform minimality of $\mathscr{X}_\Pi$ is an equivalent to $\mathscr{L}$-basis property (Nikol'skiĭ and Pavlov 1970):

$$\mathscr{X}_\Pi \in (LB) \Leftrightarrow \mathscr{X}_\Pi \in (UM) \Leftrightarrow \sigma \in (C).$$

In the vector situation, uniform minimality also implies an $\mathscr{L}$-basis property for dim $\mathfrak{N} < \infty$ (see Section 2; also Treĭl' 1986).

1.3.12. $\mathscr{L}$-basis property of simple fraction families and Carleson constant

If $\{x_\lambda\}_{\lambda \in \sigma} \in (LB)$, then, according to Theorem I.2.1, the operator of the problem of moments of family $\{x_\lambda\}_{\lambda \in \sigma}$ is an isomorphism of spaces $\bigvee_{\lambda \in \sigma} x_\lambda$ and $\ell^2 \colon \sum_{\lambda \in \sigma} |(f, x_\lambda)|^2 \asymp \|f\|^2_{H^2_+}$. Let us use the values for the constants in these inequalities expressed in terms of the Carleson constant $\delta(\sigma)$ (Nikol'skiĭ 1980: lecture VII, sec. 2). For $f \in \bigvee_{\lambda \in \sigma} x_\lambda$, inequalities are valid

$$\frac{\delta^2}{32[1 + 2\log(1/\delta)]} \|f\|^2_{H^2_+} \leq \sum_{\lambda \in \sigma} |(f, x_\lambda)|^2 \leq 32\left(1 + 2\log\frac{1}{\delta}\right)\|f\|^2_{H^2_+}. \quad (21)$$

1.3.13. Carleson–Newman condition and simple fractions

Inequalities (21) imply that for $\sigma \in (CN)$ an estimate holds

$$\sum_{\lambda \in \sigma} |(f, x_\lambda)|^2 \prec \|f\|^2_{H^2_+}, \qquad f \in H^2_+. \quad (22)$$

It so happens that the converse assertion is also true; that is,

$$\sigma \in (CN) \Leftrightarrow \sum_{\lambda \in \sigma} |(f, x_\lambda)|^2 \prec \|f\|^2_{H^2_+}, \qquad f \in H^2_+.$$

Later, we will need the following assertion.

Lemma II.1.21. *Let set σ belong to a strip $|\operatorname{Im} k| \leq$ const and be a unification of a finite number of separable sets. If vectors $\eta_\lambda \in \mathfrak{N}$ ($\lambda \in \sigma$, dim $\mathfrak{N} < \infty$) are also bounded in the norm, then for any function $f \in L^2(0, T; \mathfrak{N})$ inequality holds*

$$\sum_{\lambda \in \sigma} |(f, \eta_\lambda e^{i\lambda t})_{L^2(0, T; \mathfrak{N})}|^2 \leq C_T \|f\|^2_{L^2(0, T; \mathfrak{N})}.$$

PROOF. Without sacrificing generality, one may assume that set σ is situated in the strip $0 < c \le \text{Im } k \le C < \infty$. Indeed, mapping $\varphi(t) \mapsto e^{-\alpha t} \varphi(t)$ is an isomorphism in $L^2(0, T; \mathfrak{N})$ that transforms function $\eta_\lambda e^{i\lambda t}$ into function $\eta_\lambda e^{i(\lambda + i\alpha)t}$, and therefore the shift of the spectrum $\sigma \mapsto \sigma + i\alpha$ in $\mathbb{C}_+$ may only change constant C_T.

Now exponentials $e^{i\lambda t}$ belong to space $L^2(0, \infty)$ and are almost normed

$$\|e^{i\lambda t}\|_{L^2(0, \infty)} = \frac{1}{\sqrt{2 \text{ Im } \lambda}} \asymp 1.$$

Hence, setting $f = 0$ for $t > T$, we have

$$\sum_{\lambda \in \sigma} |(f, \eta_\lambda e^{i\lambda t})_{L^2(0, T; \mathfrak{N})}|^2 \asymp \sum_{\lambda \in \sigma} \left| \left(f, \frac{\eta_\lambda e^{i\lambda t}}{\sqrt{2 \text{ Im } \lambda}} \right)_{L^2(0, \infty; \mathfrak{N})} \right|^2.$$

Let us move to the Fourier representation and verify an inequality

$$\sum_{\lambda \in \sigma} |(\tilde{f}, \eta_\lambda x_\lambda)_{H_+^2}|^2 \prec \|\tilde{f}\|_{H_+^2}^2, \qquad \tilde{f} \in H_+^2(\mathfrak{N})$$

from which the statement of the lemma follows.

We use $\tilde{f}_p(k)$ to denote coordinate functions of vector function $f(k)$ in the orthonormal basis $\{\xi_p^0\}$ of space $\mathfrak{N}$. From the explicit expression for a scalar product (see Subsection 1.1.3), we derive

$$\sum_{\lambda \in \sigma} |(\tilde{f}, \eta_\lambda x_\lambda)_{H_+^2(\mathfrak{N})}|^2 = \sum_{\lambda \in \sigma} |(\langle \tilde{f}(\cdot), \eta_\lambda \rangle, x_\lambda)_{H_+^2(\mathbb{C})}|^2$$

$$= 4\pi \sum_{\lambda \in \sigma} \text{Im } \lambda |\langle \tilde{f}(\lambda), \eta_\lambda \rangle_{\mathfrak{N}}|^2 \prec \sum_{\lambda \in \sigma} \langle\!\langle \tilde{f}(\lambda) \rangle\!\rangle^2$$

$$= \sum_{p=1}^{\dim \mathfrak{N}} \sum_{\lambda \in \sigma} |f_p(\lambda)|^2 = \sum_{p=1}^{\dim \mathfrak{N}} \sum_{\lambda \in \sigma} (4\pi \text{ Im } \lambda)^{-1} |(f_p, x_\lambda)_{H_+^2}|^2.$$

The conditions of the lemma imply that $\sigma \in (CN)$. Applying estimate (22), we then find

$$\sum_{p=1}^{\dim \mathfrak{N}} \sum_{\lambda \in \sigma} |(\tilde{f}_p, x_\lambda)_{H_+^2}^2| \prec \sum_{p=1}^{\dim \mathfrak{N}} \|\tilde{f}_p\|_{H_+^2(\mathbb{C})}^2 = \|\tilde{f}\|_{H_+^2(\mathfrak{N})}^2.$$

The lemma is proved.

1.3.14. The inequalities for exponential sums on an interval

Proposition II.1.22 (Meyer 1985). Let $T > 0$, $\{\omega_n\}_{n \in \mathbb{Z}}$ be a strictly increasing sequence of real numbers, and $\{p_n\}_{n \in \mathbb{Z}}$ be a nonnegative sequence *such that $p_{-n} = p_n$ and $\omega_{-n} = -\omega_n$ for $n \in \mathbb{N}$. Then the inequality*

$$\left\| \sum_{n \in \mathbb{Z}} a_n p_n \exp(i\omega_n t) \right\|^2_{L^2(0,T)} \le C_T \sum_{n \in \mathbb{Z}} |a_n|^2$$

is valid with some constant $C_T < \infty$ if and only if

$$\sup_{n \in \mathbb{Z}} \sum_{n \le \omega_r < n+1} p_r^2 < \infty.$$

1.3.15. Property of functions from space K_a

Lemma II.1.23. *Let Θ be an ESF, $a > 0$, and let the embedding $\Theta^{-1}K_a \subset H^2_+(\mathfrak{N})$ take place. Then $\Theta = const.$*

PROOF. A consequence of formula (7) is that functions

$$(k - \bar{\lambda})^{-1}[1 - \exp(ia(k - \bar{\lambda}))]\eta$$

lie in space K_a for all $\lambda \in \mathbb{C}_+$, $\eta \in \mathfrak{N}$. If vector η runs through an orthonormal basis in $\mathfrak{N}$, for the elements $[\Theta^{-1}(k)]_{ij}$ of matrix function $\Theta^{-1}(k)$ we get inclusions

$$[\Theta^{-1}(k)]_{ij} \frac{(1 - e^{ia(k-\bar{\lambda})})}{(k - \bar{\lambda})} \in H^2_+; \qquad i, j = 1, 2, \ldots, \dim \mathfrak{N}.$$

Together with the estimate (1), the inclusions provide

$$|[\Theta^1(k)]_{ij}| \frac{1}{1 + |k|} \prec 1, \qquad \operatorname{Im} k \ge \varepsilon > 0,$$

which in turn implies that function $\det \Theta^{-1}(k)$, with $\operatorname{Im} k \to \infty$, is not growing faster than a polynomial.

Function $\det \Theta(k)$ is, by definition, an entire inner scalar function. That is, for some $\beta \ge 0$, an equality $\det \Theta(k) = c \, e^{ik\beta}$, $|c| = 1$, holds. From the representation of ESF (see Subsection 1.2.4), one easily deduces that Θ is a unitary constant (a constant unitary matrix) if and only if $\beta = 0$. Evidently, function $\exp(-ik\beta)$ may not grow faster in $\mathbb{C}_+$ than a polynomial except when $\beta = 0$, which proves the lemma.

1.3.16. Isomorphisms transferring exponentials into exponentials

In the space $L^2(0, \infty; \mathfrak{N})$, the following mapping is an isomorphism:

$$f(t) \overset{\mathfrak{U}_\alpha}{\mapsto} e^{i\alpha t} f(t); \qquad \alpha \in \mathbb{R}, \ \mathfrak{U}_\alpha e^{i\lambda t} \eta_\lambda = e^{i(\lambda + \alpha)t} \eta_\lambda. \tag{23}$$

In the space $\mathscr{F}^{-1}K_\Theta$, particularly in $L^2(0, T; \mathfrak{N})$ for $\Theta = \exp(ikT)$, there is also an isomorphism

$$f(t) \overset{\tilde{\mathfrak{U}}_\alpha}{\mapsto} e^{-\alpha t} f(t), \qquad \alpha \in \mathbb{R}, \ \tilde{\mathfrak{U}}_\alpha e^{i\lambda t} \eta_\lambda = e^{i(\lambda + i\alpha)t} \eta_\lambda. \tag{24}$$

Mappings (23) and (24) shift the spectrum σ of the family of exponentials $\{\eta_\lambda e^{i\lambda t}\}_{\lambda \in \sigma}$ and (along with Theorem I.1.25 and Remark I.1.27) can be used to identify their basis, uniform minimality, minimality, and W-linear independence properties.

1.4. Entire functions

We now present some information on entire scalar functions of the exponential type. These functions are of interest because they play an important role in the investigation of minimality and basis properties of families of exponentials (see Section 3). Demonstrations of the given assertions may be found in Boas (1954), Levin (1956), and Nikol'skiĭ (1969).

1.4.1. Entire functions of the exponential type

Entire function f is called an entire function of the exponential type if there exist constants C and T such that $|f(k)| \le C e^{T|k|}$ for all $k \subset \mathbb{C}$. The smallest of constants T is said to be the exponential type of the function.

1.4.2. Growth indicator and indicator diagram

2π-periodic function on $\mathbb{R}$ defined by the equality

$$h_f(\varphi) = \limsup_{r \to \infty} \frac{1}{r} \log|f(r\, e^{i\varphi})|$$

is called a growth indicator of exponential-type function f. The indicator diagram of function f is a convex set G_f such that

$$h_f(\varphi) = \sup_{k \in G_f} \mathrm{Re}(k\, e^{-i\varphi}).$$

Example II.1.24. Let $f = e^{i\alpha k} + e^{-i\beta k}$; $\alpha, \beta > 0$. Then

$$h_f(\varphi) = \begin{cases} \beta \sin \varphi, & \varphi \in [0, \pi] \\ -\alpha \sin \varphi, & \varphi \in [-\pi, 0] \end{cases}$$

and $G_f = [-i\alpha, i\beta]$.

1.4.3. Functions of the Cartwright class

Entire function f of the exponential type is said to be a function of the Cartwright class if

$$\int_{\mathbb{R}} \frac{\max\{\log|f|, 0\}}{1 + x^2} \, dx < \infty.$$

In particular, function f of the exponential type satisfying condition

$$\int_{\mathbb{R}} \frac{|f(x)|^2}{1 + x^2} \, dx < \infty$$

belongs to the Cartwright class.

The indicator diagram of a Cartwright-class function is an interval $[i\alpha, i\beta]$, $\alpha \le \beta$, of the imaginary axis. Its length is called the width of the indicator diagram.

1.4.4. Zeros of Cartwright-class functions

Proposition II.1.25. Let f be a Cartwright-class function with the width T of the indicator diagram, and let σ be the set of zeros of f, where any zero is simple and $f(0) \ne 0$. Then

(a)
$$\lim_{r \to +\infty} \frac{n_{\pm}(r)}{r} = \frac{T}{2\pi},$$

where

$$n_+(r) := \text{card}\{\lambda \in \sigma \,\|\lambda| < r, \text{Re } \lambda \ge 0\},$$

$$n_-(r) := \text{card}\{\lambda \in \sigma \,\|\lambda| < r, \text{Re } \lambda < 0\};$$

(b) *a limit exists*

$$f_0(k) = \text{p.v.} \prod_{\lambda \in \sigma} (1 - k/\lambda) = \lim_{r \to \infty} \prod_{|\lambda| < r, \lambda \in \sigma} (1 - k/\lambda);$$

(c) *indicator diagram f_0 is $[-iT/2, iT/2]$, and for some α and β we have*

$$f(k) = f_0(k) \exp(ik\alpha + \beta), \qquad \alpha \in \mathbb{R};$$

(d) *f has a factorization*

$$f(k) = \exp(-ik\alpha_+)\pi^+(k)f_e^+(k) = \exp(ik\alpha_-)\pi^-(k)f_e^-(k),$$

in which $f_e^{\pm}$ are outer functions in $\mathbb{C}_{\pm}$, respectively, $\pi^{\pm}$ are Blaschke products (in $C_{\pm}$), $\alpha_{\pm} \in \mathbb{R}$, and $\alpha_+ + \alpha_- = T$.

1.4.5. Summability and growth of functions on the real axis

Proposition II.1.26 (Nikol'skiĭ 1969). *Let f be a Cartwright-class function and $f \in L^p(\mathbb{R})$, $p > 1$. Then*

$$f(k) \xrightarrow[k \to \pm\infty]{} 0.$$

1.4.6. Functions of the sine type

Definition II.1.27. An entire function f of the exponential type is said to be of the sine type (STF) if

(a) the zeros of f lie in a strip $\{k \in \mathbb{C} \mid |y| \le h, k = x + iy\}$ for some $h > 0$;
(b) there is $y_0 \in \mathbb{R}$ such that $|f(x + iy_0)| \asymp 1$ holds for $x \in \mathbb{R}$. Note that (a) and (b) imply $|f(x + iy_1)| \asymp 1$ for any y_1, $|y_1| > h$.

Proposition II.1.28 (Levin 1961). *If f is a sine-type function, then its set of zeros is a finite unification of separable sets.*

2. Vector exponential families on the semiaxis

In this section we consider a family of subspaces $\mathscr{X}_\Pi = \{\mathscr{X}_\lambda\}_{\lambda \in \sigma}$, where $\mathscr{X}_\lambda$ are subspaces of simple vector fractions $\mathscr{X}_\lambda = (k - \bar{\lambda})^{-1}\mathfrak{N}_\lambda$; $\mathfrak{N}_\lambda$ are subspaces of an auxiliary Hilbert space $\mathfrak{N}$, $\mathfrak{N}_\lambda = \delta_\lambda \mathfrak{N}$; and δ_λ is an orthoprojector in $\mathfrak{N}$ (dimension of the auxiliary Hilbert space $\mathfrak{N}$ can be infinite). For a finite dimensional case, we produce the necessary and sufficient conditions of minimality (Blaschke condition) and of $\mathscr{L}$-basis property for family $\mathscr{X}_\Pi$. Recall from Subsection 1.1.5 that the Fourier transform transfers the fraction family $\mathscr{X}_\Pi$ into family $\{\exp(-i\bar{\lambda}t)\mathfrak{N}_\lambda\}_{\lambda \in \sigma}$ of vector exponentials, which is more frequently met in the applications; this explains the section title.

Although it is quite natural for the Blaschke condition to be required if $\mathscr{X}_\Pi$ is to be minimal in a finite-dimensional case, the condition has not yet been demonstrated in the literature.

The basis property of vector exponential families was first studied by Nikol'skiĭ and Pavlov (1970), who established that decomposition into an asymptotically orthogonal series was a sufficient condition. It has also been established that the spectrum of an $\mathscr{L}$-basis family can be decomposed in not more than dim $\mathfrak{N}$ Carlesonian sets (Vasyunin 1976). Treĭl' (1986) has proved that for a finite dimensional case, as well as for the scalar one, uniform minimality of family $\mathscr{X}_\Pi$ implies the basis property.

We first prove simple facts concerning the relationship between the properties of scalar and vector families of a general form (not necessarily of exponentials).

In the space $\mathfrak{N}$ we choose an orthonormal basis $\{\xi_p^0\}_{p=1}^{\dim \mathfrak{N}}$. Let $\mathfrak{H}$ be some Hilbert space (the space of scalar functions). Let $\mathfrak{H}(\mathfrak{N})$ denote the space of $\mathfrak{N}$-valued functions $F = \sum_p f_p \xi_p^0$, $f_p \in \mathfrak{H}$, with scalar product

$$(F, G)_{\mathfrak{H}(\mathfrak{N})} = \sum_{p=1}^{\dim \mathfrak{N}} (f_p, g_p)_{\mathfrak{H}}, \qquad G = \sum_{p=1}^{\dim \mathfrak{N}} g_p \xi_p^0, \quad g_p \in \mathfrak{H}.$$

Space $\mathfrak{H}(\mathfrak{N})$ is commonly said to be a tensor product of $\mathfrak{H}$ and $\mathfrak{N}$: $\mathfrak{H}(\mathfrak{N}) = \mathfrak{H} \otimes \mathfrak{N}$.

Lemma II.2.1. *Let $\{f_j\}_{j=1}^{\infty}$ be a family of nonzero elements of $\mathfrak{H}$ and $\{\mathfrak{N}_j\}_{j=1}^{\infty}$ be a family of nonzero subspaces in $\mathfrak{N}$, with $\mathcal{M}$ and $\mathcal{N}$ subsets of $\mathbb{N}$. Then*

(a) $\quad \varphi_{\mathfrak{H}(\mathfrak{N})}\left(\bigvee_{m \in \mathcal{M}} f_m \mathfrak{N}, \bigvee_{n \in \mathcal{N}} f_n \mathfrak{N} \right) = \varphi_{\mathfrak{H}}\left(\bigvee_{m \in \mathcal{M}} f_m, \bigvee_{n \in \mathcal{N}} f_n \right),$

(b) $\quad \varphi_{\mathfrak{H}(\mathfrak{N})}\left(\bigvee_{m \in \mathcal{M}} f_m \mathfrak{N}_m, \bigvee_{n \in \mathcal{N}} f_n \mathfrak{N}_n \right) \geq \varphi_{\mathfrak{H}}\left(\bigvee_{m \in \mathcal{M}} f_m, \bigvee_{n \in \mathcal{N}} f_n \right),$

(c) $\quad \varphi_{\mathfrak{H}(\mathfrak{N})}\left(\bigvee_{m \in \mathcal{M}} f_m \mathfrak{N}_m, \bigvee_{n \in \mathcal{N}} f_n \mathfrak{N}_n \right) \geq \varphi_{\mathfrak{H}}\left(\bigvee_{m \in \mathcal{M}} \mathfrak{N}_m, \bigvee_{n \in \mathcal{N}} \mathfrak{N}_n \right).$ (1)

PROOF.

(a) From Lemma I.1.4(c) one derives, by expanding vectors η_j over an orthonormal basis in $\mathfrak{N}$, $\eta = \sum_p \eta_j^p \xi_p^0$,

$$\sin^2 \varphi_{\mathfrak{H}(\mathfrak{N})}\left(\bigvee_{m \in \mathcal{M}} f_m \mathfrak{N}, \bigvee_{n \in \mathcal{N}} f_n \mathfrak{N} \right)$$

$$= \inf_{\eta_m \in \mathfrak{N}, \tilde{\eta}_n \in \mathfrak{N}, \, \|\sum_m f_m \eta_m\| = 1} \left\| \sum_{m \in \mathcal{M}} f_m \eta_m - \sum_{n \in \mathcal{N}} f_n \tilde{\eta}_n \right\|_{\mathfrak{H}(\mathfrak{N})}^2$$

$$= \inf_{\sum_{p=1}^{\infty} q_p^2 = 1} \sum_{p=1}^{\infty} \inf_{\|\sum f_m \eta_m^p\|^2 = q_p, \, \eta_m^p \in \mathbb{C}, \, \tilde{\eta}_n^p \in \mathbb{C}} \left\| \sum_{m \in \mathcal{M}} f_m \eta_m^p - \sum_{n \in \mathcal{N}} f_n \tilde{\eta}_n^p \right\|^2$$

$$= \inf_{\sum_{p=1}^{\infty} q_p^2 = 1} \sum_{p=1}^{\infty} q_p^2 \sin^2 \varphi_{\mathfrak{H}}\left(\bigvee_{m \in \mathcal{M}} f_m, \bigvee_{n \in \mathcal{N}} f_n \right)$$

$$= \sin^2 \varphi_{\mathfrak{H}}\left(\bigvee_{m \in \mathcal{M}} f_m, \bigvee_{n \in \mathcal{N}} f_n \right).$$ (2)

(b) In view of Lemma I.1.4(c), the value

$$\sin^2 \varphi_{\mathfrak{H}(\mathfrak{N})} \left(\bigvee_{n \in \mathcal{M}} f_m \mathfrak{N}_m, \bigvee_{n \in \mathcal{N}} f_n \mathfrak{N}_n \right)$$

may be written as an expression similar to (2), but with the additional condition $\eta_m \in \mathfrak{N}_m$, $m \in \mathcal{M}$; $\tilde{\eta}_n \in \mathfrak{N}_n$, $n \in \mathcal{N}$. Since the infimum is taken over a smaller set, this, along with (a), proves (b).

(c) Inequality (1) is demonstrated along the same lines as those in (b), since subspaces $\mathfrak{N}_j$ and elements f_j enter (1) symmetrically. The lemma is proved.

Corollary II.2.2.

(a) *An almost normed family of scalar functions $\{f_j\} \subset \mathfrak{H}$ possesses any of the following properties: minimality, uniform minimality, and the $\mathcal{L}$-basis property if and only if the family of subspaces $\{f_j \mathfrak{N}\} \subset \mathfrak{H}(\mathfrak{N})$ has the same property.*

(b) *Let the family of subspaces $\{f_j \mathfrak{N}\}$ have some of the following properties: minimality, uniform minimality, and the $\mathcal{L}$-basis. Then the family of subspaces $\{f_j \mathfrak{N}_j\}$, where $\mathfrak{N}_j$ are nonzero subspaces of $\mathfrak{N}$, has the same property.*

These statements are implied by the fact that each of the family's properties may be expressed in terms of positiveness or uniform boundedness away from zero of the angles between subspaces (see Chapter I, Subsections 1.2, 1.3).

Let us introduce a family $\Xi = \{f_j \eta_j\}_{j=1}^{\infty}$ of vector functions $f_j \in \mathfrak{H}$, $\eta_j \in \mathfrak{N}$, and a family of vectors $\Xi_{\mathfrak{N}} = \{\eta_j\}_{j=1}^{\infty}$; Γ and $\Gamma_{\mathfrak{N}}$ denote Gram matrices of families Ξ and $\Xi_{\mathfrak{N}}$, respectively. The following assertion allows one to arrive at an estimate of the Gram matrix norm of family Ξ by the norm of $\Gamma_{\mathfrak{N}}$.

Lemma II.2.3.

$$(\Gamma c, c)_{\ell^2} \leq 2 \inf_{j \in \mathbb{N}} \| f_j \|_{\mathfrak{H}}^2 (\Gamma_{\mathfrak{N}} c, c)_{\ell^2}$$

$$+ 2 \langle\langle \Gamma_{\mathfrak{N}} \rangle\rangle \| c \|_{\ell^2}^2 \sup_{j,n} \| f_j - f_n \|_{\mathfrak{H}}^2, \qquad c \in \ell^2.$$

PROOF. Let us take an arbitrary element f_{j_0}. Then for any finite element

$c \in \ell^2$ we have

$$(\Gamma c, c)_{\ell^2} = \left\| \sum_{j=1}^{\infty} c_j f_j \eta_j \right\|_{\mathfrak{H}(\mathfrak{N})}^2 = \left\| \sum_{j=1}^{\infty} c_j f_{j_0} \eta_j + \sum_{j=1}^{\infty} c_j (f_j - f_{j_0}) \eta_j \right\|_{\mathfrak{H}(\mathfrak{N})}^2$$

$$\leq 2 \left\| \sum_{j=1}^{\infty} c_j f_{j_0} \eta_j \right\|_{\mathfrak{H}(\mathfrak{N})}^2 + 2 \left\| \sum_{j=1}^{\infty} c_j (f_j - f_{j_0}) \eta_j \right\|_{\mathfrak{H}(\mathfrak{N})}^2 \qquad (3)$$

The first term may be estimated as

$$\left\| \sum_{j=1}^{\infty} c_j f_{j_0} \eta_j \right\|_{\mathfrak{H}(\mathfrak{N})}^2 = \left\| f_{j_0} \left(\sum_{j=1}^{\infty} c_j \eta_j \right) \right\|_{\mathfrak{H}(\mathfrak{N})}^2 = \| f_{j_0} \|_{\mathfrak{H}}^2 \left\langle\!\!\left\langle \sum_{j=1}^{\infty} c_j \eta_j \right\rangle\!\!\right\rangle^2$$

$$= \| f_{j_0} \|_{\mathfrak{H}}^2 (\Gamma_{\mathfrak{N}} c, c)_{\ell^2}.$$

In order to estimate the second one in (3), choose an orthonormal basis $\{\zeta_n^0\}_{n=1}^{\infty}$ in space $\mathfrak{H}$ and expand all the elements $g_j = f_j - f_{j_0}$ over it

$$f_j - f_{j_0} = \sum_{n=1}^{\infty} g_j^{(n)} \zeta_n^{(0)}.$$

Then inequalities are true

$$\left\| \sum_{j=1}^{\infty} c_j \eta_j (f_j - f_{j_0}) \right\|_{\mathfrak{H}(\mathfrak{N})}^2 = \sum_{n=1}^{\infty} \left\langle\!\!\left\langle \sum_{j=1}^{\infty} c_j \eta_j g_j^{(n)} \right\rangle\!\!\right\rangle^2$$

$$= \sum_{n=1}^{\infty} \langle \Gamma_{\mathfrak{N}} \{ c_j \eta_j g_j^{(n)} \}, \{ c_j \eta_j g_j^{(n)} \} \rangle$$

$$\leq \sum_{n=1}^{\infty} \langle\!\langle \Gamma_{\mathfrak{N}} \rangle\!\rangle \sum_{j=1}^{\infty} |c_j g_j^{(n)}|^2$$

$$= \langle\!\langle \Gamma_{\mathfrak{N}} \rangle\!\rangle \sum_{j=1}^{\infty} \| g_j \|_{\mathfrak{H}}^2 |c_j|^2$$

$$\leq \sup_{j,n} \| f_n - f_j \|_{\mathfrak{H}}^2 \langle\!\langle \Gamma_{\mathfrak{H}} \rangle\!\rangle \| c \|_{\ell^2}^2.$$

The lemma is proved.

2.1. Minimality

Here we prove that for $\dim \mathfrak{N} < \infty$ the Blaschke condition on the spectrum $\sigma \subset \mathbb{C}_+$ (see Subsection 1.2.2, condition (B)) is necessary for the minimality in $L^2(0, \infty; \mathfrak{N})$ of the family of vector exponential subspaces

$$\mathscr{E} = \{\exp(-i\bar{\lambda}t)\mathfrak{N}_\lambda\}_{\lambda \in \sigma}, \qquad \sigma \subset \mathbb{C}_+, \mathfrak{N}_\lambda \subset \mathfrak{N}.$$

Obviously, this cannot take place for dim $\mathfrak{N} = \infty$, since the minimality and even orthogonality of $\mathscr{E}$ may be gained only because of subspaces $\mathfrak{N}_\lambda$ without any restrictions on the family spectrum.

Theorem II.2.4. If dim $\mathfrak{N} < \infty$, *then* $\mathscr{E} \in (M)$ *if and only if* $\sigma \in (B)$.

PROOF. If $\sigma \in (B)$, scalar family $\mathscr{E}_{sc} = \{\exp(-i\bar{\lambda}t)\}_{\lambda \in \sigma}$ is minimal, since family $\{(k - \bar{\lambda})^{-1}\}_{\lambda \in \sigma}$ related to $\mathscr{E}_{sc}$ by the Fourier transform, is minimal (see Subsection 1.3.10). By Corollary 2, $\mathscr{E} \in (M)$.

Let $\mathscr{E} \in (M)$. We pass to the family $\hat{\mathscr{X}}$ in $H^2(\mathfrak{N})$ related to family $\mathscr{E}$ by the Fourier and U mappings (see Subsections 1.1.2–1.1.5):

$$\hat{\mathscr{X}} = \{\hat{\mathscr{X}}_\mu\}_{\mu \in \hat{\sigma}}, \qquad \hat{\mathscr{X}}_\mu = (1 - \bar{\mu}z)^{-1}\mathfrak{N}_\mu,$$

$$\mu = \omega(\lambda), \qquad \hat{\sigma} := \{\omega(\lambda)\}_{\lambda \in \sigma}.$$

Let $\hat{\mathscr{X}}' = \{\hat{\mathscr{X}}'_\mu\}_{\mu \in \hat{\sigma}}$ denote the family that is biorthogonal to $\hat{\mathscr{X}}$ and lies in subspace $G := \bigvee \hat{\mathscr{X}}$.

Subspace $\hat{\mathscr{X}}_\mu$ consists of eigenfunction of the operator $\mathfrak{S}^*$ (adjoint to shift operator $\mathfrak{S}$) corresponding to eigenvalue $\bar{\mu}$ (see Definition 1.12 and Subsections 1.3.2–1.3.4). Indeed, for $f(z) = (1 - \bar{\mu}z)^{-1}\eta$, $\eta \in \mathfrak{N}$, we have

$$(\sigma^* f(z)) = \frac{f(z) - f(0)}{z} = \frac{\bar{\mu}\eta}{(1 - \bar{\mu}z)} = \bar{\mu}f(z).$$

It follows that G is the discrete spectrum subspace of model operator $\mathfrak{S}^*|_G$ and thus G is invariant under operator $\mathfrak{S}^*$. Therefore, Proposition 1.15 provides us with a representation

$$G = H^2(\mathfrak{N}) \ominus \mathscr{S}H^2(\mathfrak{N}_1)$$

in which $\mathfrak{N}_1$ is a subspace in $\mathfrak{N}$, $\mathscr{S}(z): \mathfrak{N}_1 \to \mathfrak{N}$ is an analytical in $\mathbb{D}$ operator function, being isometric for almost all $z \in \mathbb{T}$.

We now exploit formula (15) in Section II.1 for orthoprojector P_G

$$P_G = I|_{H^2(\mathfrak{N})} - \mathscr{S}\hat{P}_+\mathscr{S}^*.$$

Here $\hat{P}_+$ is the orthoprojector from $L^2(\mathbb{T}, \mathfrak{N}_1)$ to $H^2(\mathfrak{N}_1)$ while $\mathscr{S}$ and $\mathscr{S}^*$ are multiplication operators by $\mathscr{S}(\cdot)$ and $\mathscr{S}^*(\cdot)$, respectively. Since $\hat{x}_\mu$ is the eigenfunction of operator $\sigma^*|_G$, element $\hat{x}'_\mu$ of biorthogonal system

$\hat{x}'$ is the eigenfunction of adjoint operator $P_G \mathfrak{S}|_G$:

$$P_G z \hat{x}'_\mu(z) = \mu \hat{x}'_\mu(z).$$

We now prove that

$$\hat{x}'_\mu = \frac{\mathscr{S}(z)}{z - \mu} \eta^\mu, \qquad \eta^\mu \in \operatorname{Ker} \mathscr{S}(\mu), \tag{4}$$

as for an inner operator function S (as for $\mathfrak{N}_1 = \mathfrak{N}$). With the obtained representation for P_G we get the relations

$$(z - \mu)\hat{x}'_\mu = \mathscr{S}\varphi_\mu, \qquad \varphi_\mu := \hat{P}_+ \mathscr{S}^* z \hat{x}'_\mu.$$

Now, expressing $\hat{x}'_\mu$ from the first equality,

$$\hat{x}'_\mu = (z - \mu)^{-1} \mathscr{S} \varphi_\mu \tag{5}$$

and introducing it into the second one, we have

$$\varphi_\mu = \hat{P}_+ \mathscr{S}^* \frac{z}{z - \mu} \mathscr{S} \varphi_\mu = \hat{P}_+ \frac{z}{z - \mu} \varphi_\mu. \tag{6}$$

Actually, multiplication operators by scalar function $z/(z - \mu)$ and by operator function $\mathscr{S}$ commute, and for isometry $\mathscr{S}(e^{i\theta})$ formula $\mathscr{S}^*(e^{i\theta})\mathscr{S}(e^{i\theta}) = I|_{\mathfrak{N}_1}$ holds.

Let us prove that the solution φ_μ of (6) is a constant vector. Vector function $z\varphi_\mu(z)/(z - \mu)$ may be represented in the form

$$\frac{z}{z - \mu} \varphi_\mu(\mu) + \frac{z}{z - \mu} [\varphi_\mu(z) - \varphi_\mu(\mu)] = \frac{\mu\varphi_\mu(\mu)}{z - \mu} + \varphi_\mu(\mu)$$

$$+ \frac{z}{z - \mu} (\varphi_\mu(z) - \varphi_\mu(\mu)). \tag{7}$$

The second and third items in expression (7) lie in the space $H^2(\mathfrak{N}_1)$. The first one is orthogonal to $H^2(\mathfrak{N}_1)$ in $L^2(\mathbb{T}, \mathfrak{N})$, since it may be represented in the form $\sum_{n=1}^\infty c_n z^{-n}$, $c_n \in \mathfrak{N}$. Therefore (6) takes the form

$$\varphi_\mu(z) = \varphi_\mu(\mu) + \frac{z}{z - \mu} (\varphi_\mu(z) - \varphi_\mu(\mu)),$$

whence $\varphi_\mu(z) = \varphi_\mu(\mu)$. We proved the representation (4).

Since $\hat{x}'_\mu$ is analytical in $\mathbb{D}$, $\mathscr{S}(\mu)\eta^\mu = 0$. Evidently, $\mathscr{S}(z)\eta^\mu \not\equiv 0$, because $\mathscr{S}(z)$ for almost all $z \in \mathbb{T}$ is an isometry of spaces $\mathfrak{N}_1$ and $\mathfrak{N}$. Therefore, there exists a minor of the order dim $\mathfrak{N}_1$ of matrix function $\mathscr{S}(z)$ (in some basis) that is not an identical zero. This minor tends to zero in the points of spectrum σ. Because $\mathscr{S}$ is a bounded matrix function, the minor is also bounded in $\mathbb{D}$. Hence, the set of its zeros satisfies the Blaschke condition (Subsections 1.2.7, 1.2.3, and 1.2.2). The theorem is proved.

Remark II.2.5. From the theorem it follows that for a minimal family of exponentials or simple fractions (in the case dim $\mathfrak{N} < \infty$) there exists a BPP generated by family $\{x_\lambda \mathfrak{N}_\lambda\}_{\lambda \in \sigma}$ (see Subsections 1.2.11 and 1.3.5), and $\bigvee_{\lambda \in \sigma} \{x_\lambda \mathfrak{N}_\lambda\} = K_\Pi$.

2.2. The basis property

In a scalar case, the $\mathscr{L}$-basis property of an exponential family or a simple fraction family is equivalent to its uniform minimality and is expressed via Carleson condition (C) on the family spectrum (Subsections 1.3.7, 1.3.11). From Corollary 2, one thus finds that the Carleson condition is sufficient for a vector family of exponentials or corresponding simple fractions to be an $\mathscr{L}$-basis. In a vector case, however, the Carleson condition is no longer necessary for an $\mathscr{L}$-basis property of a family.

Example II.2.6. Let σ_1 and σ_2 be two close Carlesonian sets in the sense that $\sigma_1 \cup \sigma_2 \notin (C)$. For instance, one may set $\sigma_1 = \{n + i\}_{n \in \mathbb{N}}$, $\sigma_2 = \{n + i + 1/n\}_{n \in \mathbb{N}}$. Let $\mathfrak{N}_1$ and $\mathfrak{N}_2$ be orthogonal in $\mathfrak{N}$ subspaces. Then

$$\{x_\lambda \mathfrak{N}_1\}_{\lambda \in \sigma_1} \cup \{x_\lambda \mathfrak{N}_2\}_{\lambda \in \sigma_2} \in (LB).$$

For dim $\mathfrak{N} < \infty$, it seems natural that the spectrum of an $\mathscr{L}$-basis family cannot be too "thick." Indeed, the following facts take place.

Proposition II.2.7 (Vasyunin 1976; Nikol'skiĭ 1980: lecture VII, corollary 2). *Let family $\mathscr{E} = \{\exp(i\lambda t)\mathfrak{N}_\lambda\}_{\lambda \in \sigma}$ form an $\mathscr{L}$-basis in space $L^2(0, \infty; \mathfrak{N})$, dim $\mathfrak{N} < \infty$. Then its spectrum σ is a unification of not more than dim $\mathfrak{N}$ Carlesonian sets.*

Proposition II.2.8 (Treĭl' 1986). *Let family $\{\delta_\lambda\}_{\lambda \in \sigma}$ of orthogonal projectors δ_λ in $\mathfrak{N}$ be relatively compact (any infinite sequence contains a converging subsequence), and let $\mathscr{E}$ have the form $\mathscr{E} = \{\exp(i\lambda t)\delta_\lambda \mathfrak{N}\}_{\lambda \in \sigma}$. Then family $\mathscr{E}$ is an $\mathscr{L}$-basis if it is uniformly minimal.*

If dim $\mathfrak{N} < \infty$, then family $\{\delta_\lambda\}$ is always relatively compact. So, Proposition 8 extends the equivalence of the basis property and uniform ·minimality to a finite dimensional case. For dim $\mathfrak{N} = \infty$ one should not, apparently, expect such an equivalence to be present. Consider the following example.

Example II.2.9. Let dim $\mathfrak{N} = \infty$ and set $\xi_n = \xi_n^0 + \xi_1^0$ where $\{\xi_n^0\}_{n \in \mathbb{N}}$ is an orthonormal basis in $\mathfrak{N}$. Then (see Example I.1.18) family $\{\xi_n\}_{n=2}^\infty$ is a uniform minimal one but not an $\mathscr{L}$-basis one in $\mathfrak{N}$.

Let us take a sequence of points λ_n that converge to a point $\lambda_0 \in \mathbb{C}_+$ and consider a family of vector simple fractions $\{x_{\lambda_n} \xi_n\}_{n=2}^\infty$. By the force of Corollary 2, this family is uniformly minimal. In order to demonstrate that it is not an $\mathscr{L}$-basis, we take a sequence $\{c_n\}$ of positive numbers belonging to $\ell^2 \setminus \ell^1$. Since simple fractions x_{λ_n} converge in H_+^2 to function x_{λ_n}, and since vector series

$$\sum_{n=2}^\infty c_n \xi_n = \sum_{n=2}^\infty c_n \xi_n^0 + \left(\sum_{n=2}^\infty c_n \right) \xi_1^0$$

diverges, it is not difficult to show that series $\sum_{n=2}^\infty c_n x_{\lambda_n} \xi_n$ also diverges in space $H_+^2(\mathfrak{N})$. From the Bari theorem (Proposition I.1.17) it then follows that $\{x_{\lambda_n} \xi_n\} \notin (LB)$.

We will need a test for a family of simple fractions to be a basis. First we give another definition.

Definition II.2.10 (Nikol'skiĭ 1980: lecture VII, sec. 3). Hyperbolic metric in $\mathbb{C}_+$ is a function $\rho(\lambda, \mu)$ such that

$$\tanh\tfrac{1}{2}\rho(\lambda, \mu) = R(\lambda, \mu) = \left| \frac{\lambda - \mu}{\lambda - \bar{\mu}} \right|,$$

so

$$\rho(\lambda, \mu) = \log \frac{1 + R(\lambda, \mu)}{1 - R(\lambda, \mu)}.$$

Let $D_\rho(\lambda, r)$ be "a circle" of radius r and center λ in this metrics.

Let $\sigma \subset \mathbb{C}_+$. Denote $G_m(r)$, $m = 1, 2, \ldots$, connected components of set $G(r)$, with $G(r)$ being a unification in $\lambda \in \sigma$ of the circles $D_\rho(\lambda, r)$. Write $\Lambda_m(r)$ for the set of points from σ lying in $G_m(r)$: $\Lambda_m(r) = \sigma \cap G_m(r)$ and

$L_m(r)$ for subspaces of simple fractions with spectra $\Lambda_m(r)$:

$$L_m(r) := \bigvee_{\lambda \in \Lambda_m(r)} (k - \bar{\lambda})^{-1}.$$

Proposition II.2.11 (Vasyunin 1977; Nikol'skiĭ 1980: lecture IX, sec. 5). *Let set σ be a unification of N Carlesonian sets σ_j. Then*

(a) *for any $r > 0$, the family of subspaces $\{L_m(r)\}$ forms an $\mathscr{L}$-basis;*
(b) *for $r < r_0 := 1/(2N) \min_j \inf_{\lambda \neq \mu; \, \lambda, \mu \in \sigma_j} \rho(\lambda, \mu)$, the dimension of $L_m(r)$ is not more than N (card $\Lambda_m(r) \leq N$).*

The proof of assertion (a) is presented in Vasyunin (1977) and Nikol'skiĭ (1980) for only $r = r_* := r_0/16$. However, it is true for smaller r as well. For $r > r_*$ we have

$$G_m(r_*) = \bigcup_{\lambda \in \Lambda_m(r_*)} D_\rho(\lambda, r_*) \subset \bigcup_{\lambda \in \Lambda_m(r_*)} D_\rho(\lambda, r).$$

It is clear that the latter unification is connected and so for any m, $G_m(r_*)$ lies in some component $G_n(r)$. Therefore, subspace $L_n(r)$ either coincides with the subspace $L_m(r_*)$ or contains this subspace along with other subspaces $L_{m_j}(r_*)$:

$$L_n(r) = \bigvee_j L_{m_j}(r_*).$$

Since the family of subspaces $\{L_m(r_*)\}$ constitutes an $\mathscr{L}$-basis, the same is true for the family of subspaces $\{L_n(r)\}$.

In other words, assertion (a) means the following. Under the conditions $\sigma \in (CN)$, $\sigma \notin (C)$, the simple fraction family $\{x_\lambda\}_{\lambda \in \sigma}$ is not an $\mathscr{L}$-basis in H_+^2. However, joining simple fractions corresponding to "close" points of the spectrum in groups, one finds the subspaces spanned over those groups of fractions to form the $\mathscr{L}$-basis.

Proposition 11 enables one to obtain a criterion for a family of vector exponentials to be an $\mathscr{L}$-basis, if σ satisfies the Carleson–Newman condition (CN). (The condition was formulated in Ivanov 1985.)

Theorem II.2.12. *If $\sigma \in (CN)$, then the family $\mathfrak{X}_\Pi = \{(k - \bar{\lambda})^{-1} \mathfrak{N}_\lambda\}_{\lambda \in \sigma}$ forms an $\mathscr{L}$-basis if and only if for some $r > 0$ an inequality holds*

$$\inf_m \min_{\lambda \in \Lambda_m(r)} \varphi_\mathfrak{N}\left(\mathfrak{N}_\lambda, \bigvee_{\mu \in \Lambda_m(r), \mu \neq \lambda} \mathfrak{N}_\mu\right) > 0. \qquad (8)$$

Roughly speaking, condition (8) means that in each group Λ_m the family of subspaces $\{\mathfrak{N}_\lambda\}_{\lambda \in \Lambda_m}$ is an $\mathscr{L}$-basis in $\mathfrak{N}$, and this property is "uniform" in groups (i.e., with respect to m).

PROOF. According to Proposition 11, scalar subspace family

$$\left\{ \bigvee_{\lambda \in \Lambda_m(r)} (k - \bar{\lambda})^{-1} \right\}_m$$

is an $\mathscr{L}$-basis. The same is true for family $\{\mathscr{L}_m\}$ (Corollary 2a)

$$\mathscr{L}_m := \bigvee_{\lambda \in \Lambda_m(r)} (k - \bar{\lambda})^{-1} \mathfrak{N}_\lambda.$$

Let $\mathscr{P}_m$ be a projector in $H^2_+(\mathfrak{N})$ on $\mathscr{L}_m$ parallel to $\bigvee_{n \neq m} \mathscr{L}_n$. The $\mathscr{L}$-basis property of $\{\mathscr{L}_m\}$ is equivalent, in views of Proposition I.1.24(b), to the estimates

$$\|\varphi\|^2 \asymp \sum_m \|\mathscr{P}_m \varphi\|^2, \qquad \varphi \in \bigvee \mathscr{L}_m = \bigvee \mathscr{X}_\Pi. \tag{9}$$

Let us write $\mathscr{P}_{m,\lambda}$ for the projector in $\mathscr{L}_m$ on $(k - \bar{\lambda})^{-1}\mathfrak{N}_\lambda$ parallel to

$$\bigvee_{\mu \in \Lambda_m(r), \mu \neq \lambda} (k - \bar{\mu})^{-1} \mathfrak{N}_\mu.$$

It is evident that

$$\mathscr{P}_m = \left(\sum_{\lambda \in \Lambda_m(r)} \mathscr{P}_{m,\lambda} \right) \mathscr{P}_m$$

and that $\mathscr{P}_{m,\lambda}\mathscr{P}_m$ is the projector on $(k - \bar{\lambda})^{-1}\mathfrak{N}_\lambda$, $\lambda \in \Lambda_m(r)$, parallel to $\bigvee_{\mu \in \sigma, \mu \neq \lambda} (k - \bar{\mu})^{-1}\mathfrak{N}_\mu$. That is why the $\mathscr{L}$-basis property of family, $\mathscr{X}_\Pi$ is equivalent (Proposition I.1.24(b)) to the estimates

$$\|\varphi\|^2 \asymp \sum_{\lambda, \mu} \|\mathscr{P}_{m,\lambda}\mathscr{P}_m \varphi\|^2, \qquad \varphi \in \bigvee \mathscr{X}_\Pi. \tag{10}$$

From a comparison of (10) and (9) it follows that the $\mathscr{L}$-basis property of $\mathscr{X}_\Pi$ is equivalent to the estimate uniform in m and φ.

$$\|\varphi\|^2 \asymp \sum_{\lambda \in \Lambda_m(r)} \|\mathscr{P}_{m,\lambda}\varphi\|^2, \qquad \varphi \in \mathscr{L}_m, m \in \mathbb{N}. \tag{11}$$

Since for $r < r_0$ sets $\Lambda_m(r)$ contain not more than N points, (11) is equivalent to the estimate

$$\sup_{m} \max_{\lambda \in \Lambda_m(r)} \|\mathscr{P}_{m,\lambda}\| < \infty. \tag{12}$$

By Lemma I.1.9(a), condition (12) may be written in the form

$$\alpha(r) := \inf_{m} \min_{\lambda \in \Lambda_m(r)} \varphi_{H^2_+(\mathfrak{N})}\left(x_\lambda \mathfrak{N}_\lambda, \bigvee_{\lambda \in \Lambda_m(r), \mu \neq \lambda} x_\mu \mathfrak{N}_\mu\right) > 0. \tag{13}$$

So, we just proved the following lemma.

Lemma II.2.13. For $r < r_0$, condition (13) is equivalent to the $\mathscr{L}$-basis property of family $\mathscr{X}_\Pi$.

Let condition (8) be valid. By Lemma 1(c),

$$\varphi_{H^2_+(\mathfrak{N})}\left(x_\lambda \mathfrak{N}_\lambda, \bigvee_{\lambda \in \Lambda_m(r), \mu \neq \lambda} x_\lambda \mathfrak{N}_\lambda\right) \geq \inf_{m, \lambda} \varphi_{\mathfrak{N}}\left(\mathfrak{N}_\lambda, \bigvee_{\lambda \in \Lambda_m(r), \mu \neq \lambda} \mathfrak{N}_\lambda\right).$$

Then Lemma 13 implies that family $\mathscr{X}_\Pi$ is an $\mathscr{L}$-basis.

Suppose that condition (8) is not true for all r. Let us show that then $\mathscr{X}_\Pi \notin (LB)$. The idea of the demonstration is that for small r "almost" linear dependence of subspace family $\{\mathfrak{N}_\nu\}_{\nu \in \Lambda_m(r)}$ implies "almost" linear dependence of simple fraction subspaces $\{x_\nu \mathfrak{N}_\nu\}_{\nu \in \Lambda_m(r)}$.

So, by the force of the suggestion, for each r, $0 < r < r_0$, and every $\varepsilon > 0$, a point $\lambda \in \Lambda_m(r)$ may be found there along with vectors $\eta_\lambda \in \mathfrak{N}_\lambda$, $\langle\langle \eta_\lambda \rangle\rangle = 1$, $\eta_\mu \in \mathfrak{N}_\mu$, $\mu \in \Lambda_m(r)$, $\mu \neq \lambda$, such that

$$\left\langle\left\langle \eta_\lambda + \sum_{\mu \in \Lambda_m(r), \mu \neq \lambda} \eta_\mu \right\rangle\right\rangle < \varepsilon. \tag{14}$$

In each subspace $\mathfrak{N}_\nu$, $\nu \in \Lambda_m(r)$, let us now choose an orthonormal basis $\{\eta_\nu^{(p)}\}_{p=1}^{n_\nu}$, $n_\nu = \dim \mathfrak{N}_\nu$. We denote Gram matrices of family $\{x_\nu \eta_\nu^{(p)}\}_{p=1, \nu \in \Lambda_m(r)}^{p=n_\nu} \subset H^2_+(\mathfrak{N})$ and family $\{\eta_\nu^{(p)}\}_{p=1, \nu \in \Lambda_m(r)}^{p=n_\nu} \subset \mathfrak{N}$ by Γ and $\Gamma_{\mathfrak{N}}$, respectively.

Expand vectors η_ν over this basis

$$\eta_\nu = \sum_{p=1}^{n_\nu} c_\nu^{(p)} \eta_\nu^{(p)}$$

and consider vector $c = \{c_\nu^{(p)}\}_{p=1, \nu \in \Lambda_m(r)}^{p=n_\nu}$ (as an element of ℓ^2). Since

$$\sum_{\nu, p} |c_\nu^{(p)}|^2 = \sum_\nu \sum_p |c_\nu^{(p)}|^2 = \sum_\nu \langle\langle \eta_\nu \rangle\rangle^2,$$

we have $\|c\|_{\ell^2} \geq 1$.

Inequality (14), in terms of $\Gamma_{\mathfrak{N}}$, means

$$(\Gamma_{\mathfrak{N}}c, c) = \left\langle\!\!\left\langle \eta_\lambda + \sum_{v \in \Lambda_m(r), \, v \neq \lambda} \eta_v \right\rangle\!\!\right\rangle^2 < \varepsilon.$$

Then Lemma 3 provides an estimate

$$(\Gamma c, c) \leq 2\varepsilon + 2\|\Gamma_{\mathfrak{N}}\| \, \|c\|^2 \max_{\lambda, \, \mu \in \Lambda_m(r)} \|x_\lambda - x_\mu\|_{H_+^2}^2. \tag{15}$$

Let us now prove that for $r < r_0$

$$\|\Gamma_{\mathfrak{N}}\| \leq N, \tag{16}$$

and that

$$\max_{\lambda, \, \mu \in \Lambda_m(r)} \|x_\lambda - x_\mu\|_{H_+^2} \leq Nr. \tag{17}$$

We first check inequality (16), and for this we take an arbitrary vector $d = \{d_v^{(p)}\}_{p=1, \, v \in \Lambda_m(r)}^{p=n_v}$; then

$$\frac{(\Gamma_{\mathfrak{N}}d, d)}{\|d\|^2} = \frac{\langle\!\langle \sum_{v, p} d_v^{(p)} \eta_v^{(p)} \rangle\!\rangle^2}{\sum_{v, p} |d_v^{(p)}|^2} = \frac{\langle\!\langle \sum_{v \in \Lambda_m(r)} \tilde{\eta}_v \rangle\!\rangle^2}{\sum_{v \in \Lambda_m(r)} \langle\!\langle \tilde{\eta}_v \rangle\!\rangle^2}, \tag{18}$$

where

$$\tilde{\eta}_v := \sum_{p=1}^{n_v} d_v^{(p)} \eta_v^{(p)}.$$

An obvious inequality

$$\left\|\sum_{p=1}^{N} z_p\right\|_{\mathfrak{H}}^2 \leq N \sum_{p=1}^{N} \|z_p\|_{\mathfrak{H}}^2 \tag{19}$$

and relation (18) produce inequality (16).

To justify inequality (17), note that between two points $v, \mu \in \Lambda_m(r)$, a path exists running through the centers of circles $D_\rho(\lambda, r)$ (G_m is a connected set). There are not more than N circles; hence, $\rho(\mu, v) \leq Nr$. Using as well the estimate from Lemma 1.22, we arrive at the inequality (17).

The proved inequalities (16) and (17) produce

$$(\Gamma c, c) \leq 2(\varepsilon + N^3 r^2)\|c\|_{\ell^2}^2.$$

Since the numbers ε and r may be infinitesimal, the last inequality is incompatible with the $\mathcal{L}$-basis property of family $\mathscr{X}_\Pi$: if $\mathscr{X}_\Pi \in (LB)$, then by Proposition I.1.20(b) operator Γ has a bounded inverse. This completes the proof of the theorem.

In a finite dimensional situation, one can relax the requirements of Theorem 12 exploiting Proposition 7 to some extent.

Corollary II.2.14. If dim $\mathfrak{N} < \infty$, *then* $\mathscr{X}_\Pi \in (LB)$ *if and only if for some r condition (8) holds.*

The exponential families arising in control problems for string net (see Chapter VII) have asymptotically serial structure. In this case, the convenient sufficient conditions of basis property may be given.

Definition II.2.15. Family $\{(k - \bar{\lambda}_n)^{-1}\delta_{\lambda_n}\mathfrak{N}\}_{\lambda_n \in \sigma}$, *where* δ_{λ_n} *are ortho-projectors in* $\mathfrak{N}$, *is said to be a Carleson series* $C(\delta)$, *if* $\sigma \in (C)$ *and* $\delta_{\lambda_n} \xrightarrow[n \to \infty]{} \delta$ *for some orthoprojector* δ *in* $\mathfrak{N}$.

Corollary II.2.16. Let $\mathscr{X}_\Pi$ *be a unification of Carleson series* $C(\delta_j)$ *and family* $\{\delta_j\mathfrak{N}\}$ *of subspaces be an* $\mathscr{L}$-*basis. Then* $\mathscr{X}_\Pi \in (LB)$.

In the following examples the spectrum lies in a strip parallel to the real axis and the hyperbolic metric is equivalent to the Euclidean one.

Example II.2.17. Set

$$\mathscr{X}_\Pi := \{(k - 2n + i)^{-1}\delta_1\mathfrak{N}\}_{n \in \mathbb{Z}} \cup \{(k - 2n + 1 + i)^{-1}\delta_2\mathfrak{N}\}_{n \in \mathbb{Z}}$$

$$\cup \{(k - n - \gamma_n + i)^{-1}\delta_3\mathfrak{N}\}_{n \in \mathbb{Z}}, \qquad 0 < |\gamma_n| \le \gamma < 1/2.$$

Here, δ_1, δ_2, and δ_3 are distinct one-dimensional projectors in $\mathbb{C}^2$, $\mathscr{X}_\Pi$ is a unification of three Carleson series in $\mathbb{C}^2$, and Corollary 16 is inapplicable. Nevertheless, the family spectrum is a unification of only two Carlesonian sets $\sigma^{(1)} = \{n + i\}_{n \in \mathbb{Z}}$ and $\sigma^{(2)} = \{n + \gamma_n + i\}_{n \in \mathbb{Z}}$. For $r < 1/2 - \gamma$, not more than two points fall in $\Lambda_m(r)$, and not more than one from each of the sets $\sigma^{(1)}$ and $\sigma^{(2)}$. Therefore the angles in (8) take only one of three values

$$\varphi(\delta_1\mathfrak{N}, \delta_2\mathfrak{N}), \qquad \varphi(\delta_1\mathfrak{N}, \delta_3\mathfrak{N}) \quad \text{and} \quad \varphi(\delta_2\mathfrak{N}, \delta_3\mathfrak{N})$$

and are bounded away from zero. Hence, $\mathscr{X}_\Pi \in (LB)$.

In Chapter VII, which focuses on the controllability of a system of strings, a family also appears with a number of Carleson series greater than the dimension of space.

Example II.2.18. Set

$$\mathscr{X}_\Pi = \{(k - n + i)^{-1}\eta_n^{(1)}\}_{n\in\mathbb{Z}} \cup \{(k - n - \gamma_n + i)^{-1}\eta_n^{(2)}\}_{n\in\mathbb{Z}},$$

where $0 < |\gamma_n| \le \gamma < 1/2$ and

$$\eta_n^{(1)} = \begin{pmatrix} \sin\alpha_n \\ \cos\alpha_n \end{pmatrix}, \qquad \eta_n^{(2)} = \begin{pmatrix} \sin(\alpha_n + \beta) \\ \cos(\alpha_n + \beta) \end{pmatrix}, \qquad \alpha_n \in \mathbb{R}, \ \beta \in (0, \pi).$$

The behavior of vector $\eta_n^{(1)}$ is disordered, but for any n

$$\varphi(\eta_n^{(1)}, \eta_n^{(2)}) = \beta$$

and $\mathscr{X}_\Pi \in (LB)$.

3. Families of vector exponentials on an interval

In this section we consider families of projections of simple vector fractions

$$P_\Theta \mathscr{X}_\Pi = \{P_\Theta (k - \bar\lambda)^{-1}\mathfrak{N}_\lambda\}_{\lambda\in\sigma}$$

for dim $\mathfrak{N} < \infty$. Here, Θ is an ESF (an entire singular inner operator function in $\mathbb{C}_+$) $\sigma \subset \mathbb{C}_+$ is the family spectrum satisfying the Blaschke condition (B), and $\mathfrak{N}_\lambda$ are subspaces of $\mathfrak{N}$, dim $\mathfrak{N}_\lambda =: n_\lambda$. We provide a simple fraction family with subscript Π corresponding to BPP Π generated by the family: σ is the set of semisimple zeros of Π and Ker $\Pi^*(\lambda) = \mathfrak{N}_\lambda$.

Family $\mathscr{X}_\Pi$ is a Fourier image of exponential family $\mathscr{E} = \{e^{-i\bar\lambda t}\mathfrak{N}_\lambda\} \subset L^2(0, \infty; \mathfrak{N})$, and for $\Theta = \exp(ikT)$ family $P_\Theta\mathscr{X}_\Pi$ is a Fourier image of family $\mathscr{E}_T = \{\chi_T e^{-i\bar\lambda t}\mathfrak{N}_\lambda\}_{\lambda\in\sigma} \subset L^2(0, T, \mathfrak{N})$ (characteristic function χ_T of interval $(0, T)$ is usually omitted). Everywhere in this section, dim $\mathfrak{N} = N < \infty$ and orthonormal basis $\{\xi_j^0\}_{j=1}^N$ is specified in $\mathfrak{N}$. At times it may seem more convenient to pass from the family of subspaces $\mathscr{X}_\Pi$ to the family of elements choosing an orthonormal basis $\{\eta_\lambda^{(j)}\}$, $j = 1, \dots, n_\lambda$, in $\mathfrak{N}_\lambda$ for each λ. The family of normalized elements $\{x_\lambda\eta_\lambda^{(j)}\}_{j,\lambda}$, where

$$x_\lambda(k) = \sqrt{\frac{\mathrm{Im}\,\lambda}{\pi}}\,\frac{1}{k - \bar\lambda} \in H_+^2, \qquad \lambda \in \mathbb{C}_+, \ \|x_\lambda\|_{H_+^2} = 1,$$

is denoted by $\mathscr{X}_\Pi$, along with the family of subspaces.

3.1. Geometrical indications of minimal and basis properties

Lemma II.3.1. For any ESF Θ and each BPP Π, there may be found an ESF $\tilde\Theta$ and BPP $\tilde\Pi$ such that

$$\Pi\tilde\Theta = \Theta\tilde\Pi =: S \tag{1}$$

in a way that

$$K_s = K_\Pi \oplus \Pi K_{\tilde{\Theta}} = K_\Theta \oplus \Theta K_{\tilde{\Pi}} \qquad (2)$$

and

$$K_\Pi \bigvee K_\Theta = K_s \qquad (3)$$

hold.

PROOF. Let a number α be large enough for function

$$\Theta_1(k) = \exp(ik\alpha)\Theta^{-1}(k)$$

to be bounded in $\mathbb{C}_+$ (and thus to be an ESF). Define functions Θ_2 (an ESF) and $\tilde{\Pi}$ (a BPP) from the factorization problem

$$\Theta_1\Pi = \tilde{\Pi}\Theta_2 \qquad (4)$$

and set $\tilde{\Theta} := \Theta_2^{-1} \exp(ik\alpha)$.

Functions $\tilde{\Theta}$ and $\tilde{\Pi}$ satisfy an equation

$$\Pi\tilde{\Theta} = \Theta\tilde{\Pi}$$

and to check (1) it remains to show that $\tilde{\Theta}$ is an ESF. One may see from the construction that $\tilde{\Theta}$ is an entire function unitary-valued on the real axis. It is an ESF if it is bounded in $\mathbb{C}_+$. From (4) we obtain

$$\Theta_2^{-1} = \Pi^{-1}\Theta_1^{-1}\tilde{\Pi},$$

which implies

$$\tilde{\Theta} = \Pi^{-1}\Theta\tilde{\Pi}.$$

This relation may be written as

$$\tilde{\Theta} = \hat{\Pi}\Theta\tilde{\Pi}/\det \Pi, \qquad (5)$$

where $\hat{\Pi}$ is the matrix function composed by cofactors of matrix Π. Since $\det \hat{\Pi} = (\det \Pi)^{N-1}$, $\hat{\Pi}$ is a BPP. Take an arbitrary element φ_{ij} of matrix $\hat{\Pi}\Theta\tilde{\Pi}$, which is evidently bounded in $\mathbb{C}_+$. As $\tilde{\Theta}$ is an entire function, φ_{ij} has to contain an inner factor $\det \Pi$:

$$\varphi_{ij} = f_{ij} \det \Pi$$

and f_{ij} is an analytical and is bounded in the $\mathbb{C}_+$ function. From formula (5) it follows that f_{ij} is an element of the matrix $\tilde{\Theta}$.

To verify equalities (2), we exploit the following: if $\mathfrak{H}_1$ is a subspace of Hilbert space $\mathfrak{H}$ and $\mathfrak{H}_2$ is a subspace of $\mathfrak{H}_1$, then

$$\mathfrak{H} \ominus \mathfrak{H}_2 = [\mathfrak{H} \ominus \mathfrak{H}_1] \oplus [\mathfrak{H}_1 \ominus \mathfrak{H}_2]. \qquad (6)$$

Using this, we find that

$$K_s = H_+^2(\mathfrak{N}) \ominus \Pi\tilde{\Theta}H_+^2(\mathfrak{N}) = [H_+^2(\mathfrak{N}) \ominus \Pi H_+^2(\mathfrak{N})]$$

$$\oplus \Pi[H_+^2(\mathfrak{N}) \ominus \tilde{\Theta}H_+^2(\mathfrak{N})].$$

Similarly, replacing Π by Θ and $\tilde{\Pi}$ by $\tilde{\Theta}$, we get

$$K_s = H_+^2(\mathfrak{N}) \ominus \Theta\tilde{\Pi}H_+^2(\mathfrak{N}) = [H_+^2(\mathfrak{N}) \ominus \Theta H_+^2(\mathfrak{N})]$$

$$\oplus \Theta[H_+^2(\mathfrak{N}) \ominus \tilde{\Pi}H_+^2(\mathfrak{N})]$$

and formulas (2) are proved.

Let us demonstrate that (3), (2) imply $K_\Pi \bigvee K_\Theta \subset K_s$. Since any subspace of the form K_s is invariant under action of the operator $\mathfrak{S}^*$, then, according to Proposition 1.16, subspace $K_\Pi \bigvee K_\Theta$ has the form K_{s_1}, where S_1 is an inner operator function dividing S from the left:

$$S = S_1 S_2 \tag{7}$$

(S_2 is an inner operator function in $\mathbb{C}_+$.) Since subspaces K_Π and K_Θ lie in $K_{s_1} = K_\Pi \bigvee K_\Theta$, the same assertion implies that Π and Θ are divisors of S:

$$S_1 = \Pi S_\Pi = \Theta S_\Theta, \tag{8}$$

where S_Π and S_Θ are inner functions.

Now formulas (1), (7), and (8) produce the equalities

$$S_\Pi S_2 = \tilde{\Theta}, \qquad S_\Theta S_2 = \tilde{\Pi}. \tag{9}$$

Let us show that the equalities are correct only if $S = $ const (i.e., the greatest common divisor of BPP $\tilde{\Pi}$ and ESF $\tilde{\Theta}$ is trivial). Consider the determinant of the first equality in (9). Taking into account that the determinant of an ESF is by definition of the form $c \exp(ik\beta)$, $\beta \geq 0$, $|c| = 1$, we have

$$\det S_\Pi \det S_2 = c \exp(ik\beta). \tag{10}$$

Similarly, taking the determinant of the second equality in (9), we get

$$\det S_\Theta \det S_2 = b, \tag{11}$$

with b being a Blaschke product. From (10) it follows that a scalar function $\det S$ has no zeros, while (11) implies that it has no singular cofactor as well. Therefore, $\det S = $ const and hence $S = $ const. The lemma is proved.

Lemma II.3.2. For an ESF Θ and a BPP Π, the relations

$$K_\Theta \cap K_\Pi^\perp = \Theta H^2_-(\mathfrak{N}) \cap \Pi H^2_+(\mathfrak{N}),$$

$$\varphi(K_\Theta, K_\Pi^\perp) = \varphi(\Theta H^2_-(\mathfrak{N}), \Pi H^2_+(\mathfrak{N})),$$

$$K_\Theta^\perp \cap K_\Pi = \Theta H^2_+(\mathfrak{N}) \cap \Pi H^2_-(\mathfrak{N}),$$

$$\varphi(K_\Theta^\perp, K_\Pi) = \varphi(\Theta H^2_+(\mathfrak{N}), \Pi H^2_-(\mathfrak{N}))$$

are true with subspaces $K_\Pi^\perp$ and $K_\Theta^\perp$ being orthogonal complements in subspace $K_\Pi \bigvee K_\Theta$ to K_Π and K_Θ, respectively.

PROOF. For any inner operator function K_{s_0} in $\mathbb{C}_+$, an equality holds:

$$S_0 H^2_-(\mathfrak{N}) = K_{s_0} \oplus H^2_-(\mathfrak{N}). \tag{12}$$

Indeed, the operator of multiplication by S_0 is unitary in $L^2(\mathbb{R}; \mathfrak{N})$ and that is why

$$S_0 L^2(\mathbb{R}; \mathfrak{N}) = L^2(\mathbb{R}; \mathfrak{N}).$$

Representing the space $L^2(\mathbb{R}; \mathfrak{N})$ as a sum of Hardy spaces for the upper and lower half-planes, we obtain

$$S_0 H^2_+(\mathfrak{N}) \oplus S_0 H^2_-(\mathfrak{N}) = H^2_+(\mathfrak{N}) \oplus H^2_-(\mathfrak{N}). \tag{13}$$

If one now subtracts subspace $S_0 H^2_+(\mathfrak{N})$ from both sides of the equality, one arrives at (12), which implies the useful relation

$$K_{s_0} = S_0 H^2_-(\mathfrak{N}) \ominus H^2_-(\mathfrak{N}).$$

It is then easy to check that the right-hand side coincides with $S_0 H^2_-(\mathfrak{N}) \cap H^2_+(\mathfrak{N})$. So we have

$$K_{s_0} = S_0 H^2_-(\mathfrak{N}) \ominus H^2_-(\mathfrak{N}) = S_0 H^2_-(\mathfrak{N}) \cap H^2_+(\mathfrak{N}). \tag{14}$$

Using (12) for $S_0 = \Theta$, we represent subspace $\Theta H^2_-(\mathfrak{N})$ in the form

$$\Theta H^2_-(\mathfrak{N}) = K_\Theta \oplus H^2_-(\mathfrak{N}). \tag{15}$$

Let S be an inner operator function whose existence is stated in Lemma 1. Then

$$\Pi H^2_+(\mathfrak{N}) = [\Pi H^2_+(\mathfrak{N}) \ominus S H^2_+(\mathfrak{N})] \oplus S H^2_+(\mathfrak{N})$$

$$= \Pi[H^2_+(\mathfrak{N}) \ominus \tilde{\Theta} H^2_+(\mathfrak{N})] \oplus S H^2_+(\mathfrak{N}),$$

and formula (2) provides

$$\Pi H_+^2(\mathfrak{N}) = K_\Pi^\perp \oplus S H_+^2(\mathfrak{N}). \tag{16}$$

From (15) and (16) it follows that subspaces $K_\Pi^\perp$ and K_Θ differ from $\Pi H_+^2(\mathfrak{N})$ and $\Theta H_-^2(\mathfrak{N})$ by pairwise orthogonal summands. The addition of such summands does not change both the intersection of subspaces and the angle between them.

In a similar way, from formulas

$$\Pi H_-^2(\mathfrak{N}) = K_\Pi \oplus H_-^2(\mathfrak{N}), \qquad \Theta H_+^2(\mathfrak{N}) = K_\Theta^\perp \oplus S H_+^2(\mathfrak{N})$$

one derives the second pair of the desired equalities and thus completes the proof of the lemma.

To return to the family $P_\Theta \mathscr{X}_\Pi$, we denote its elements by

$$\mathscr{X}_{\Theta, \lambda} := P_\Theta (k - \bar{\lambda})^{-1} \mathfrak{N}_\lambda$$

and reserve notations $\mathscr{X}_\Pi' = \{\mathscr{X}_\lambda'\}$ and $\mathscr{X}_{\Theta,\Pi}' = \{\mathscr{X}_{\Theta,\lambda}'\}$ for the subspace families biorthogonal to $\mathscr{X}_\Pi$ and $P_\Theta \mathscr{X}_\Pi$, respectively (if they exist). We always assume $\mathscr{X}_\Pi' \subset K_\Pi$ but do not demand the inclusion $\mathscr{X}_{\Theta,\Pi}' \subset \bigvee P_\Theta \mathscr{X}_\Pi$.

Theorem II.3.3. Let Θ be an ESF and Π be a BPP.

(a) Family $P_\Theta \mathscr{X}_\Pi$ is then complete in the space K_Θ if and only if

$$K_\Theta \cap K_\Pi^\perp = \{0\}. \tag{17}$$

(b) If family $P_\Theta \mathscr{X}_\Pi$ is minimal,

$$K_\Theta^\perp \cap K_\Pi = \{0\}. \tag{18}$$

(c) If family $P_\Theta \mathscr{X}_\Pi$ is minimal, complete in K_Θ, and $\sigma \in (CN)$, then the biorthogonal family is complete in K_Θ.

(d) The family $P_\Theta \mathscr{X}_\Pi$ is an $\mathscr{L}$-basis if and only if $\mathscr{X}_\Pi$ is an $\mathscr{L}$-basis and

$$\varphi(K_\Theta^\perp, K_\Pi) > 0. \tag{19}$$

(e) The family $P_\Theta \mathscr{X}_\Pi$ is a basis in K_Θ if and only if family $\mathscr{X}_\Pi$ is an $\mathscr{L}$-basis, and (19) and

$$\varphi(K_\Theta, K_\Pi^\perp) > 0 \tag{20}$$

hold.

PROOF.

(a) Let family $P_\Theta \mathscr{X}_\Pi$ be complete; that is, $\bigvee P_\Theta \mathscr{X}_\Pi = K_\Theta$. Then the image of operator $P_\Theta: K_\Pi \mapsto K_\Theta$ is dense in K_Θ and hence the adjoint operator, which equals $P_\Pi|_{K_\Theta}$ by formula (3) in Section I.1, is invertible, That is, there are no elements in K_Θ orthogonal to K_Π.

Conversely, if (17) holds, operator $P_\Pi|_{K_\Theta}$ is invertible. Therefore, $\bigvee P_\Theta K_\Pi = K_\Theta$. Since $K_\Pi = \bigvee \mathscr{X}_\Pi$, assertion (a) is true.

Preserving the notations of the families, let us turn now to the families of elements $\{x_\lambda \eta_\lambda^{(j)}\}$ and $\{P_\Theta x_\lambda \eta_\lambda^{(j)}\}$, $\lambda \in \sigma$, $j = 1, 2, \ldots, n_\lambda$, from the subspace ones. We note that this family of almost normed elements and the family $P_\Theta \mathscr{X}_\Pi$ of subspaces form an $\mathscr{L}$-basis (a basis) simultaneously.

(b) For family $\mathscr{X}'_\Pi$ biorthogonal to the family of simple fractions, a relation is valid

$$\bigvee \mathscr{X}'_\Pi = \bigvee \mathscr{X}_\Pi = K_\Pi$$

(see Subsection 1.3.5). That is why the conditions of Theorem I.1.31 are fulfilled. Assertion (a) of this theorem states that operator $P_\Pi|_{\mathfrak{M}}$, $\mathfrak{M} := \bigvee P_\Theta \mathscr{X}_\Pi$ has a dense image in K_Π. The same goes for the operator $P_\Pi|_{K_\Theta} \supset P_\Pi|_{\mathfrak{M}}$. Therefore adjoint operator $P_\Theta|_{K_\Pi}$ is invertible, which is just what formula (18) expressed. Assertion (b) is verified.

(c) If $\sigma \in (CN)$, then, by definition, σ is a unification of a finite number of Carlesonian sets:

$$\sigma = \bigcup_{j=1}^{M} \sigma_j, \qquad \sigma_j \in (C), j = 1, \ldots, M.$$

By Subsection 1.3.11, family $\{x_\lambda\}_{\lambda \in \sigma_j}$ forms an $\mathscr{L}$-basis. Drawing on Corollary II.2.2, we conclude that $\{x_\lambda \mathfrak{N}_\lambda\}_{\lambda \in \sigma_j} \in (LB)$, $j = 1, \ldots, M$. Assertion (c) follows from Theorem I.1.31(b).

(d) If family $P_\Theta \mathscr{X}_\Pi$ is an $\mathscr{L}$-basis, it is almost normed and uniformly minimal in K_Θ. Consequently, it is also *-uniformly minimal so that Lemma I.1.28 becomes applicable to it. By the force of this lemma, the family of elements $\mathscr{X}_\Pi$ is *-uniformly minimal and, since it is normed, $\mathscr{X}_\Pi$ is uniformly minimal. According to Proposition 2.8, uniform minimality of the simple fraction family $\{x_\lambda \eta_\lambda^{(j)}\}$ is equivalent to its $\mathscr{L}$-basis property.

From the Bari theorem (Proposition I.1.17) for $x = \sum_{\lambda, j} c_\lambda^{(j)} x_\lambda \eta_\lambda^{(j)}$, one has

$$\|x\|_{H^2_+(\mathfrak{N})}^2 \asymp \sum_{\lambda, j} |c_\lambda^{(j)}|^2. \tag{21}$$

By the condition, family $P_\Theta \mathscr{X}_\Pi$ is an $\mathscr{L}$-basis. The same theorem gives

$$\sum_{\lambda,j} |c_\lambda^{(j)}|^2 \asymp \left\| \sum_{\lambda,j} c_\lambda^{(j)} P_\Theta x_\lambda \eta_\lambda^{(j)} \right\|_{H^2_+(\mathfrak{N})}^2 = \|P_\Theta x\|_{H^2_+(\mathfrak{N})}^2. \tag{22}$$

Comparing estimates (21) and (22), we find that

$$\|P_\Theta x\|_{H^2_+(\mathfrak{N})}^2 \asymp \|x\|_{H^2_+(\mathfrak{N})}^2.$$

This means that operator $P_\Theta|_{K_\Pi}$ is an isomorphism on its image. Then Lemma I.1.10(a) implies (19).

Conversely, let $\mathscr{X}_\Pi \in (LB)$ and inequality (19) be true. Then by the same lemma, $P_\Theta|_{K_\Pi}$ is an ismorphism on its image. Hence, $P_\Theta \mathscr{X}_\Pi \in (LB)$.
(e) Family $P_\Theta \mathscr{X}_\Pi$ is a basis in K_Θ if and only if $P_\Theta \mathscr{X}_\Pi \in (LB)$ and $\bigvee P_\Theta \mathscr{X}_\Pi = K_\Theta$. In accordance with what has been proved, this is equivalent to the following: $\mathscr{X}_\Pi \in (LB)$, $P_\Theta|_{K_\Pi}$ is an isomorphism on its image, and $\text{Cl}\, P_\Theta K_\Pi = K_\Theta$. These properties of operator $P_\Theta|_{K_\Pi}$ are equivalent to the fact that the operator is an isomorphism between subspaces K_Π and K_Θ. From Lemma I.1.10(c) we now obtain assertion (e). The theorem is proved.

For a family Ξ of general type (not necessarily exponential), a family Ξ' biorthogonal to it may be incomplete in $\bigvee \Xi$ (Example I.1.18). Assertion (c) of the theorem attests that for a family of vector exponentials under the condition $\sigma \in (CN)$ the completeness of the biorthogonal family takes place. This is proved at the abstract level and with the use of only the completeness of the family biorthogonal to an exponential family on a semiaxis. For a scalar case, completeness of the biorthogonal family has been proved (Young 1980) without any condition on the family spectrum by the methods of the theory of functions of the complex variable.

3.2. Minimal vector families and the generating function

We now introduce one of the main subjects of the exponential families theory: the generating function containing all the information about family $P_\Theta \mathscr{X}_\Pi$; in particular, the zeros of its determinant coincide with the spectrum σ of the family.

Definition II.3.4. Let Θ be an ESF and Π be a BPP generated by a family of subspaces $\mathscr{X}_\Pi = \{x_\lambda \mathfrak{N}_\lambda\}$ (so that σ is the set of zeros of $\det \Pi$. The zeros are semisimple and $\text{Ker}\, \Pi^*(\lambda) = \mathfrak{N}_\lambda$).
Entire strong operator-function F is said to be a generating function

(GF) of family $P_\Theta \mathscr{X}_\Pi = \{\mathscr{X}_{\Theta,\lambda}\}_{\lambda \in \sigma}$, $\mathscr{X}_{\Theta,\lambda} = P_\Theta x_\lambda \mathfrak{N}_\lambda$, if it has a factorization

$$F = \Pi F_e^+ = \Theta F_e^- \qquad (23)$$

in which $F_e^\pm$ are outer operator-functions in $\mathbb{C}_\pm$.

From the definition of a GF it follows that $\operatorname{Ker} F^*(\lambda) = \mathfrak{N}_\lambda$. Set $\mathfrak{N}^\lambda := \operatorname{Ker} F(\lambda)$; obviously, $\dim \mathfrak{N}^\lambda = \dim \mathfrak{N}_\lambda =: n_\lambda$.

Let us impose two restrictions on functions Θ and Π:

(1) For some $\rho > 0$, ESF Θ is divisible by function $e^{ik\rho}$ (i.e., function $\Theta_\rho(k) = \exp(-ik\rho)\Theta(k)$ is an inner one in $\mathbb{C}_+$). This condition is equivalent to the fact that Θ_ρ is bounded in $\mathbb{C}_+$

$$\langle\langle e^{-ik\rho} \Theta(k)\rangle\rangle \le 1, \qquad k \in \mathbb{C}_+. \qquad (24)$$

(2) The spectrum, say σ, of family $\mathscr{X}_\Pi$ satisfies a condition

$$\inf_{\lambda \in \sigma} \operatorname{Im} \lambda > 0. \qquad (25)$$

For an exponential family $\mathscr{E}$ (related to $P_\Theta \mathscr{X}_\Pi$ by the Fourier transform), the first limitation means that $\mathscr{E}$ is considered in subspace $\mathscr{F}^{-1}K_\Theta$ containing the space $L^2(0, \rho; \mathfrak{N})$. In fact, in view of the formula (6) and factorization $\Theta = \exp(ik\rho)\Theta_\rho$, we have

$$K_\Theta = K_\rho \oplus e^{ik\rho} K_{\Theta_\rho}$$

and $\mathscr{F}^{-1}K_\rho = L^2(0, \rho; \mathfrak{N})$. The second restriction actually means only the boundedness from below of $\operatorname{Im} \lambda$, $\lambda \in \sigma$, since it may always be fulfilled with the help of a shift $\sigma \mapsto \sigma + ic$ (see Subsection 1.3.6).

When the theory of exponentials is used in control problems, condition (1) implies that a control effect is acting at least during a fixed period of time ρ at any point where it is "applied."

Theorem II.3.5. Let F be a GF for family $P_\Theta \mathscr{X}_\Pi$. Then $P_\Theta \mathscr{X}_\Pi \in (M)$ and the family of subspaces $\mathscr{X}'_{\Theta,\Pi} = \{\mathscr{X}'_{\Theta,\lambda}\}_{\lambda \in \sigma}$ biorthogonal to $P_\Theta \mathscr{X}_\Pi$ has the form

$$\mathscr{X}'_{\Theta,\lambda} = (k - \lambda)^{-1} F(k) \mathfrak{N}^\lambda. \qquad (26)$$

We first prove an auxiliary assertion.

Lemma II.3.6. Let a family of subspaces $\{\mathscr{Y}_\lambda\}_{\lambda\in\sigma}$, $\mathscr{Y}_\lambda \subset K_\Theta$, satisfy conditions

$$\mathscr{Y}_\lambda \perp P_\Theta \mathscr{X}_\mu, \qquad \mu \neq \lambda, \tag{27}$$

$$P_{\mathscr{X}_{\Theta,\lambda}} \mathscr{Y}_\lambda = \mathscr{X}_{\Theta,\lambda}, \tag{28}$$

where $P_{\mathscr{X}_{\Theta,\lambda}}$ is the orthoprojector on subspace $\mathscr{X}_{\Theta,\lambda}$. Then $\{P_\Theta \mathscr{X}_\Pi\} \in (M)$ and $\{\mathscr{Y}_\lambda\}$ is a biorthogonal family.

PROOF OF THE LEMMA. Let us turn from the subspace family to the family of elements $\{x_{\Theta,\lambda}^{(j)}\}$, $\lambda \in \sigma$, $j = 1, 2, \ldots, n_\lambda$, $x_{\Theta,\lambda}^{(j)} := P_\Theta x_\lambda \eta_\lambda^{(j)}$, where vector family $\{\eta_\lambda^{(j)}\}$ is orthonormalized in $\mathfrak{N}_\lambda$ for any λ. Equality (28) reveals that for any j there exists a solution $\mathscr{y}_\lambda^{(j)}$ to the equation

$$P_{\mathscr{X}_{\Theta,\lambda}} \mathscr{y}_\lambda^{(j)} = x_{\Theta,\lambda}^{(j)} / \|x_{\Theta,\lambda}^{(j)}\|^2.$$

Then for $\lambda \neq \mu$ condition (27) implies

$$(\mathscr{y}_\lambda^{(j)}, x_{\Theta,\mu}^{(r)})_{H_+^2(\mathfrak{N})} = 0, \qquad \lambda \neq \mu. \tag{29}$$

For $\lambda = \mu$ as a consequence of the choice of family $\{\mathscr{y}_\lambda^{(j)}\}$, one finds that

$$(\mathscr{y}_\lambda^{(j)}, x_{\Theta,\lambda}^{(r)}) = (x_{\Theta,\lambda}^{(j)}, x_{\Theta,\lambda}^{(r)}) / \|x_{\Theta,\lambda}^{(j)}\|^2 = \delta_r^j. \tag{30}$$

From (29) and (30) it follows that the family of elements $\{\mathscr{y}_\lambda^{(j)}\}_{\lambda\in\sigma}^{j=1,\ldots,n_\lambda}$ is biorthogonal to $\{x_{\Theta,\lambda}^{(j)}\}$. The lemma is proved.

PROOF OF THE THEOREM. Factorization of a GF F gives rise to the fact that F is a strong H_+^2 function while $\Theta^{-1}F$ is a strong H_-^2 function. So, by (23), (26), and Definition II.1.8, we have

$$\mathscr{X}'_{\Theta,\lambda} \subset H_+^2(\mathfrak{N}), \qquad \Theta^{-1}\mathscr{X}'_{\Theta,\lambda} \subset H_-^2(\mathfrak{N}).$$

Identity (15) shows that subspace $\mathscr{X}'_{\Theta,\lambda}$ lies in K_Θ. We now check conditions (27) and (28) for family $\mathscr{X}'_{\Theta,\lambda}$. Taking arbitrary vectors $\eta^\lambda \in \mathfrak{N}^\lambda$ and $\eta \in \mathfrak{N}$ and calculating the integral with the help of residues, for $v \in \mathbb{C}_+$ we get

$$((k-\lambda)^{-1}F(k)\eta^\lambda, P_\Theta(k-\bar{v})^{-1}\eta)_{H_+^2(\mathfrak{N})} = ((k-\lambda)^{-1}F(k)\eta^\lambda, (k-\bar{v})^{-1}\eta)_{H_+^2(\mathfrak{N})}$$

$$= \int_{\mathbb{R}} \frac{\langle F(k), \eta^\lambda \rangle}{(k-\lambda)(k-v)} \, dk = 2\pi i \begin{cases} \dfrac{\langle F(\lambda)\eta^\lambda, \eta \rangle}{\lambda - v} + \dfrac{\langle F(v)\eta^\lambda, \eta \rangle}{v - \lambda}, & \lambda \neq v, \\[2ex] \langle F'(v)\eta^\lambda, \eta \rangle, & \lambda = v, \end{cases}$$

$$= 2\pi i \begin{cases} \dfrac{\langle F(v)\eta^\lambda, \eta \rangle}{v - \lambda}, & \lambda \neq v, \\[2ex] \langle F'(\lambda)\eta^\lambda, \eta \rangle, & \lambda = v \end{cases} \tag{31}$$

(we have exploited the equality $F(\lambda)\eta^\lambda = 0$ here and $F'(\lambda) := (d/dk)F(k)|_{k=\lambda}$).
If one sets $v = \mu \in \sigma$ and takes vector $\eta = \eta_\mu \in \mathfrak{N}_\mu$, then from (31) and the
equality $F^*(\mu)\eta_\mu = 0$ one derives

$$\left(\frac{F(k)\eta^\lambda}{k-\lambda}, P_\Theta \frac{\eta_\mu}{k-\bar\mu}\right)_{H^2_+(\mathfrak{N})} = 2\pi i \begin{cases} 0, & \mu \neq \lambda, \\ \langle F'(\mu)\eta^\mu, \eta_\mu\rangle, & \mu = \lambda. \end{cases} \tag{32}$$

Thus, condition (27) for family $\{\mathscr{X}'_{\Theta,\lambda}\}$ is verified. Let us check condition
(28). Subspaces $\mathscr{X}'_{\Theta,\lambda}$ and $\mathscr{X}_{\Theta,\lambda}$ are of the same dimension, since dim $\mathfrak{N}^\lambda =$
dim $\mathfrak{N}_\lambda$. Therefore, it suffices to show that the projector on the subspace
$\mathscr{X}_{\Theta,\lambda}$ is invertible over $\mathscr{X}'_{\Theta,\lambda}$. Let us assume the opposite, namely,
that a nonzero vector $\eta^\lambda \in \mathfrak{N}^\lambda$ may be found such that

$$\frac{F(k)\eta^\lambda}{k-\lambda} \perp \mathscr{X}_{\Theta,\lambda} = P_\Theta x_\lambda \mathfrak{N}_\lambda.$$

Then (32) implies

$$F'(\lambda)\eta^\lambda \in \mathfrak{N}^\perp_\lambda := \mathfrak{N} \ominus \mathfrak{N}_\lambda. \tag{33}$$

Since $\mathfrak{N}^\perp_\lambda = [\text{Ker } F^*(\lambda)]^\perp = F(\lambda)\mathfrak{N}$, relation (33) may be written as
$F'(\lambda)\eta^\lambda = F(\lambda)\eta$ for some $\eta \in \mathfrak{N}$. This equality contradicts the fact that
the zeros of function F are semisimple (because the zeros of Π are
semisimple and the $F^\pm_e$ zeros are absent). Hence, the conditions of
Lemma 6 are satisfied, and family $\{\mathscr{X}'_{\Theta,\lambda}\}$ is biorthogonal to $P_\Theta\mathscr{X}_\Pi$. The
theorem is proved.

Remark II.3.7. The proof of Theorem 5 does not use all the information
about a GF. It is not difficult to find out that family $P_\Theta\mathscr{X}_\Pi$ tends to be
minimal if an entire matrix-function $\tilde F$ may be found such that

(a) functions $\tilde F$ and $\Theta^{-1}\tilde F$ are strong $H^2_\pm$ functions, respectively, and
(b) function F has semisimple zeros in the points of the spectrum σ, and
Ker $\tilde F^*(\lambda) = \mathfrak{N}_\lambda, \lambda \in \sigma$.

In such a case, a biorthogonal family is described by formula (26), in
which F has to be replaced by $\tilde F$. In particular, $\tilde F$ may have zeros outside
of the spectrum σ, and the factorization of $\tilde F$ in $\mathbb{C}_+$ may contain an ESF.

Let us give an expression for the biorthogonal family of elements via
the GF.

Corollary II.3.8. Let F be a GF of the family $\{P_\Theta x_\lambda \eta_\lambda\}_{\lambda \in \sigma}, \langle\langle\eta_\lambda\rangle\rangle = 1$.

Then biorthogonal family $\{x'_{\Theta,\lambda}\}$ *is given by*

$$x'_{\Theta,\lambda} = \frac{1}{2i\sqrt{\pi \operatorname{Im}\lambda}\,\langle F'(\lambda)\eta^\lambda, \eta_\lambda\rangle}\,\frac{F(k)}{k-\lambda}\,\eta^\lambda, \tag{34}$$

where $\eta^\lambda \in \operatorname{Ker} F(\lambda)$, $\langle\langle\eta^\lambda\rangle\rangle = 1$.

Relation (34) follows from the equality (32).

Theorem II.3.9. Suppose $P_\Theta \mathscr{X}_\Pi \in (M)$. *Then*

(a) *entire operator function* F, $F \not\equiv 0$, *may be found such that functions* $\Pi^{-1}F$ *and* $\Theta^{-1}F$ *are strong* H_+^2 *function and* H_-^2 *function, respectively. Along with it there exists a nonzero minor of* F *vanishing in the points of the spectrum;*

(b) *if family* $P_\Theta \mathscr{X}_\Pi$ *is complete in subspace* K_Θ *and* $\sigma \in (CN)$, *then a GF of the family exists.*

The idea of the proof for a scalar situation is in the following. One takes an arbitrary element $x'_{\Theta,\mu}$ of a biorthogonal to $P_\Theta \mathscr{X}_\Pi$ family, sets $f(k) := (k-\mu)x'_{\Theta,\mu}(k)$, and checks that this function satisfies the requirements of the theorem.

In a vector situation, in order to construct an operator function one needs several vector functions of the form $(k-\lambda_j)^{-1}x'_{\Theta,\lambda_j}(k)$. So, we start the proof with the choice of those functions.

Denote a family of subspaces biorthogonal to $P_\Theta \mathscr{X}_\Pi$ as $\mathscr{X}'_{\Theta,\Pi} = \{\mathscr{X}'_{\Theta,\lambda}\}_{\lambda\in\sigma}$ and choose in every subspace $\mathscr{X}'_{\Theta,\lambda}$ an orthonormal basis $\{x'^{(r)}_{\Theta,\lambda}\}_{r=1}^{n_\lambda}$. We write $M(k)$ for the dimension in $\mathfrak{N}$ of the linear span of all the vectors $x'^{(r)}_{\Theta,\lambda}(k)$, $\lambda\in\sigma$, $r=1,\dots,n_\lambda$. Set $M = \max_{k\in\mathbb{C}\setminus\sigma} M(k)$ and suppose that this maximum is attained on the vector family,

$$\{x'^{(r_j)}_{\Theta,\lambda}(k_0)\}_{j=1}^M.$$

In what follows we assume that all the points λ_j are different and omit superscript (r_j). The general case can be studied similarly.

Define operator function F by its value at a point k on the vectors $\eta \in \mathfrak{N}$ in the following way:

$$F(k)\eta = \sum_{j=1}^M (k-\lambda_j)\langle\eta, \xi_j^0\rangle x'_{\Theta,\lambda_j}(k). \tag{35}$$

(family $\{\xi_j^0\}$ is an orthonormal basis in $\mathfrak{N}$).

Since at k_0 vectors $x'_{\Theta,\lambda_j}(k_0)$ are linearly independent, $F \neq 0$. All the elements of the space K_Θ, particularly x'_{Θ,λ_j} are entire vector functions (Subsection 1.3.1). Therefore F is also an entire function.

Let us demonstrate that F is a strong H^2_+ function while $\Theta^{-1}F$ is a strong H^2_- one. We first verify F to be a strong function. From the definition of F one derives

$$\frac{\langle\langle F(k)\rangle\rangle}{1+|k|} \leq \sum_{j=1}^{M} \frac{|k|+|\lambda_j|}{1+|k|} \langle\langle x'_{\Theta,\lambda_j}(k)\rangle\rangle \prec \sum_{j=1}^{M} \langle\langle x'_{\Theta,\lambda_j}(k)\rangle\rangle.$$

Elements x'_{Θ,λ_j} of space K_Θ lie in $L^2(\mathbb{R}, \mathfrak{N})$. Hence, F is a strong function. Let us take some vector $\eta \in \mathfrak{N}$, then vector function

$$\frac{F(k)}{k+i}\eta = \sum_{j=1}^{M} \frac{k-\lambda_j}{k+i} \langle \eta, \xi_j^0 \rangle x'_{\Theta,\lambda_j}(k)$$

belongs to the subspace $H^2_+(\mathfrak{N})$, because elements x'_{Θ,λ_j} belong to it while factor $(k-\lambda_j)/(k+i)$ is analytic and bounded in the upper half-plane. Therefore, F is a strong H^2_+ function.

From formula (15) we conclude that functions $\Theta^{-1}x'_{\Theta,\lambda_j}$ lie in the space $H^2_-(\mathfrak{N})$. So, as in the case of the upper half-plane, for any $\eta \in \mathfrak{N}$ vector function

$$\frac{\Theta^{-1}F(k)}{k-i}\eta = \sum_{j=1}^{M} \frac{k-\lambda_j}{k-i} \langle \eta, \xi_j^0 \rangle \Theta^{-1}x'_{\Theta,\lambda_j}(k)$$

lies in $H^2_-(\mathfrak{N})$. Thus, $\Theta^{-1}F$ is a strong H^2_- function.

Let us establish that

$$\operatorname{Ker} F^*(\lambda) \supset \mathfrak{N}_\lambda. \tag{36}$$

As is easily seen from the definition of F, operator $F^*(k)$ is defined by the formula

$$F^*(k)\eta = \sum_{j=1}^{M} \overline{(k-\lambda_j)} \langle \eta, x'_{\Theta,\lambda_j}(k) \rangle \xi_j^0. \tag{37}$$

Take vector $\eta_\lambda \in \mathfrak{N}_\lambda$ and let $\mu \in \sigma$. Then the biorthogonality of the families $P_\Theta \mathscr{X}_\Pi$ and $\mathscr{X}'_{\Theta,\Pi\lambda}$ implies

$$\delta_\mu^\lambda = (P_\Theta x_\lambda \eta_\lambda, x'_{\Theta,\mu}) = 2\pi i \sqrt{\frac{\operatorname{Im}\lambda}{\pi}} \langle \eta_\lambda, x'_{\Theta,\mu}(\lambda) \rangle. \tag{38}$$

Let λ not coincide with any of the numbers λ_j, $j = 1, 2, \ldots, N$. Then by the force of (38) for $\eta \in \mathfrak{N}_\lambda$, all the scalar products in (37) tend to zero.

If λ does coincide with some λ_j, then the jth summand is nullified due to the factor $\overline{(k - \lambda_j)}$, while the other summands are vanishing again by the force of (38). Thus, function $\Pi^{-1}F$ has no poles in $\mathbb{C}_+$ and is a strong H^2_+ function, since F is such a function.

To prove assertion (a) of the theorem, it remains only to check that for $\lambda \in \sigma$, rank $F(\lambda) < M$. Assume that at some point $\lambda \in \sigma$ rank $F(\lambda) = M$. From the definition of F it follows immediately that λ differs from the points $\lambda_1, \ldots, \lambda_M$. Let us choose vector $\eta_\lambda \in \mathfrak{N}_\lambda$, which is not orthogonal to $x'_{\Theta, \lambda}(\lambda)$. Vector family $\{x'_{\Theta, \lambda_j}(\lambda)\}_{j=1}^M$ is then orthogonal to η_λ by the relation (38).

Since

$$\dim \operatorname{Lin}\{x'_{\Theta, \lambda_j}(\lambda)\}_{j=1}^M = \operatorname{rank} F(\lambda) = M,$$

$$\dim \operatorname{Lin}(\{x'_{\Theta, \lambda_j}(\lambda)\}_{j=1}^M \cup x'_{\Theta, \lambda}(\lambda)) = M + 1.$$

The last equality contradicts the definition of number M, which proves assertion (a).

(b) Let us show that for the complete family $P_\Theta \mathscr{X}_\Pi$ under Carleson–Newman condition ($\sigma \in (CN)$), the family of subspaces $\{x'_{\Theta, \lambda}(k_0)\}_{\lambda \in \sigma}$ is complete in $\mathfrak{N}$, that is, that $M = N$. Indeed, if $M < N$, vector η_0 may be found orthogonal to all the vectors $x'_{\Theta, \lambda}(k_0)$, which are the elements of subspace $\mathscr{X}'_{\Theta, \lambda}(k_0)$. Therefore (see (38)), vector function $P_{\Theta_0} x_k \eta_0$ is orthogonal to the biorthogonal family $\mathscr{X}'_{\Theta, \Pi}$. On the other hand, this vector function is not an identical zero in view of condition (24) imposed on Θ. We have just found a discrepancy with the completeness of family $\mathscr{X}'_{\Theta, \Pi}$, following from Theorem II.3.3(c).

So $M = N$ and the constructed matrix function F is not degenerate at k_0. Since $\Pi^{-1}F$ is a strong H^2_+ function, it has a factorization

$$\Pi^{-1}F = \Pi_+ \Theta_+ F_e^+ \tag{39}$$

(Π_+ is a BPP, Θ_+ is an ESF). In the lower half-plane, function $\Theta^{-1}F$ has a factorization

$$\Theta^{-1}F = \Pi_-^{-1}\Theta_-^{-1}F_e^- \tag{40}$$

(Π_- is a BPP, Θ_- is an ESF). Formulas (39) and (40) take into account the fact that F is an entire operator function and hence has no nonentire singular factor.

For the desired factorization of F, it remains only to establish the absence of the "extra" zeros of F – the absence of $\Pi_\pm$ cofactors – and the triviality of the "extra" singular cofactors $\Theta_\pm$, that is, the equalities $\Theta_\pm = \text{const}$.

Let us demonstrate that an extra zero of F enables one to present a nonzero element g orthogonal to the complete family $P_\Theta \mathcal{X}_\Pi$ in K_Θ. We first check that $\det F$ does not tend to zero outside of σ. Let $v \in \mathbb{C} \backslash \sigma$ and $\eta \in \mathfrak{N}$. Suppose $F(v)\eta = 0$. Set $g(k) = F(k)\eta/(k - v)$. This vector function is an entire one and by the force of assertion (a) lies in the subspace $H_+^2(\mathfrak{N}) \cap \Theta H_-^2(\mathfrak{N})$ equal to K_Θ (formula (14)). Equalities (31) imply that $g \perp P_\Theta \mathcal{X}_\Pi$. Since family $P_\Theta \mathcal{X}_\Pi$ is complete in K_Θ, $g = 0$, that is, $\eta = 0$. Therefore, F has no zeros outside the spectrum σ.

In a scalar situation, it would now be sufficient to demonstrate that function f (a scalar analog of operator function F) has no multiple zeros. But if λ is such a multiple zero, then evidently $f(\lambda) = f'(\lambda) = 0$ and formula (32) would show function $g(k) = f(k)/(k - \lambda)$ (lying in K_Θ) to be orthogonal to the complete family $P_\Theta \mathcal{X}_\Pi$.

In a vector situation, the construction of such an element g is more complicated. Let us specify a point $\lambda \in \sigma$ and first verify the equality

$$\text{Ker } F^*(\lambda) = \mathfrak{N}_\lambda. \tag{41}$$

Suppose that it is not valid. That is, by the force of (36),

$$\dim \text{Ker } F^*(\lambda) > \dim \mathfrak{N}_\lambda. \tag{42}$$

Let subspace $\text{Ker } F(\lambda)$ be denoted by $\mathfrak{N}^\lambda$. Then assumption (42) may be written as

$$\dim \mathfrak{N}^\lambda > \dim \mathfrak{N}_\lambda,$$

since

$$\dim \text{Ker } F(\lambda) = \dim \text{Ker } F^*(\lambda).$$

Consider a subspace

$$\mathcal{Y}_\lambda := F(k)\mathfrak{N}^\lambda/(k - \lambda),$$

which lies in K_Θ and $\mathcal{Y}_\lambda \perp P_\Theta \mathcal{X}_\mu$ for $\mu \neq \lambda$. Since

$$\dim \mathcal{Y}_\lambda = \dim \mathfrak{N}^\lambda > \dim \mathfrak{N}_\lambda = \dim P_\Theta \mathcal{X}_\lambda,$$

a nonzero element g may be found in $\mathcal{Y}_\lambda$ orthogonal to $P_\Theta \mathcal{X}_\lambda$ and, hence the complete family $P_\Theta \mathcal{X}_\Pi$. The contradiction proves equality (41) to be correct.

Now, to verify the triviality of the factors Π_+ and Π_- in factorizations (39) and (40), one has only to show that the zeros of F are semisimple.

Let λ not be a semisimple zero of function F. Then, by definition, vectors $\eta_1^\lambda \in \mathfrak{N}^\lambda$, $\eta_1^\lambda \neq 0$, and $\eta \in \mathfrak{N}$ may be found such that

$$F'(\lambda)\eta_1^\lambda + F(\lambda)\eta = 0. \tag{43}$$

Let us seek an element g of the form

$$g(k) = \frac{\psi(k)}{(k-\lambda)^2}, \qquad \psi(k) = F(k)[\eta_1^\lambda + (k-\lambda)(\eta + \eta^\lambda)],$$

which is orthogonal to family $P_\Theta K_\Pi$. Here, vector $\eta^\lambda \in \mathfrak{N}^\lambda$ is "a free parameter." From relation (43) it follows that ψ has a second-order zero at the point λ, and hence $g \in K_\Theta$, since, as it was shown, F is an H_+^2 function, and $\Theta^{-1}F$ is an H_-^2 one. Let $\mu \neq \lambda$, $\mu \in \sigma$. Function $\langle \psi(k), \eta_\mu \rangle$ for $\eta_\mu \in \mathfrak{N}_\mu$ tends to zero at $k = \mu$. Therefore

$$\left(g, P_\Theta \frac{\eta_\mu}{k - \bar\mu} \right)_{H_+^2(\mathfrak{N})} = \int_{\mathbb{R}} \frac{\langle \psi(k), \eta_\mu \rangle}{(k-\lambda)^2(k-\mu)} \, dk$$

$$= 2\pi i \left(\operatorname*{res}_{k=\lambda} \frac{\langle \psi(k), \eta_\mu \rangle}{(k-\lambda)^2(k-\mu)} + \operatorname*{res}_{k=\mu} \frac{\langle \psi(k), \eta_\mu \rangle}{(k-\lambda)^2(k-\mu)} \right) = 0.$$

We only have to show that under some choice of vector η^λ, function $g(k)$ is orthogonal to subspace $P_\Theta \mathfrak{X}_\lambda$. Take an arbitrary element $P_\Theta (k-\bar\lambda)^{-1}\eta_\lambda$ of $P_\Theta \mathfrak{X}_\lambda$. Then

$$\left(g(k), P_\Theta \frac{\eta_\lambda}{k - \bar\lambda} \right) = 2\pi i \operatorname*{res}_j (k-\lambda)^{-3} h(k) = 2\pi i h''(\lambda),$$

where $h(k) := \langle \psi(k), \eta_\lambda \rangle$. For g to be orthogonal to $P_\Theta \mathfrak{X}_\Pi$, $h''(\lambda) = 0$ is required for any $\eta_\lambda \in \mathfrak{N}_\lambda$. Since

$$h''(\lambda) = \langle F''(\lambda)\eta_1^\lambda + 2F'(\lambda)[\eta + \eta^\lambda], \eta_\lambda \rangle,$$

η^λ has to be chosen from the condition

$$\tilde\eta + 2F'(\lambda)\eta^\lambda \perp \mathfrak{N}_\lambda, \qquad \tilde\eta := F''(\lambda)\eta_1^\lambda + 2F'(\lambda)\eta$$

so that it is enough to solve the equation

$$P_{\mathfrak{N}_\lambda} F'(\lambda)\eta^\lambda = -P_{\mathfrak{N}_\lambda}\tilde\eta. \tag{44}$$

Lemma II.3.10. *Operator $P_{\mathfrak{N}_\lambda}F'(\lambda)$ maps $\mathfrak{N}^\lambda$ on $\mathfrak{N}_\lambda$.*

If we consider the lemma to be provisionally proved, we see that there is a solution $\eta^\lambda \in \mathfrak{N}_\lambda$ for equation (44), and thus a nonzero element $g \in K_\Theta$ is found that is orthogonal to $P_\Theta \mathfrak{X}_\Pi$.

PROOF OF LEMMA 10. According to (41) the dimensions of the subspaces $\mathfrak{N}^{\lambda}$ and $\mathfrak{N}_{\lambda}$ coincide. Therefore the operator $P_{\mathfrak{N}_{\lambda}}F'(\lambda)|_{\mathfrak{N}^{\lambda}}$ is an isomorphism if it is nondegenerate. Let

$$P_{\mathfrak{N}_{\lambda}}F'(\lambda)\eta_0^{\lambda} = 0, \qquad \eta_0^{\lambda} \in \mathfrak{N}^{\lambda}. \tag{45}$$

Set

$$g = F(k)\eta_0^{\lambda}/(k - \lambda).$$

Then, by (32) and (45), $g \perp P_{\Theta}\mathscr{X}_{\Pi}$, from which $\eta_0^{\lambda} = 0$. The lemma is proved.

We have demonstrated that in the factorizations (39) and (40), BPP factors $\Pi_{\pm}$ are absent. Let us show that ESF Θ_+ is a unitary constant. Define ESF $\tilde{\Theta}$ and BPP $\tilde{\Pi}$ from the factorization condition

$$\Pi\Theta_+ = \tilde{\Theta}_+\tilde{\Pi}. \tag{46}$$

The biorthogonal to the $P_{\Theta}\mathscr{X}_{\Pi}$ family of subspaces, in view of Remark 7, is described by the formula

$$\mathscr{X}'_{\Theta,\lambda} = (k - \lambda)^{-1}F(k)\mathfrak{N}^{\lambda}.$$

Identity (46) then implies $\mathscr{X}'_{\Theta,\lambda} \in \tilde{\Theta}_+ H^2_+(\mathfrak{N})$ and, hence,

$$\bigvee \mathscr{X}'_{\Theta,\lambda} \subset \tilde{\Theta}_+ H^2_+(\mathfrak{N}).$$

Since family $P_{\Theta}\mathscr{X}_{\Pi}$ is complete in K_{Θ} and $\sigma \in (CN)$, by Theorem 3(c) we obtain

$$K_{\Theta} \subset \tilde{\Theta}_+ H^2_+(\mathfrak{N}).$$

From the condition on Θ (see formula (24)), we conclude that $K_{\rho} \subset \tilde{\Theta}_+ H^2_+(\mathfrak{N})$ and thus $\tilde{\Theta}_+^{-1}K_{\rho} \subset H^2_+(\mathfrak{N})$. Lemma 1.23 shows that $\tilde{\Theta}_+ = \text{const}$. Then ESF Θ_+ is constant as well.

ESF Θ_- is proved to be a constant in a similar manner. Indeed, factorization (40) (accounting for the absence of Π_-) implies the inclusion

$$\Theta^{-1}\mathscr{X}'_{\Theta,\lambda} \subset \Theta_-^{-1}H^2_-(\mathfrak{N}).$$

Family $\{\mathscr{X}'_{\Theta,\lambda}\}$ is complete in K_{Θ}. Therefore, family $\{\Theta^{-1}\mathscr{X}'_{\Theta,\lambda}\}$ is complete in the space $\Theta^{-1}K_{\Theta}$ equal to $H^2_+(\mathfrak{N}) \ominus \Theta^{-1}H^2_-(\mathfrak{N})$ by the force of (14). (This subspace is an analog to subspace K_{Θ} for the lower half-plane.) So we get an embedding

$$\Theta_-^{-1}H^2_+(\mathfrak{N}) \supset H^2_+(\mathfrak{N}) \ominus \Theta^{-1}H^2_-(\mathfrak{N}).$$

Function Θ^{-1} is inner in $\mathbb{C}_-$ and is divisible by $\exp(-ik\rho)$. Hence

$$H^2_-(\mathfrak{N}) \ominus \Theta^{-1}H^2_-(\mathfrak{N}) \supset H^2_-(\mathfrak{N}) \ominus e^{-ik\rho}H^2_-(\mathfrak{N}).$$

If one applies to the embedding

$$H^2_-(\mathfrak{N}) \supset \Theta_-[H^2_-(\mathfrak{N}) \ominus e^{-ik\rho} H^2_-(\mathfrak{N})]$$

Lemma 1.23 formulated for the lower half-plane, one gets the triviality of ESF Θ_-. Thus the theorem is proved completely.

Remark II.3.11. If family $P_\Theta \mathscr{X}_\Pi$ is not complete in subspace K_Θ, then there may be no GF at all. As an example, let us take in $L^2(0, 2\pi)$ family $\mathscr{E}$ of "harmonics" $e^{-t} e^{int}$ with positive numbers, $n \in \mathbb{N}$.

The family $P_\Theta \mathscr{X}_\Pi$ of simple fraction projections corresponding to family $\mathscr{E}$ is of the form $\{P_{2\pi}(k - n + i)^{-1}\}_{n \in \mathbb{N}}$. If one takes a family of such fractions for all integer $n \in \mathbb{Z}$, it will have a GF $f(k) = e^{i\pi k} \sin \pi(k - i)$. However, for family $P_{2\pi} \mathscr{X}_\Pi$ there is no GF. Actually, in the opposite case a Cartwright-class function would be found with the set of zeros $\{n + i\}_{n \in \mathbb{N}}$. But this is impossible: the zeros of a Cartwright-class function are "equally" divided between the right and left half-planes.

3.3. Hilbert operator and exponential families on an interval

Let $W(x)$ be a nonnegative matrix function, $x \in \mathbb{R}$. Further, let L^2_W denote a space of vector functions squarely integrable over $\mathbb{R}$:

$$L^2_W = \left\{ f \mid \|f\|^2_{L^2_W} := \int_{\mathbb{R}} \langle f(x), W(x)f(x) \rangle \, dx < \infty \right\}.$$

In this section, we relate the basis property of a family of exponentials (a family of simple fractions $P_\Theta \mathscr{X}_\Pi$) with the boundedness of the Hilbert operator $\mathscr{H}$ (see Subsection I.1.5) in the space L^2_W with $W = F^*F$, where F is a GF of the family.

Introduce the sets $\mathscr{D}_\pm$ lying in $H^2_\pm(\mathfrak{N})$, respectively. $\mathscr{D}_+$ is a linear span (not a closure) of elements

$$\left(\frac{k - i}{k + i} \right)^m \frac{1}{k + i} \eta, \qquad \eta \in \mathfrak{N}, \; m = 0, 1, \ldots. \tag{47}$$

Similarly,

$$\mathscr{D}_- = \operatorname*{Lin}_{\eta \in \mathfrak{N}, m = 0, 1, \ldots} \left\{ \left(\frac{k + i}{k - i} \right)^m \frac{1}{k - i} \eta \right\}.$$

Elements (47) of family $\mathscr{D}_+$ are the images of the elements $z^m \eta \in H^2(\mathfrak{N})$ under the mapping of Hardy space in $\mathbb{D}$ on Hardy space in $\mathbb{C}_+$ (up to a constant factor). Hence, Cl $\mathscr{D}_+ = H_+^2(\mathfrak{N})$ and, similarly, Cl $\mathscr{D}_- = H_-^2(\mathfrak{N})$. From the definition, it follows that if $f \in \mathscr{D}_+ + \mathscr{D}_-$, then it satisfies the estimate

$$|f(k)| \leq C_f(1 + |k|)^{-1}, \qquad k \in \mathbb{R}. \tag{48}$$

Let F be a GF of family $P_\Theta \mathscr{X}_\Pi$ (thus, we assume $P_\Theta \mathscr{X}_\Pi$ to be minimal; see Theorem 5) and $\mathbb{F}$ be a multiplication by F operator. Since F is a strong function, operator $\mathbb{F}$ does not lead out the functions satisfying (48) of $L^2(\mathbb{R}, \mathfrak{N})$. Therefore operator

$$\mathscr{P} := \mathbb{F} P_- \mathbb{F}^{-1}$$

is properly defined on a (linear) set $\mathbb{F}(\mathscr{D}_+ + \mathscr{D}_-)$, where P_- is the ortho-projector from $L^2(\mathbb{R}, \mathfrak{N})$ on $H_-^2(\mathfrak{N})$.

Lemma II.3.12. Operator $\mathscr{P}$ is continuously extendable up to the skew projector

$$\tilde{\mathscr{P}} := \mathscr{P}_{\Theta H_-^2}^{\|\Pi H_+^2}$$

if and only if family $P_\Theta \mathscr{X}_\Pi$ is complete in the subspace K_Θ.

PROOF. Take an arbitrary element g from the domain of $\mathscr{P}$: $g = Fu_+ + Fu_-$, $u_+ \in \mathscr{D}_+$, $u_- \in \mathscr{D}_-$. Then

$$\mathscr{P}g = \mathbb{F} P_-(u_+ + u_-) = \mathbb{F} u_-.$$

Therefore

$$\mathscr{P}|_{\mathbb{F}\mathscr{D}_+} = 0, \qquad \mathscr{P}|_{\mathbb{F}\mathscr{D}_-} = I.$$

The first of the relations implies operator $\mathscr{P}|_{\mathbb{F}\mathscr{D}_+}$ to be continuously extended by zero on the set Cl $\mathbb{F}\mathscr{D}_+$. From the second one, it follows that operator $\mathscr{P}|_{\mathbb{F}\mathscr{D}_-}$ is continued by the identity operator to the set Cl $\mathbb{F}\mathscr{D}_-$. Let us show that $\mathscr{P}$ can be continued to the set Cl $\mathbb{F}\mathscr{D}_+ +$ Cl $\mathbb{F}\mathscr{D}_-$ if and only if

$$\text{Cl } \mathbb{F}\mathscr{D}_+ \cap \text{Cl } \mathbb{F}\mathscr{D}_- = \{0\}. \tag{49}$$

In fact, if (49) is not true, sequences $u_n^+ \in \mathscr{D}_+$, $u_n^- \in D_-$ may be found such that

$$u_n^+ \to u, \qquad u_n^- \to u, \qquad u \neq 0.$$

Then $\mathscr{P}u_n^+ = 0$, $\mathscr{P}u_n^- = u_n^- \to u$ and the continuation of $\mathscr{P}$ is not possible. If (49) is true, continuation of $\mathscr{P}$ is obviously possible.

Let us now show that

$$\text{Cl } \mathbb{F}\mathscr{D}_+ = \Pi H_+^2(\mathfrak{N}), \qquad \text{Cl } \mathbb{F}\mathscr{D}_- = \Theta H_-^2(\mathfrak{N}). \tag{50}$$

We use factorization of the GF F. In the upper half-plane $F = \Pi F_e^+$. Multiplication by Π is a unitary operator; that is why

$$\mathrm{Cl}\ \mathbb{F}\mathcal{D}_+ = \Pi\ \mathrm{Cl}\ F_e^+ \mathcal{D}_+.$$

By an outer function property (Subsection 1.2.8), $\mathrm{Cl}\ F_e^+ \mathcal{D}_+ = H_+^2(\mathfrak{N})$, and the first of equalities (50) is justified. In $\mathbb{C}_-$, operator function F has a factorization of the form $F = \Theta F_e^-$. Therefore, as in the arguments above, we prove the second equality from (50).

So, we have proved that the condition

$$\Pi H_+^2(\mathfrak{N}) \cap \Theta H_-^2(\mathfrak{N}) = \{0\} \tag{51}$$

is equivalent to the property of operator $\mathscr{P}$ to be extendable up to a skew projector $\mathscr{P} = \mathscr{P}_{\Theta H_-^2(\mathfrak{N})}^{\|\Pi H_+^2(\mathfrak{N})}$ defined over a set

$$\Pi H_+^2(\mathfrak{N}) + \Theta H_-^2(\mathfrak{N}). \tag{52}$$

To complete the proof of the lemma, one only has to note that (51) is equivalent to the completeness of family $P_\Theta \mathscr{X}_\Pi$ by the force of Theorem 3(a) and Lemma 2.

Lemma II.3.13.
(a) *Operator $\mathscr{P}$ is continuously extendable to the whole space $L^2(\mathbb{R}, \mathfrak{N})$ if and only if*

$$\varphi(\Theta H_-^2(\mathfrak{N}), \Pi H_+^2(\mathfrak{N})) > 0. \tag{53}$$

(b) *Inequality (53) implies inequality*

$$\varphi(\Theta H_+^2(\mathfrak{N}), \Pi H_-^2(\mathfrak{N})) > 0. \tag{54}$$

PROOF. First, we demonstrate that

$$\mathrm{Cl}\ (\Pi H_+^2(\mathfrak{N}) + \Theta H_-^2(\mathfrak{N})) = L^2(\mathbb{R}, \mathfrak{N}). \tag{55}$$

Here we exploit the minimality of family $P_\Theta \mathscr{X}_\Pi$ induced by the existence of the GF (Theorem 5). Assume (55) to be invalid. Then a nonzero element may be found in the intersection of orthogonal (in $L^2(\mathbb{R}, \mathfrak{N})$) complements to the subspaces $\Pi H_+^2(\mathfrak{N})$ and $\Theta H_-^2(\mathfrak{N})$. By formulas (13), those orthogonal complements are equal to $\Pi H_-^2(\mathfrak{N})$ and $\Theta H_+^2(\mathfrak{N})$, respectively. But Theorem 3(b) and Lemma 2 pronounce the intersection of these subspaces to be trivial. So, formula (55) is proved.

(a) Let (53) be true. Then by Lemma I.1.9(b)

$$\Pi H_+^2(\mathfrak{N}) + \Theta H_-^2(\mathfrak{N}) = \mathrm{Cl}(\Pi H_+^2(\mathfrak{N}) + \Theta H_-^2(\mathfrak{N})),$$

and by (55)

$$\Pi H_+^2(\mathfrak{N}) + \Theta H_-^2(\mathfrak{N}) = L^2(\mathbb{R}, \mathfrak{N}).$$

Under condition (53), relation (51) is obviously true, and according to Lemma 12, operator $\mathscr{P}$ is extendable up to a skew projector $\mathscr{P}^{\|\Pi H_+^2(\mathfrak{N})}_{\Theta H_-^2(\mathfrak{N})}$ defined over the whole space $L^2(\mathbb{R}, \mathfrak{N})$. On the other hand, if $\mathscr{P}$ is extendable in such a manner, then, as it was established in Lemma 12, the continuation is the skew projector $\mathscr{P}^{\|\Pi H_+^2(\mathfrak{N})}_{\Theta H_-^2(\mathfrak{N})}$ and by Lemma I.1.9(a) the angle between $\Pi H_+^2(\mathfrak{N})$ and $\Theta H_-^2(\mathfrak{N})$ is positive.

(b) Under condition (53) and taking into account (55), we can use Lemma I.1.9(c) and obtain

$$\varphi((\Theta H_-^2(\mathfrak{N}))^\perp, (\Pi H_+^2(\mathfrak{N}))^\perp) > 0.$$

In view of (13), we complete the proof of the lemma.

We now prove one of the main results of our book. Recall that GF of family $P_\Theta \mathscr{X}_\Pi$ is determined by functions Θ and Π. It is the same for the family of subspaces and for the family of elements.

Theorem II.3.14. Let F be a GF of family $P_\Theta \mathscr{X}_\Pi$. Then

(a) *Operator $P_\Theta|_{K_\Pi}$ is an isomorphism of spaces K_Π and K_Θ if and only if the Hilbert operator is bounded in the space $L^2_{F^*F}$.*

(b) *The family $P_\Theta \mathscr{X}_\Pi$ of elements forms a basis in the space K_Θ if and only if family $\mathscr{X}_\Pi$ forms a basis in K_Π and the Hilbert operator is bounded in the space $L^2_{F^*F}$.*

PROOF.

(a) By Lemma I.1.10(b), operator $P_\Theta|_{K_\Pi}$ is an isomorphism of the subspaces K_Θ and K_Π if and only if

$$\varphi(K_\Theta^\perp, K_\Pi) > 0, \qquad \varphi(K_\Theta, K_\Pi^\perp) > 0. \tag{56}$$

By Lemma 2, these inequalities are equivalent to (53) and (54). Thus, by Lemma 13(b), inequality (53) is a necessary and sufficient condition for operator $P_\Theta|_{K_\Pi}$ to be isomorphic. But (53) is equivalent to the boundedness of the skew projector $\tilde{\mathscr{P}}$. By Lemma 12, $\tilde{\mathscr{P}}$ coincides with operator $\mathscr{P}$ on a dense set. Let us use the expression for the Riesz projector P_- via the Hilbert operator $\mathscr{H}$, that is, formula (2) in Section II.1. In this way, we find that $\mathscr{P}$ is bounded (i.e., $P_\Theta|_{K_\Pi}$ is isomorphic) if and only if operator $\mathbb{F}\mathscr{H}\mathbb{F}^{-1}$ is bounded in $L^2(\mathbb{R}, \mathfrak{N})$.

Let us now demonstrate the latter operator to be isometric (unitary equivalent) to the Hilbert operator in the weight space L^2_{F*F}. To do this, we introduce an isometry

$$\mathbb{F}^{-1}: L^2 \mapsto L^2_{F*F}$$

by the rule $(\mathbb{F}^{-1}u)(x) = F^{-1}(x)u(x)$. The fact is that indeed an isometry follows from the identity

$$\|\mathbb{F}^{-1}u\|^2_{L^2_{F*F}} = \int_\mathbb{R} \langle F^{-1}u, (F^*F)F^{-1}u \rangle \, dx = \int_\mathbb{R} \langle u, u \rangle \, dx = \|u\|^2_{L^2(\mathbb{R},\mathfrak{N})}.$$

Operator $\mathbb{F}\mathscr{H}\mathbb{F}^{-1}$ is then isometric to operator $\mathbb{F}^{-1}(\mathbb{F}\mathscr{H}\mathbb{F}^{-1})\mathbb{F} = \mathscr{H}$ in L^2_{F*F}. Assertion (a) is proved.

(b) If operator $\mathscr{H}$ is bounded in L^2_{F*F}, projector $P_\Theta|_{K_\Pi}$ is an isomorphism in view of (a). Then family $P_\Theta\mathscr{X}_\Pi$, which is the image of family $\mathscr{X}_\Pi$, is a basis in K_Θ if $\mathscr{X}_\Pi$ is a basis in K_Π.

Conversely, if a family of elements $P_\Theta\mathscr{X}_\Pi$ forms a basis in K_Θ, then by Theorem 3(e) $\mathscr{X}_\Pi \in (LB)$, and (56) holds. The latter inequality is equivalent to the boundedness of $\mathscr{H}$ in L^2_{F*F}.

The theorem is proved.

Remark II.3.15. Let a condition of the semisimplicity of zeros not be imposed on a BPP Π. In order for family $P_\Theta\mathscr{X}_\Pi$ to form a basis, it is necessary to consider multiple fractions (or function $t^r \exp(-i\bar{\lambda}t)\eta$ for the exponential family). Nevertheless, if there is a function F with a factorization of the form (23), then it is true that assertion (a) of Theorem 14 remains as before. Indeed, in the proof we do not consider zeros of F to be semisimple.

3.4. Indications for Hilbert operator to be bounded

Definition II.3.16. The Hilbert transform of function $v \in L^\infty(\mathbb{R})$ is function

$$\tilde{v}(x) = \frac{1}{\pi} \, \text{p.v.} \int_\mathbb{R} \left(\frac{1}{x-t} + \frac{t}{1+t^2} \right) v(t) \, dt.$$

Note that if $v \in L^\infty \cap L^2$, then $\tilde{v}$ differs from $\mathscr{H} v$ only by a constant. The term $t/(1 + t^2)$ is added inside the parentheses for convergence; if the integral converges without it, the term produces a constant. It is useful to note that the Schwarz formula

$$V(k) = \frac{1}{\pi i} \int_{\mathbb{R}} \left(\frac{1}{t - k} - \frac{t}{1 + t^2} \right) v(t) \, dt$$

restores analytic function V in $\mathbb{C}_+$ by the trace of its real part v on $\mathbb{R}$: $V(x) = v(x) + i\tilde{v}(x)$ under the condition Im $V(i) = 0$.

For a scalar case, two criteria are known for the Hilbert operator to be bounded in space L_w^2 with the weight w.

Proposition II.3.17 (Helson and Szego 1960: Hunt, Muckenhoupt, and Weeden 1973; Nikol'skiĭ 1980; Garnett 1981: chap. VI, th. 1.2). *The following conditions are equivalent:*

(a) *the Hilbert operator is bounded in the space L_w^2, $w(x) \geq 0$,*
(b) *Helson–Szego condition: functions v, $u \in L^\infty(\mathbb{R})$, $\|v\|_{L^\infty} < \pi/2$, may be found such that $w = \exp(u + \tilde{v})$,*
(c) *Muckenhoupt condition:*

$$\sup \left(\frac{1}{|I|} \int_I w(x) \, dx \cdot \frac{1}{|I|} \int_I \frac{1}{w(x)} \, dx \right) < \infty, \tag{A_2}$$

where sup *is taken over all the intervals $I = (\alpha, \beta) \subset \mathbb{R}$.*

Muckenhoupt condition (A_2) is given in Hunt, Muckenhoupt, and Weeden (1973) for operator $\mathscr{H}$ to be bounded in L_w^2 for $p = 2$.

Let us provide some elementary information about condition (A_2).

Lemma II.3.18. Let $f_1, f_2 \in (A_2)$. *Then*

(a) $f_1 + f_2 \in (A_2)$,
(b) $f(x) := \max\{f_1(x), f_2(x)\} \in (A_2)$.

PROOF.

(a) By the definition of the Muckenhoupt condition there exists constant c such that for any interval $I = (\alpha, \beta)$

$$\int_I f_j(x) \, dx \leq c|I|^2 \left[\int_I f_j^{-1}(x) \, dx \right]^{-1}, \qquad j = 1, 2.$$

Hence,

$$\frac{1}{|I|} \int_I (f_1 + f_2)\, dx \, \frac{1}{|I|} \int_I (f_1 + f_2)^{-1}\, dx$$

$$\leq c \left\{ \frac{1}{\int_I f_1^{-1}\, dx} + \frac{1}{\int_I f_2^{-1}\, dx} \right\} \int_I (f_1 + f_2)^{-1}\, dx$$

$$= c \frac{\int_I (f_1 + f_2)^{-1}\, dx}{\int_I f_1^{-1}\, dx} + c \frac{\int_I (f_1 + f_2)^{-1}\, dx}{\int_I f_2^{-1}\, dx} \leq c + c = 2c,$$

since

$$\frac{1}{f_1 + f_2} \leq \frac{1}{f_j}, \qquad j = 1, 2.$$

(b) Note that

$$\tfrac{1}{2}(f_1(x) + f_2(x)) \leq f(x) \leq f_1(x) + f_2(x).$$

Assertion (b) now follows from (a) and the definition of Muckenhoupt condition (A_2).

The lemma is proved.

Proposition II.3.19 (Garnett 1981). *Let $w \in (A_2)$. Then*

$$\int_{\mathbb{R}} \frac{w(x)}{1 + x^2}\, dx < \infty.$$

Example II.3.20. Let us check that function $|x|^\alpha$ satisfies (A_2) if and only if $|\alpha| < 1$.

Let $|\alpha| < 1$. Because function $w(x) = |x|^\alpha$ is even, we may confine ourselves to intervals of two kinds: $0 < a < b$ and $a < 0 < b$ with $b \geq |a|$. For an interval of the first kind we have

$$\int_a^b w\, dx \asymp b^{\alpha+1} - a^{\alpha+1} = a^{\alpha+1}(c^{\alpha+1} - 1), \qquad c := b/a,$$

and

$$\int_a^b w^{-1}\, dx \asymp b^{-\alpha+1} - a^{-\alpha+1} = a^{-\alpha+1}(c^{-\alpha+1} - 1).$$

Therefore

$$\frac{1}{|I|^2} \int_I w\, dx \int_I w^{-1}\, dx \asymp \frac{(c^{\alpha+1} - 1)(c^{-\alpha+1} - 1)}{(c - 1)^2} =: \varphi_1(c).$$

It is not difficult to show function φ_1 to be bounded on the semiaxis $[1, \infty)$.

On the intervals $(-|a|, b)$, $b \geq |a|$, of the second kind we have for $c := b/|a|$

$$\frac{1}{|I|^2} \int_I w \, dx \int_I w^{-1} \, dx \asymp \frac{1}{(b + |a|)^2} (b^{\alpha+1} + |a|^{\alpha+1}) \cdot (b^{-\alpha+1} + |a|^{-\alpha+1})$$

$$= \frac{(c^{\alpha+1} + 1)(c^{-\alpha+1} + 1)}{(c + 1)^2} =: \varphi_2(c).$$

Function φ_2 is bounded on $[1, \infty)$.

If $|\alpha| \geq 1$, then

$$\int_{\mathbb{R}} \frac{w(x)}{1 + x^2} \, dx = \infty.$$

Therefore, the condition of Proposition 19, necessary for (A_2), is not satisfied.

In a vector situation a necessary and sufficient condition for the Hilbert operator to be bounded in the space with matrix weight $W(x) \geq 0$ is not known. Let us prove several results in this direction.

Theorem II.3.21.

(1) *Hilbert operator $\mathcal{H}$ is bounded in L^2_W if and only if it is bounded in $L^2_{(W)^{-1}}$.*

(2) *If $\mathcal{H}$ is bounded in L^2_W then*
   (a) *for any vector $\eta \in \mathfrak{N}$, $\eta \neq 0$,*

$$\langle W(\cdot)\eta, \eta \rangle \in (A_2) \tag{57}$$

   (b) $\langle\langle W(\cdot)\rangle\rangle \in (A_2)$.
(3) *If*

$$W(k) \asymp \langle\langle W(k)\rangle\rangle I_{\mathfrak{N}}, \qquad k \in \mathbb{R}, \tag{58}$$

then any of the conditions 2(a) and 2(b) are sufficient for operator $\mathcal{H}$ to be bounded in L^2_W.

PROOF. Operator $\mathcal{H}$ in L^2_W is unitary equivalent to operator $W^{1/2} \mathcal{H} W^{-1/2}$ in L^2 (see the proof of Theorem 14). The adjoint to the latter operator is $-W^{-1/2} \mathcal{H} W^{1/2}$. (The Hilbert operator is antiselfadjoint, $\mathcal{H}^* = -\mathcal{H}$, as is seen from the formula for the Riesz projector P_+.) And operator $W^{-1/2} \mathcal{H} W^{1/2}$ is unitary equivalent to operator $\mathcal{H}$ in $L^2_{(W)^{-1}}$.

(2a) For any $\eta \in \mathcal{N}$, $\langle\langle \eta \rangle\rangle = 1$, subspace

$$L_\eta := \{u \in L_W^2 \mid u(k) = f(k)\eta, f(k) \text{ is a scalar function}\}$$

of vectors spanned on η, is invariant under the action of $\mathcal{H}$, and $\mathcal{H}|_{L_\eta}$ is isometric to the "scalar" Hilbert operator in L_w^2 with weight $w = \langle W(\cdot)\eta, \eta \rangle$. From the boundedness of $\mathcal{H}$ in L_W^2 and the criterion for the scalar Hilbert operator to be bounded, condition (57) follows.

(2b) As has been proved, for any vector ξ_j^0 of an orthonormal basis in $\mathfrak{N}$, function $\langle W(\cdot)\xi_j^0, \xi_j^0 \rangle$ satisfies the Muckenhoupt condition. From Lemma 18(b) we then have

$$w_{\max}(\cdot) := \max_j \langle W(\cdot)\xi_j^0, \xi_j^0 \rangle \in (A_2). \qquad (59)$$

Let us check that

$$w_{\max}(k) \leq \langle\langle W(k) \rangle\rangle \leq N w_{\max}(k). \qquad (60)$$

The left inequality is obvious. To prove the right one we denote $g(k)$ the normed eigenvector of (matrix) operator $W(k)$, corresponding to its largest eigenvalue $\langle\langle W(k) \rangle\rangle$. Then, expanding basis vectors ξ_j^0 of $\mathfrak{N}$ over the basis of $W(k)$ eigenvectors, we arrive at

$$\langle W(k)\xi_j^0, \xi_j^0 \rangle \geq \langle\langle W(k) \rangle\rangle |\langle \xi_j^0, g(k) \rangle|^2. \qquad (61)$$

Since

$$1 = \langle\langle g(k) \rangle\rangle = \sum_{j=1}^N |\langle \xi_j^0, g(k) \rangle|^2,$$

$$\max_j |\langle \xi_j^0, g(k) \rangle|^2 \geq N^{-1}. \qquad (62)$$

Inequalities (61) and (62) imply the right one in (60). Now, (59) provides us with 2(b).

(3) Under condition (58), the metrics in spaces L_W^2 and $L_{\langle\langle W \rangle\rangle I}^2$ are equivalent. If the Muckenhoupt condition is fulfilled by function $\langle\langle W \rangle\rangle$, then Hilbert operator is bounded in $L_{\langle\langle W \rangle\rangle I}^2$. Hence, it is bounded also in $L_{w_{\max}}^2$ (function $w_{\max}$ is determinate in (59)). From inequalities (60) it follows that functions $w_{\max}$ and $\langle\langle W \rangle\rangle$ both satisfy the Muckenhoupt condition or both do not. Since from condition 2(a) and Lemma 18 it follows that $w_{\max} \in (A_2)$, the theorem is proved.

Corollary II.3.22. Let a GF of family $P_\Theta \mathscr{X}_\Pi$ *satisfy inequalities*

$$\langle\langle F(k) \rangle\rangle \prec 1, \qquad \langle\langle F(k)^{-1} \rangle\rangle \succ 1, \, k \in \mathbb{R}. \qquad (63)$$

Then $P_\Theta|_{K_\Pi}$ *is an isomorphism.*

Indeed, under condition (63), $F^*(k)F(k) \asymp I_{\mathfrak{R}}$, and one should only apply Theorem 21.

4. Minimality and the basis property of scalar exponential families in $L^2(0, T)$

The tests showing an exponential family to be minimal or to form a basis are formulated in a simpler and more convenient way for scalar applications as compared with the vector situation. Questions surrounding the theory of scalar exponential families are discussed in the literature (Paley and Wiener 1934; Levin 1956; Hrushchev, Nikol'skiĭ, and Pavlov 1981; Schwartz 1959; Young 1980). In this section, we consider several results that are significant for what follows. We demonstrate only a few of them since they are fairly well known.

Throughout the section, except in a few stipulated cases, we consider family $\mathscr{E}_T = \{e^{i\lambda_n t}\}_{n \in \mathbb{Z}}$ in the space $L^2(0, T)$. The family spectrum – that is, the set $\{\lambda_n\}$ – is denoted by σ. We assume that

$$\sup_{n \in \mathbb{Z}} |\operatorname{Im} \lambda_n| < \infty; \quad \lambda_n \neq \lambda_m; \quad n \neq m \qquad \operatorname{Re} \lambda_n \le \operatorname{Re} \lambda_{n+1}.$$

4.1. Minimality

Without sacrificing generality, one may consider (see Subsection 1.3.15) that $\sigma \in \mathbb{C}_+$.

Theorem II.4.1. Family $\mathscr{E}_T$ *is minimal in* $L^2(0, T)$ *if and only if an entire function* f *of the exponential type may be found with the indicator diagram of width not greater than* T, *which tends to zero on* σ *and satisfies the condition*

$$\int_{-\infty}^{\infty} \frac{|f(x)|^2}{1 + x^2} \, dx < \infty.$$

This assertion can be found in the literature (Paley and Wiener 1934) or can be proved easily enough in a manner similar to that of Theorem 3.5

(see also Remark 3.7 and Theorem 3.9(a)). For the demonstration we introduce simple fraction family

$$\mathscr{X}_T = \{P_T(k - \bar{\lambda})^{-1}\}_{\lambda \in \sigma}$$

with P_T being the orthoprojector in H^2_+ on $K_T := H^2_+ \ominus e^{ikT}H^2_+$. Family $\mathscr{E}_T$ is related to $\mathscr{X}_T$ by an isomorphism. In fact, the mapping $g(t) \mapsto \overline{g(t)}$ turns $\mathscr{E}_T$ into family $\{e^{-i\lambda t}\}_{\lambda \in \sigma}$, which is transformed into $\mathscr{X}_T$ by the inverse Fourier transform (constant factor $1/\sqrt{2\pi}$ is omitted).

Let family $\mathscr{E}_T$ be minimal in $L^2(0, T)$. Then family $\mathscr{X}_T$ is minimal in K_T. Take an arbitrary element $x'_\lambda(k)$ of the family biorthogonal to $\mathscr{X}_T$. One easily sees that function $f(k) = (k - \lambda)x'_\lambda(k)$ satisfies the conditions of the theorem.

Conversely, let f satisfy the conditions. Then it belongs to the Cartwright class (see Subsection 1.4.3) and allows factorization (Proposition 1.25)

$$f = e^{-ik\alpha_+}\pi_+ f^+_e = e^{ik\alpha_-} f^-_e.$$

The assumptions made provide $\alpha_+ + \alpha_- \leq T$ and $\pi_+(\lambda) = 0$ for $\lambda \in \sigma$. Set $f_0(k) = f(k) e^{ik\alpha_+}$ and let r_λ denote the multiplicity of the root of $f_0(k)$ at $k = \lambda$. Let us check the family

$$\left\{\frac{r_\lambda!}{2\pi i f_0^{(r_\lambda)}(\lambda)} \cdot \frac{f_0(k)}{(k - \lambda)^{r_\lambda}}\right\}_{\lambda \in \sigma}$$

to be biorthogonal to $\mathscr{X}_T$ (compare with Theorems 3.5 and 3.9). For $\mu \in \sigma$, $\mu \neq \lambda$, we have

$$\left(\frac{f_0(k)}{(k - \lambda)^{r_\lambda}}, \frac{1}{k - \bar{\mu}}\right) = 2\pi i \operatorname*{res}_{k=\mu} \frac{f_0(k)}{(k - \lambda)^{r_\lambda}(k - \mu)} = 0.$$

At the same time,

$$\left(\frac{f_0(k)}{(k - \lambda)^{r_\lambda}}, \frac{1}{k - \bar{\lambda}}\right) = 2\pi i \operatorname*{res}_{k=\lambda} \frac{f_0(k)}{(k - \lambda)^{r_\lambda + 1}} = 2\pi i \frac{f_0^{(r_\lambda)}(\lambda)}{r_\lambda!}.$$

The theorem is proved.

The following assertion contains a test for a family $\mathscr{E}_T$ to be nonminimal. Set

$$n(r) = \operatorname{card}\{\lambda \in \sigma \mid |\lambda| < r\}.$$

Corollary II.4.2. *If*

$$\limsup_{r \to \infty} \frac{n(r)}{r} > \frac{T}{\pi}, \tag{1}$$

then family $\mathscr{E}_T$ is not minimal in $L^2(0, T)$.

PROOF. Let $\mathscr{E}_T$ be minimal and f be the function described in Theorem 1. Let us write $\tilde{\sigma}$ for the set of zeros of f and $\tilde{n}(r) := \mathrm{card}\{\lambda \in \tilde{\sigma} \mid |\lambda| < r\}$. Proposition 1.25 implies

$$\lim_{r \to \infty} \tilde{n}(r)/r \leq \frac{T}{\pi}.$$

But along with it $\tilde{\sigma} \supset \sigma$, and therefore $\tilde{n}(r) \geq n(r)$, which contradicts (1).

4.2. The basis property of family $\{e^{i\lambda_n t}\}$

The problem of how to describe the Riesz bases of exponentials has a long history. The first such effort seems to have been made by R. Paley and N. Wiener (1934), who proved that $\{e^{i\lambda_n t}\}$ is a Riesz basis in $L^2(0, 2\pi)$ if $\sup_{n \in \mathbb{Z}} |\lambda_n - n| = d < \pi^{-2}$. R. J. Duffin and J. J. Eachus (1942) demonstrated the same statement to be valid for $d < \pi^{-1} \log 2$. The problem was solved in these terms by M. I. Kadets (1964), who proved the result for $d < 1/4$. The examples in Ingham (1934) and Levinson (1940) show that Kadet's result cannot be improved.

All the investigations just mentioned rest on the fact that $\{e^{i\lambda_n t}\}$ is close to the orthonormal basis $\{e^{int}\}$. An alternative approach to the description of Riesz bases of exponentials in $L^2(0, T)$ was developed by B. Ya. Levin (1961). Entire function F of the exponential type, whose set of zeros coincides with σ and the indicator diagram is a segment $[-iT, 0]$ of the imaginary axis, plays an important role in this approach. If F is a Cartwright-class function, it allows an explicit representation of the form (see Subsection 1.4.4)

$$F(z) = e^{izT/2} \text{ p.v.} \prod_{n \in \mathbb{Z}} (1 - z/\lambda_n).$$

(In the case where one of the numbers λ_n equals zero, the corresponding factor $1 - z/\lambda_n$ is replaced by z.)

If $\sigma \in \mathbb{C}_+$ and the condition

$$\int_{-\infty}^{\infty} \frac{|F(x)|^2}{1 + x^2} \, dx < \infty \tag{2}$$

holds, F is a generating function (GF) for a simple fraction family $\{P_T(k - \bar{\lambda}_n)^{-1}\}_{n \in \mathbb{Z}}$ in space K_T in the sense of Definition 3.4.

The foregoing provides the grounds to call an entire function of the exponential type with simple zeros $\{\lambda_n\}$ and with indicator diagram $[-iT, 0]$ a generating function of family $\{e^{i\lambda_n t}\}$ in $L^2(0, T)$. This definition is broader than the one introduced in Section 3 since it does not assume

$\mathscr{E}_T$ to be minimal. That is, it does not assume condition (2) to be valid. A broader definition of a generating function is more convenient in this section, and we conserve notation GF for it.

B. Ya. Levin (1961) and V. D. Golovin (1964) were the first to formulate a generating function for the basis property of exponential families.

Proposition II.4.3. Let a GF of family $\mathscr{E}_T$ be a sine-type function (STF, see Subsection 1.4.6), and let the set σ be separable. That is, $\inf_{n \neq m}|\lambda_n - \lambda_m| > 0$. Then $\mathscr{E}_T$ is a Riesz basis in $L^2(0, T)$.

From a contemporary point of view, this result is just a simple combination of Theorem 3.14(b) and Corollary 3.22. Note that the separability of σ is necessary for $\mathscr{E}_T$ to be uniformly minimal.

V. E. Katsnelson (1971) generalized Kadet's theorem for the case of zeros of an STF.

Proposition II.4.4. Let a GF of family $\mathscr{E}_T$ be an STF, and $\{\delta_n\}_{n \in \mathbb{Z}}$ be a bounded sequence of complex numbers such that

$$|\text{Re } \delta_n| \leq d \inf_{n \neq m} |\text{Re}(\lambda_m - \lambda_n)|, \qquad d < 1/4, \; n \in \mathbb{Z}. \tag{3}$$

If set $\{\lambda_n + \delta_n\}$ is separable, then family $\{e^{i(\lambda_n + \delta_n)t}\}$ is a Riesz basis in $L^2(0, T)$.

This theorem was strengthened by Avdonin (1974a, 1974b), who replaced condition (3) on sequence $\{\delta_n\}$ by an analog valid "in the mean." To formulate the result, we need the following definition.

Definition II.4.5. Let $\sigma = \{\lambda_n\}_{n \in \mathbb{Z}}$, $\sup_{n \in \mathbb{Z}}|\text{Im } \lambda_n| < \infty$, and $\{\alpha_j\}_{j \in \mathbb{Z}} \subset \mathbb{R}$ be a growing sequence such that $\sup_{j \in \mathbb{Z}} l_j < \infty$, $l_j := \alpha_{j+1} - \alpha_j$. Decomposition

$$\sigma = \bigcup_{j \in \mathbb{Z}} \sigma_j, \qquad \sigma_j := \{\lambda_n \mid \alpha_j \leq \text{Re } \lambda_n < \alpha_{j+1}\},$$

is said to be a Λ-decomposition of set σ.

Proposition II.4.6 (Avdonin 1974a). *Replace condition (3) of Proposition 4 by the requirement*

$$\left| \sum_{n: \; \lambda_n \in \sigma_j} \text{Re } \delta_n \right| \leq dl_j, \qquad d < 1/4, \; j \in \mathbb{Z}, \tag{4}$$

to some Λ-decomposition of set σ. Then Proposition 4 remains valid.

One can easily check that conditions (4) are equivalent to the inequality

$$\limsup_{r \to +\infty} \sup_{x \in \mathbb{R}} \frac{\Delta_x(r)}{2r} < 1/4, \qquad \Delta_x(r) := \sum_{x - r < \operatorname{Re} \lambda_n < x + r} \operatorname{Re} \delta_n.$$

Avdonin (1974a) also presented the following generalization of the Levin–Golovin theorem.

Proposition II.4.7. Assume that

(a) *set* $\{\lambda_n\}$ *is separable and* $|\operatorname{Im} \lambda_n| \leq h < \infty$, $n \in \mathbb{Z}$;
(b) *generating function F of family* $\mathcal{E}_T$ *satisfies a condition*

$$|F(x + iy_0)| \asymp |x|^\alpha, \qquad x \in \mathscr{R}$$

with some $\alpha \in (-1/2, 1/2)$ *and some* y_0, $|y_0| > h$. *Then* $\mathcal{E}_T$ *is a Riesz basis in* $L^2(0, T)$.

Avdonin (1974a) demonstrated a similar statement for functions of more general form than $|x|^\alpha$; he also showed that $|F(x + iy_0)|$ may be strongly oscillating.

Propositions 3, 4, 6, and 7 are proved by means of the theory of functions of complex variables and by some infinite product estimates. A criterion for the Riesz basis property of family $\mathcal{E}_T$ was obtained by B. S. Pavlov (1979) with the help of a "geometric" approach: for $\operatorname{Im} \lambda_n \geq c > 0$ exponential family $\{e^{i\lambda_n t}\}_{n \in \mathbb{Z}}$ forms a basis in $L^2(0, T)$ if and only if it is an $\mathscr{L}$-basis in $L^2(0, \infty)$ and the projector from $\bigvee_{L^2(0, \infty)} \{e^{i\lambda_n t}\}$ on $L^2(0, T)$ is an isomorphism. This approach for vector families is explained in Section 3. Accounting for an isomorphism $f(t) \mapsto f(t) e^{-\alpha t}$ of the space $L^2(0, T)$ (see Subsection 1.3.16), the necessary and sufficient conditions by which a family of exponentials constitutes a basis may be formulated as follows.

Theorem II.4.8 (Pavlov 1979). Family $\{e^{i\lambda_n t}\}_{n \in \mathbb{Z}}$ *forms a Riesz basis in* $L^2(0, T)$ *if and only if*

(a) *set* $\{\lambda_n\}$ *is separable and* $|\operatorname{Im} \lambda_n| \leq h < \infty$, $n \in \mathbb{Z}$;
(b) *for some* y_0, $|y_0| > h$, *function* $w(x)$,

$$w(x) := |F(x + iy_0)|^2, \qquad F(z) := \text{p.v.} \prod_{m \in \mathbb{Z}} (1 - z/\lambda_n),$$

satisfies *Muckenhoupt condition* (A₂) *or the Helson–Szego condition* (*see Proposition* II.3.17) *while the width of function F indicator diagram equals T.*

The following relation is useful in studying the basis property (see, e.g., Avdonin 1974a)

$$|F(x + iy_0)| \asymp |F(x + iy_1)|, \qquad x \in \mathbb{R},$$

for any y_1 such that $\inf_{n \in \mathbb{Z}} |\text{Im } \lambda_n - y_1| > 0$. Therefore condition (b) of Theorem 8 implies $|F(x + iy_1)|^2 \in (A_2)$. In applying this relation, it is convenient to assume that $\text{Im } \lambda_n \geq c > 0$ and to establish the inclusion

$$|F(x)|^2 \in (A_2).$$

Propositions 3 and 7 now become simple consequences of the fact that function $w(x)$ obeying the estimates

$$w(x) \asymp |x|^\alpha, \qquad x \in \mathbb{R}, \alpha \in (-1, 1)$$

satisfies condition (A₂) (see Example 3.20).

Some other known indications for a basis property of $\mathscr{E}_T$ can be derived from Theorem 8 (Hrushchev, Nikol'skiĭ, and Pavlov 1981), and a criterion for it can be obtained in terms of the function $N_\sigma(x)$,

$$N_\sigma(x) := \begin{cases} \text{card}\{\lambda_n \mid 0 \leq \text{Re } \lambda_n \leq x\}, & x \geq 0, \\ -\text{card}\{\lambda_n \mid x \leq \text{Re } \lambda_n < 0\}, & x < 0. \end{cases}$$

In particular, if $\{e^{i\lambda_n t}\}$ is a Riesz basis in $L^2(0, T)$, then function $\psi(x) := N_\sigma(x) - (T/2\pi)x$ belongs to the BMO class. Recall that the BMO class consists of locally summable functions f on the real axis for which

$$\sup_{I \in \mathscr{J}} \frac{1}{|I|} \int_I |f(x) - f_I| \, dx < \infty, \qquad f_I := \frac{1}{|I|} \int_I f(x) \, dx.$$

Here $\mathscr{J}$ is the set of all intervals of $\mathscr{R}$.

Corollary II.4.9. *If family $\{e^{i\lambda_n t}\}$ forms a Riesz basis in $L^2(0, T)$, then*

$$\sup_{n \in \mathbb{Z}} |\lambda_{n+1} - \lambda_n| < \infty. \tag{5}$$

Indeed, if (5) is not valid, then for any $m \in \mathbb{N}$ an interval I_m of the length m may be found on which function $N_\sigma(x)$ is constant. Then one easily checks that

$$\frac{1}{m}\int_{I_m}|\psi(x) - \psi_{I_m}|\,dx = \frac{T}{2\pi}\frac{m}{4}$$

and, hence, $\psi \notin \text{BMO}$.

Remark II.4.10. With the help of Pavlov's approach, a necessary and sufficient condition was obtained (Hrushchev, Nikol'skiĭ, and Pavlov 1981) for family

$$\{\sqrt{|\text{Im}\,\lambda_n|}\,e^{i\lambda_n t}\}_{n\in\mathbb{Z}}$$

with semibounded spectrum ($\inf_{n\in\mathbb{Z}}\text{Im}\,\lambda_n > -\infty$ or $\sup_{n\in\mathbb{Z}}\text{Im}\,\lambda_n < \infty$) as a Riesz basis in $L^2(0, T)$. The constraint for the spectrum to be semibounded was removed by Minkin (1991).

Let us prove the stability of both minimal and basis properties of exponential families under the perturbation of a finite set of their elements.

Lemma II.4.11. Let $\mathcal{N}$ be an arbitrary finite set of integers,

$$\{\mu_n\}_{n\in\mathcal{N}} \cap \{\lambda_n\}_{n\in\mathbb{Z}\setminus\mathcal{N}} = \varnothing, \qquad \mu_n \neq \mu_m, n \neq m.$$

Then the replacement of $\{\lambda_n\}_{n\in\mathcal{N}}$ by $\{\mu_n\}_{n\in\mathcal{N}}$ violates neither the minimal nor basis property of family $\mathcal{E}_T$.

PROOF. Without a loss of generality, we are able to consider that

$$\inf_{n\in\mathbb{Z}}\text{Im}\,\lambda_n > 0, \qquad \inf_{n\in\mathcal{N}}\text{Im}\,\mu_n > 0.$$

Let $\mathcal{E}_T$ be a Riesz basis in $L^2(0, T)$ and $F(z)$ be its GF. By Theorem 8, $|F(x)|^2 \in (A_2)$. Function

$$\tilde{F}(z) := F(z)\sum_{n\in\mathcal{N}}\frac{z - \mu_n}{z - \lambda_n}$$

is a generating function for family

$$\tilde{\mathcal{E}}_T := \{e^{i\lambda_n t}\}_{n\in\mathbb{Z}\setminus\mathcal{N}} \cup \{e^{i\mu_n t}\}_{n\in\mathcal{N}}.$$

It is easy to see that

$$\left|\prod_{n\in\mathcal{N}}\frac{x - \mu_n}{x - \lambda_n}\right| \asymp 1, \qquad x \in \mathbb{R}.$$

That is why $|\tilde{F}(x)|^2 \in (A_2)$, and, hence, family $\tilde{\mathcal{E}}_T$ is a Riesz basis.

The same arguments prove the implication

$$\mathscr{E}_T \in (M) \Rightarrow \tilde{\mathscr{E}}_T \in (M).$$

One only has to refer to the function introduced in Theorem 1 instead of the GF F.

Theorem 8 and Corollary 9 enable one to prove that the proximity of the set $\{\lambda_n\}$ to the zeros of an STF is not only sufficient but also necessary for family $\mathscr{E}_T$ to form a basis.

Proposition II.4.12 (Avdonin and Joó 1988). *Family $\{e^{i\lambda_n t}\}_{n\in\mathbb{Z}}$ forms a Riesz basis in $L^2(0, T)$ if and only if*

(a) *set $\{\lambda_n\}$ is separable and $\sup_{n\in\mathbb{Z}}|\text{Im }\lambda_n| < \infty$;*
(b) *there exists an STF with an indicator diagram of width T and a set of zeros $\{\mu_n\}$ such that for some $d \in (0, 1/4)$*

$$d\,\text{Re}(\lambda_{n-1} - \lambda_n) \leq \text{Re}(\mu_n - \lambda_n) \leq d\,\text{Re}(\lambda_{n+1} - \lambda_n), \qquad n \in \mathbb{Z}.$$

The latter inequalities may be replaced by the equivalent ones:

$$\hat{d}\,\text{Re}(\mu_{n-1} - \mu_n) \leq \text{Re}(\lambda_n - \mu_n) \leq \hat{d}\,Re(\mu_{n+1} - \mu_n), \quad \hat{d} \in (0, 1/4).$$

Proposition 12 leads directly to an assertion about the stability of the basis of exponentials under small variations of its spectrum.

Proposition II.4.13. *If family $\{e^{i\mu_n t}\}_{n\in\mathbb{Z}}$ constitutes a Riesz basis in $L^2(0, T)$, then there exists $\varepsilon > 0$ such that for any sequence $\{\lambda_n\}_{n\in\mathbb{Z}}$ satisfying $|\lambda_n - \mu_n| < \varepsilon$, family $\{e^{i\lambda_n t}\}_{n\in\mathbb{Z}}$ is also a Riesz basis in $L^2(0, T)$.*

In Section 5 (Theorem 5.5) a similar statement is proved for a family of vector exponentials.

Remark II.4.14. Since a perturbation of a finite number of exponentials does not violate the basis property (Lemma 11), conditions $|\lambda_n - \mu_n| < \varepsilon$ have to hold only for large $|n|$.

The following simple consequence of Proposition 12 concerns the basis property conservation under the variation of $\text{Im }\lambda_n$.

Corollary II.4.15. *Let family $\{e^{i\lambda_n t}\}_{n\in\mathbb{Z}}$ form a Riesz basis in $L^2(0, T)$ and let $\{\delta_n\}$ be a bounded sequence of real numbers. If set $\{\lambda_n + i\delta_n\}$ is separable, then family $\{\exp(i(\lambda_n + i\delta_n)t)\}_{n\in\mathbb{Z}}$ is a Riesz basis in $L^2(0, T)$.*

The geometric approach turned out to be useful for studying not only exponential families but also other function families. Bases from reproducing kernels were considered by Hrushchev, Nikol'skiĭ, and Pavlov (1981) and bases from Mittag–Leffler functions by Gubreev (1987) and Hrushchev (1987).

4.3. Basis subfamilies of $\{e^{i\lambda_n t}\}$

The following result is useful for describing properties of reachability sets for the systems of hyperbolic type, but it is of interest for other reasons as well.

Theorem II.4.16. If family $\{e^{i\lambda_n t}\}_{n \in \mathbb{Z}}$ forms a Riesz basis in $L^2(0, T)$, then for any $T' \in (0, T)$ there exists a subfamily $\mathscr{E}' \subset \mathscr{E}_T$ constituting a Riesz basis in $L^2(0, T')$.

Avdonin (1977b: p. 98) discusses this statement for the case $\lambda_n = T/(2\pi)n + o(1)$, as does Ivanov (1983c). Avdonin, Horvath, and Joó (1989, see also Theorem 25 below) prove the existence of a basis for the uniform density of the distribution of subfamily and Avdonin (1991) proves Theorem 16 in the given formulation. All these authors exploit the conditions for $\mathscr{E}_T$ as a basis stated in Proposition 6.

To prove Theorem 16, we start by clarifying uniformity in the distribution of STF zeros.

Theorem II.4.17. Let F be an STF with the indicator diagram of the width T and the set of zeros $\{\lambda_n\}_{n \in \mathbb{Z}}$. Set

$$N(x, r) := \operatorname{card}\{\lambda_n \mid x \le \operatorname{Re} \lambda_n < x + r\}, \qquad x \in \mathbb{R}, r > 0.$$

Then

$$\frac{N(x, r)}{r} \to \frac{T}{2\pi} \qquad (r \to \infty) \tag{6}$$

uniformly relative to $x \in \mathbb{R}$.

PROOF. Let us consider the types of F in the upper and lower half-planes to be $T/2$. This may always be achieved by multiplying by $e^{i\gamma z}$, where $\gamma \in [-T/2, T/2]$. Let $h := \sup_{n \in \mathbb{Z}} |\operatorname{Im} \lambda_n|, H > h$. In the half-plane $\operatorname{Im} z > H$, choose a single-valued continuous branch of $\arg F(z)$. Levin and Ostrovskiĭ

(1979: lemma 3) have shown that

$$\arg F(r + iH) - \arg F(iH) = -\pi N(0, r) + \mathcal{O}(1).$$

The same arguments lead to the equality

$$\arg F(x + r + iH) - \arg F(x + iH) = -\pi N(x, r) + \mathcal{O}(1). \qquad (7)$$

Here, $\mathcal{O}(1)$ denotes a bounded function on r and x. Set

$$\Phi(z) = cF(x + iH)\, e^{iTz/2},$$

with the constant $c \in \mathbb{C}$ chosen in such a way that $\Phi(i) > 0$. From formula (7) we get

$$\arg \Phi(x + r) - \arg \Phi(x) = -\pi N(x, r) + rT/2 + \mathcal{O}(1).$$

On the other hand, since function $\log \Phi(z)$ is bounded in $\mathbb{C}_+$ and $\operatorname{Im} \Phi(i) = 0$, functions $\arg \Phi(x)$ and $\log \Phi(x)$ are related by (see Hrushchev, Nikol'skiĭ, and Pavlov 1981: p. 234)

$$\arg \Phi(x) = \tilde{\varphi}(x) := \frac{1}{\pi} \, \text{p.v.} \int_{-\infty}^{\infty} \left[\frac{1}{x - t} + \frac{t}{1 + t^2} \right] \varphi(t)\, dt,$$

$$\varphi(x) := \log |\Phi(x)|,$$

we assume that $\arg \Phi(i) = 0$. Therefore, to prove the theorem it is enough to show that

$$\frac{1}{r} \, \text{p.v.} \int_{-\infty}^{\infty} \left[\frac{1}{x + r - t} - \frac{1}{x - t} \right] \varphi(t)\, dt \xrightarrow[r \to \infty]{} 0$$

uniformly in $x \in \mathbb{R}$.

Since F is an STF, function $\varphi(x)$ is bounded over the real axis. From Levin and Ostrovskiĭ (1979: lemma 3) it follows also that its derivative $\varphi'(x)$ is bounded on $\mathbb{R}$. Therefore (see, for instance, Avdonin and Joó 1988: p. 8),

$$\text{p.v.} \int_{t \in B(x, r)} \left[\frac{1}{x + r - t} - \frac{1}{x - t} \right] \varphi(t)\, dt = \mathcal{O}(1),$$

where

$$B(x, r) = \{t \in \mathbb{R} \mid |x + r - t| < 1\} \cup \{t \in \mathbb{R} \mid |x - t| < 1\}.$$

The fact that

$$\int_{\mathbb{R} \setminus B(x, r)} \frac{|\varphi(t)|\, dt}{|x - t|\,|x + r - t|} \xrightarrow[r \to \infty]{} 0$$

uniformly in $x \in \mathbb{R}$ follows immediately from the boundedness of φ. Theorem 17 is proved.

Theorem II.4.18. *Let* $\{\lambda_n\}_{n \in \mathbb{Z}}$ *be a sequence of complex numbers such that*

$$\sup_{n \in \mathbb{Z}} |\operatorname{Im} \lambda_n| < \infty, \qquad \inf_{n \neq m} |\lambda_m - \lambda_n| > 0$$

and let corresponding function $N(x, r)$ *satisfy condition* (6) *uniformly in* $x \in \mathbb{R}$. *Then for any* $T' \in (0, T)$, *family* $\{e^{i\lambda_n t}\}_{n \in \mathbb{Z}}$ *contains a subfamily* $\mathcal{E}'$ *that forms a Riesz basis in* $L^2(0, T')$.

PROOF. We construct family $\mathcal{E}'$ as a perturbation of the orthogonal in $L^2(0, T)$ basis $\{e^{i(2\pi/T')mt}\}_{m \in \mathbb{Z}}$. Let $I = [\alpha, \beta)$ be an arbitrary interval of the real axis. Set

$$M_I = \left\{ m \in \mathbb{Z} \,\Big|\, \frac{2\pi}{T'} m \in I \right\}, \qquad N_I = \{ n \in \mathbb{Z} \mid \operatorname{Re} \lambda_n \in I \},$$

$$m_* = \min\{m \mid m \in M_I\}, \qquad n_* = \min\{n \mid n \in N_I\},$$

$$m^* = \max\{m \mid m \in M_I\}, \qquad n^* = \max\{n \mid n \in N_I\},$$

and construct mapping $Q_*: M_I \mapsto \mathbb{Z}$ as follows:

$$Q_*(m_*) = n_*, \qquad Q_*(m_* + 1) = n_* + 1, \qquad Q_*(m_* + 2) = n_* + 2, \ldots.$$

Sequence $\{\delta_m'\}$ is specified for $m \in M_I$ by the equalities

$$\delta_m' = \lambda_{Q_*(m)} - \frac{2\pi}{T'} m.$$

Similarly, we define the mapping $Q^*: M_I \mapsto \mathbb{Z}$

$$Q^*(m^*) = n^*, \qquad Q^*(m^* - 1) = n^* - 1, \qquad Q^*(m^* - 2) = n^* - 2, \ldots;$$

and the sequence $\{\delta_m''\}$

$$\delta_m'' = \lambda_{Q^*(m)} - \frac{2\pi}{T'} m, \qquad m \in M_I.$$

According to the condition of the theorem, the density of set $\{Re\, \lambda_n\}$ is more than the density of $\{2\pi m/T'\}$. Therefore, if interval I is large enough, it can be proved that the inequalities are valid

$$-c \leq \sum_{m \in M_I} \operatorname{Re} \delta_m' \leq 0, \qquad 0 \leq \sum_{m \in M_I} \operatorname{Re} \delta_m'' \leq c. \tag{8}$$

Mappings Q_* and Q^* can be constructed for any interval. Since condition (6) is uniform in $x \in \mathbb{R}$, inequalities (8) are valid for intervals that are long enough, and constant c depends only on the length.

Accurate calculations show that for the validity of (8) it is sufficient if the length R of I satisfies the inequality

$$R \geq \frac{2T}{T - T'} \cdot r_1.$$

Here, r_1 is found from the conditions

$$\left| \frac{1}{r} N(x, r) - \frac{T}{2\pi} \right| < \delta \cdot \frac{T - T'}{2\pi}, \tag{9}$$

for $r > r_1 > 2\pi/T$, $x \in \mathbb{R}$, for some $\delta \in (0, 1)$.

Let us now take any interval I_1 of length lR, $l \in \mathbb{N}$. Using mappings Q and Q^* on subintervals of length R and taking into account (8), we can construct mapping $Q: M_{I_1} \mapsto \mathbb{Z}$ such that corresponding sequence $\{\delta_m\}$

$$\delta_m = \lambda_{Q(m)} - \frac{2\pi}{T'} m, \qquad m \in M_{I_1}$$

satisfies the condition

$$\left| \sum_{m \in M_{I_1}} \operatorname{Re} \delta_m \right| \leq c = \frac{c}{lR} |I_1|. \tag{10}$$

Let us take l to be large so as to provide $c/lR < 1/4$. Since inequality (10) holds for any interval of the form $[jlR, (j + 1)lR)$, $j \in \mathbb{Z}$, by the force of Proposition 6 family $\{\exp(i\lambda_{Q(m)}t)\}_{m \in \mathbb{Z}}$ constitutes a Riesz basis in $L^2(0, T')$. The theorem is proved.

Remark II.4.19. The proof of Theorem 18 shows that its conditions may be sharpened in the following way. If condition (9) holds for all sufficiently large r, then there exists a subfamily of $\{e^{i\lambda_n t}\}$ forming a Riesz basis in $L^2(0, T')$.

Theorem 16 is a direct consequence of Proposition 12 and Theorems 17 and 18.

4.4. Algorithm of a basis subfamily extraction

In the optimal control problems for DPS (see Avdonin, Ivanov, and Ishmukhametov 1991), it is useful to have a simple algorithm for the extraction of a basis exponential subfamily. In contrast to the more general situation described in Theorems 16 and 18, sequence $\{\lambda_n\}$ in such cases has a known asymptotic behavior. To illustrate this, let us consider exponential family $\mathscr{E}_T = \{e^{\pm i\lambda_n t}\}_{n\in\mathbb{N}}$ with a separable sequence $\{\lambda_n\}$ satisfying an asymptotic relation

$$\lambda_n = \lambda_n^0 + o(1), \qquad \lambda_n^0 = \frac{2\pi}{T}(n - 1/2).$$

Let $T' \in (0, T)$. It is necessary to construct an algorithm of extraction of a basis in the $L^2(0, T')$ subfamily.

We try to construct such a basis as a small perturbation in the mean (in the sense of Proposition 6) of the orthogonal in $L^2(0, T')$ basis

$$\{e^{\pm i\mu_n t}\}_{m\in\mathbb{N}}, \qquad \mu_n := \frac{2\pi}{T}(m - 1/2).$$

We define sequence $\{n_m\}$ in the following manner. For each $m \in \mathbb{N}$ we find an integer n_m from the conditions

(i) $|\lambda_{n_m}^0 - \mu_m| = \min_{n\in\mathbb{N}} |\lambda_n^0 - \mu_m|$,

(ii) for the set of such m to which each of the two points of set $\{\lambda_n^0\}_{n\in\mathbb{Z}}$ closest to μ_m correspond, the signs of differences $\lambda_{n_m}^0 - \mu_m$ are interchanging.

Note that condition (ii) becomes necessary when the numbers T and T' are commensurable; that is, $T/T' \in \mathbb{Q}$.

Theorem II.4.20. If sequence $\{n_m\}$ is constructed by the algorithm (i), (ii), then family $\mathscr{E}'$ forms a Riesz basis in $L^2(0, T')$.

PROOF. By Proposition 13 and Remark 14, it suffices to show that family $\mathscr{E}'_0 = \{e^{i\lambda_{n_m}^0 t}\}$ is a Riesz basis in $L^2(0, T')$. It is convenient to perform the change of variable $t \mapsto (2\pi/T)t$ and move on to the case

$$T = 2\pi, \qquad \lambda_n^0 = n - 1/2, \qquad \mu_m = a(m - 1/2), \quad a > 1.$$

Set $\delta_m := \mu_m - \lambda_{n_m}^0$.

Consider first the situation when a is rational; that is, $a = p/q$; $p, q \in \mathbb{N}$. In accordance with Definition 5, let us construct Λ-partitioning of set

$\sigma := \{\pm \mu_m\}_{m \in \mathbb{N}}$ by setting $\alpha_j = 2jp, j \in \mathbb{Z}$. So,

$$\sigma = \bigcup_{j \in \mathbb{Z}} \sigma_j, \qquad \sigma_j = \{\mu_m \mid m = 2qj + 1, 2qj + 2, \ldots, 2q(j + 1)\}.$$

Since both sets σ_j and Λ_j^0,

$$\Lambda_j^0 = \{\lambda_m^0 \mid n = 2pj + 1, 2pj + 2, \ldots, 2p(j + 1)\}$$

are symmetric in relation to point $(2j + 1)p$, we have

$$\sum_{m: \mu_m \in \sigma_j} \delta_m = 0, \qquad j \in \mathbb{Z}.$$

Therefore, by Proposition 6, family $\mathscr{E}_0'$ is a Riesz basis in $L^2(0, T')$.

Now let a be an irrational number. In this case, one δ_m is able to write

$$\delta_m = \mu_m - [\mu_m] - \tfrac{1}{2}$$

where $[\mu_m]$ is an integer part of number μ_m (if a is rational, such a representation takes place for noninteger μ_m). Let us establish that the value

$$S_{pl} := \frac{1}{l} \sum_{m = p+1}^{p+1} (\mu_m - [\mu_m]) - \frac{1}{2} = \frac{1}{l} \sum \delta_m$$

tends to zero with $l \to \infty$ uniformly in p. Then Theorem 20 will follow from Proposition 6. In other words, roughly speaking, one has to prove that the fraction parts of sequence $\{a(m - 1/2)\}$ are $1/2$ in the mean. Polya and Szegö (1964: v. 1, part 2, chap. 4, secs. 2, 3) proved the following close result: if for sequence $\{x_m\}_{m \in \mathbb{N}}$, $x_m \in (0, 1)$, a relation

$$\frac{1}{l} \sum_{m=1}^{l} e^{2\pi i j x_m} \xrightarrow[l \to \infty]{} 0$$

holds for any integer j, $j \neq 0$, then

$$\frac{1}{l} \sum_{m=1}^{l} x_m \xrightarrow[l \to \infty]{} 1/2. \tag{11}$$

For $x_n = \mu_n - [\mu_n]$ it is not difficult to check the equalities

$$\left| \sum_{m=p+1}^{p+1} e^{2\pi i j x_m} \right| \leq d_j, \qquad j \neq 0.$$

Using this, we can prove the assertion about uniform convergence of S_{pl} to zero in the same way as (11). The theorem is proved.

4.5. *Riesz bases of elements of the form* $t^s e^{i\lambda_n t}$

When the moment method is applied to the investigation of DPS described by non-self-adjoint operators, families of functions appear (see Avdonin 1977b, 1980):

$$\tilde{\mathscr{E}}_T = \{e^{i\lambda_n t}, t e^{i\lambda_n t}, \dots, t^{m_n - 1} e^{i\lambda_n t}\}_{n \in \mathbb{Z}} \tag{12}$$

in space $L^2(0, T)$, as well as their vector analogs in $L^2(0, T; \mathbb{C}^N)$. Properties of such families may be studied by the scheme from Section 3, by means of a GF with zeros of points λ_n of multiplicity m_n. Because it sometimes requires rather cumbersome constructions to account for multiple zeros in a vector situation, we focus here on some results on the basis properties of scalar families of the form (12) and omit the proofs.

Definition II.4.21. Generating function of family $\tilde{\mathscr{E}}_T$ (if it exists) is an entire function of the exponential type with the indicator diagram $[-iT, 0]$ and the set of zeros $\{\lambda_n\}$ of corresponding multiplicity m_n.

Theorem II.4.22. Family $\tilde{\mathscr{E}}_T$ forms a Riesz basis in $L^2(0, T)$ if and only if

(a) $\sup_{n \in \mathbb{Z}} |\text{Im } \lambda_n| =: h < \infty, \sup_{n \in \mathbb{Z}} m_n < \infty, \inf_{n \neq j} |\lambda_n - \lambda_j| > 0,$

(b) *for some* $y_0 \in \mathbb{R}, |y_0| > h$, *function* $w(x) = |F(x + iy_0)|^2$ *satisfies condition* (A_2).

Here F is a GF of family $\tilde{\mathscr{E}}_T$.

The theorem is proved according to the scheme of Section 3. Sedletskiĭ (1982) shows the conditions to be sufficient for $\tilde{\mathscr{E}}_T$ to form a basis. Their necessity may be demonstrated by noting that the family $\tilde{\mathscr{E}}_T$ is almost normed if and only if

$$\sup_{n \in \mathbb{Z}} |\text{Im } \lambda_n| =: h < \infty, \qquad \sup_{n \in \mathbb{Z}} m_n < \infty.$$

The corresponding generalization of Propositions 6 and 12 reads as follows.

Theorem II.4.23. Family $\tilde{\mathscr{E}}_T$ forms a Riesz basis of $L^2(0, T)$ if and only if set $\{\lambda_n\}_{n \in \mathbb{Z}}$ is separable and an STF exists with the set of zeros $\{\mu_n\}_{n \in \mathbb{Z}}$ of multiplicity m_n such that

$$\sup_{n \in \mathbb{Z}} |\delta_n| < \infty, \qquad \limsup_{r \to +\infty} \sup_{x \in \mathbb{Z}} \frac{|\Delta_x(r)|}{2r} < \tfrac{1}{4}.$$

Here,

$$\delta_n = \lambda_n - \mu_n, \qquad \Delta_x(r) = \sum_{x - r < \operatorname{Re} \mu_n < x + r} m_n \operatorname{Re} \delta_n.$$

This theorem may be made more precise. Namely, points μ_n may be chosen in a way that for some $d \in (0, 1/4)$

$$d \operatorname{Re}(\lambda_{n-1} - \lambda_n) \le \operatorname{Re}(\mu_n - \lambda_n) \le d \operatorname{Re}(\lambda_{n+1} - \lambda_n), \qquad n \in \mathbb{Z}.$$

Avdonin, Horvath, and Joó (1989) have proved the sufficiency part of the theorem. To demonstrate the necessity, one ought to modify somewhat the proof of Proposition 12.

Theorems 23 and 17 imply the following corollary.

Corollary II.4.24. *If family $\tilde{\mathscr{E}}_T$ forms a Riesz basis in $L^2(0, T)$, then for any $\varepsilon > 0$, $r(\varepsilon) > 0$ may be found such that for all $r > r(\varepsilon)$ and all $x \in \mathbb{R}$ an inequality holds:*

$$\left| \frac{1}{r} \left(\sum_{x \le \operatorname{Re} \lambda_n < x + r} m_n \right) - \frac{T}{2\pi} \right| < \varepsilon. \tag{13}$$

From this the following analogs of Theorem 16 are obtained along the lines of the arguments of Theorem 18. For the specific case when sequence $\{\lambda_n\}$ satisfies condition (13), these results are proved in Avdonin, Horvath, and Joó (1989).

Theorem II.4.25. *Let family $\tilde{\mathscr{E}}$ form a Riesz basis in $L^2(0, T)$. Then for any $T' \in (0, T)$ there exists a subfamily $\tilde{\mathscr{E}}' \subset \tilde{\mathscr{E}}$, constituting a Riesz basis in $L^2(0, T')$. Moreover, family $\tilde{\mathscr{E}}'$ may be of the form*

$$\{e^{i\lambda_n t}, t\, e^{i\lambda_n t}, \dots, t^{m_n - 1}\, e^{i\lambda_n t}\}_{n \in \mathbb{Z}(T')},$$

where $\mathbb{Z}(T')$ is a subset of $\mathbb{Z}$.

4.6. Complementation of a basis on subinterval up to a basis on interval

Theorem II.4.26. *Let family $\mathscr{E} = \{e^{i\lambda_n t}\}_{n \in \mathbb{Z}}$ form a Riesz basis in $L^2(0, T)$. Then for any $T_1 > T$ a family $\mathscr{E}_0 = \{e^{i\mu_n t}\}_{n \in \mathbb{Z}}$ may be found such that family $\mathscr{E}_1 := \mathscr{E} \cup \mathscr{E}_0$ is a Riesz basis in $L^2(0, T_1)$.*

PROOF. Without sacrificing generality, one may consider the set $\sigma = \{\lambda_n\}$ to lie in the strip $0 < c \le \operatorname{Im} z < C$. Since $\mathscr{E}$ is a basis, then by Theorem 3.9(b) one may find GF f of family $\mathscr{E}$:

$$f(z) = b(z)f_e^+ z = e^{iTz} f_e^-(z),$$

where b is the BP constructed for set σ. By the force of Theorem 3.14(b) and Proposition 3.17, function $|f(x)|^2$ satisfies condition (A$_2$).

Set

$$\mu_n = \frac{\pi}{\beta} + i\alpha, \qquad \beta = \frac{T_1 - T}{2\pi}, \quad \alpha > C,$$

and consider family $\mathscr{E}_0 = \{e^{i\mu_n t}\}_{n \in \mathbb{Z}}$. In space $L^2(0, T_1 - T)$ this family evidently has GF f_0:

$$f_0(z) = e^{i\pi\beta z} \sin[\pi\beta(z - i\alpha)] = b_0(z)f_{e,0}^+(z) = e^{2\pi i\beta z} f_{e,0}^-(z),$$

which obeys the estimate $|f_0(x)| \asymp 1$, $x \in \mathbb{R}$.

Let us consider family $\mathscr{E}_1 = \mathscr{E} \cup \mathscr{E}_0$. In space $L^2(0, T)$ the latter family possesses GF f_1:

$$f_1 = ff_0 = bb_0 f_e^+ f_{e,0}^+ = e^{izT_1} f_e^- f_{e,0}^-.$$

From the behavior of functions f and f_0 on the real axis we conclude that $|f_1(x)|^2 \in (A_2)$.

Since $\mathscr{E}$ is a basis, set σ is separable. Then set $\sigma \cup \{\mu_n\}_{n \in \mathbb{Z}}$ is separable by its construction. Therefore, by Proposition 8 (or Theorem 3.14(b), family $\mathscr{E}_1$ is a Riesz basis in $L^2(0, T)$.

Seip (1995) investigated the more general complementation problem of an $\mathscr{L}$-basis from exponentials to a basis.

4.7. On the families of exponentials with the imaginary spectrum

Let us present several facts regarding the properties of exponential families of the form

$$\mathscr{E}_T = \{\sqrt{2\mu_n}\, e^{-\mu_n t}\}_{n \in \mathbb{N}} \subset L^2(0, T),$$

$$0 < \mu_1 < \mu_2 < \cdots, \quad \mu_n \to +\infty. \tag{14}$$

Since

$$\|e^{-\mu_n t}\|_{L^2(0, T)}^2 = \int_0^T e^{-2\mu_n t}\, dt = \frac{1}{2\mu_n}(1 - e^{-2\mu_n t}),$$

family $\mathscr{E}_T$ is almost normed.

Lemma II.4.27. If $\mathscr{E}_T \in (UM)$, then

$$\liminf_{n \to +\infty} \frac{\mu_{n+1}}{\mu_n} > 1. \tag{15}$$

PROOF. Let the lower limit in (15) equal 1; that is, a relation

$$\frac{\mu_{n_j+1}}{\mu_{n_j}} \xrightarrow[j \to \infty]{} 1 \tag{16}$$

will be valid for sequence $\{n_j\}$. Set $v_j = \mu_{n_j}$, $\tilde{v}_j = \mu_{n_j+1}$. One can easily check that

$$\cos \varphi_{L^2(0,T)}(e^{-v_j t}, e^{-\tilde{v}_j t}) = \frac{2\sqrt{v_j}\sqrt{\tilde{v}_j}}{v_j + \tilde{v}_j} \frac{1 - e^{-(v_j + \tilde{v}_j)T}}{\sqrt{1 - e^{-v_j T}}\sqrt{1 - e^{-\tilde{v}_j T}}}.$$

In view of (14) and (16), the right-hand side of the equality tends to unity when $j \to \infty$. Therefore

$$\varphi(e^{-\mu_{n_j} t}, e^{-\mu_{n_j+1} t}) \xrightarrow[j \to \infty]{} 0.$$

Hence, $\mathscr{E}_T \notin (UM)$, which completes the proof.

Proposition II.4.28 (Schwartz 1959). *If $\mathscr{E}_T \in (M)$, then the orthoprojector from $L^2(0, \infty)$ on $L^2(0, T)$ is an isomorphism of spaces*

$$\mathrm{Cl}_{L^2(0,\infty)} \mathrm{Lin}\{e^{-\mu_n t}\}_{n \in \mathbb{N}}$$

and

$$\mathrm{Cl}_{L^2(0,T)} \mathrm{Lin}\{e^{-\mu_n t}\}_{n \in \mathbb{N}}.$$

Corollary II.4.29. The norms of elements e'_n of the biorthogonal to $\mathscr{E}_T$ family satisfy the estimates

$$\|e'_n\|_{L^2(0,T)} \le C_T \|\tilde{e}'_n\|_{L^2(0,\infty)} = C_T \prod_{m \neq n} \left| \frac{\mu_m + \mu_n}{\mu_m - \mu_n} \right|,$$

in which $\{\tilde{e}'_n\}$ is a family biorthogonal to $\mathscr{E}_\infty = \{\sqrt{2\mu_n}\, e^{-\mu_n t}\}$ in space $L^2(0, \infty)$.

The estimates follow directly from Proposition 28 and formula (11) in Section II.1.

In a special, but significant case of power asymptotics of μ_n one is able to give explicit estimates for the norms of a family biorthogonal to $\mathscr{E}_\infty$.

Proposition II.4.30 (Fattorini and Russell 1971). *Let*

$$\mu_n = M(n + \beta)^\alpha, \qquad \alpha > 1, \beta \in \mathbb{R}, M > 0.$$

Then

$$\|\tilde{e}'_n\|_{L^2(0,\,\infty)} = \exp(c_\alpha n + o(n)), \qquad c_\alpha := 2 \text{ p.v.} \int_0^\infty \frac{t^{1/\alpha}}{t^2 - 1}\, dt.$$

For more general assumptions on $\{\mu_n\}$, Fattorini and Russell (1974), Avdonin (1977a, 1977b), and Hansen (1991) obtained estimates of $\|\tilde{e}'_n\|_{L^2(0,\,\infty)}$.

5. Additional information about vector exponential theory

5.1. Relationship of minimality of vector and scalar exponential families on an interval

Theorem II.5.1. Let the family of subspaces $\mathscr{E} = \{e^{i\lambda_n t}\}_{n \in \mathbb{N}}$, $\mathfrak{N}_n \subset \mathfrak{N}$, $\dim \mathfrak{N} = N$, $\lambda_n \in \mathbb{C}_+$, *be minimal in space* $L^2(0, T; \mathfrak{N})$ *and all the numbers* $\{\lambda_n\}_{n \in \mathbb{N}}$ *be different. Then scalar family* $\{e^{i\lambda_n t}\}_{n=N+1}^\infty$ *is minimal in* $L^2(0, NT)$.

PROOF. By Lemma I.1.28, family $\mathscr{E}$ is minimal in space $L^2(0, \infty; \mathfrak{N})$. According to Theorem 2.4, its spectrum satisfies the Blaschke condition. Let Π denote a BPP generated by family $\{(k - \bar{\lambda}_n)^{-1}\mathfrak{N}_n\}_{n \in \mathbb{N}}$ and apply Theorem 3.9(a) to family $P_T \mathscr{X}_\Pi$. Let F be a matrix function constructed in this theorem while $\tilde{F}$ is its identically nondegenerate $M \times M$ submatrix whose determinant turns into zero at the spectrum points. The latter calls for the following factorization

$$\tilde{F} = \tilde{\Pi} S_+ F_e^+ = e^{ikT} S_-^{-1} F_e^-, \tag{1}$$

where $S_\pm$ are inner, $F_e^\pm$ are outer functions, and $\tilde{\Pi}$ is a BPP whose spectrum coincides with set $\{\lambda_n\}_{n=1}^\infty$.

Functions $S_\pm$, in their turn, may also be presented in a factorized form

$$S_\pm = \Pi_\pm \Theta_\pm \tag{2}$$

with $\Pi_\pm$ being BPP, $\Theta_\pm$ being ESF (since $\tilde{F}$ is an entire function, it has no other singular cofactors).

Let us now pass to scalar functions, and to do this, introduce notations

$$b_\pm := \det \Pi_\pm, \qquad f_e^\pm := \det F_e^\pm, \qquad b := \det \tilde{\Pi}$$

and numbers $\alpha_\pm$ determined by

$$\det \Theta_\pm = \exp(ik\alpha_\pm).$$

(We include a unimodular constant to the other factors.) From formulas (1) and (2) giving $\tilde{F}$ factorization, we derive a factorization of scalar function $\det \tilde{F}$:

$$\det \tilde{F} = bb_+ \, e^{ik\alpha_+} f_e^+ = e^{+ikMT} b_-^{-1} e^{-ik\alpha_-} f_e^-. \tag{3}$$

For function

$$f := (\det \tilde{F}) b_- \exp(-ik\alpha_+),$$

we get out of (3)

$$f = bb_- b_+ f_e^+ = e^{ik\delta} f_e^-, \qquad \delta := MT - \alpha_- - \alpha_+ \le MT \le NT.$$

It is evident that f is an exponential-type function and the width of its indicator diagram is not greater than NT, and that f equals zero at the points of the spectrum. We now demonstrate f to have only power growth on $\mathbb{R}$, and, by dividing it by a polynomial with zeros $\lambda_1, \lambda_2, \ldots, \lambda_M$, we arrive at the function legible for the minimality criterion (Theorem 4.1).

Lemma II.5.2. For $R \ge M + 1$ function $|f(k)|^2/(1 + k^2)^R$ is summable over the real axis.

PROOF OF THE LEMMA. Since $\tilde{F}$ is a strong function according to Theorem 3.9, for any matrix element f_{ij} of matrix $\tilde{F}$, function $|f_{ij}|^2/(1 + k^2)$ is summable on $\mathbb{R}$; f_{ij} is an entire function of exponential type, and if it has no zeros, then $\log f_{ij}$ is also an entire function increasing no faster than $|k|$. So f_{ij} has to be of the form $\exp(i\alpha k + \beta)$, $\alpha \in \mathbb{R}$. In this case f_{ij} is bounded on $\mathbb{R}$. If f_{ij} is a zero of f_{ij}, then the entire function $f_{ij}/(k - \lambda_{ij})$ of the exponential type is squarely integrable, and it is bounded on $\mathbb{R}$ (Proposition 1.26). In any case,

$$|f_{ij}(k)| \prec 1 + |k|, \qquad k \in \mathbb{R}.$$

Therefore, $|\det \tilde{F}(k)| \prec (1 + k^2)^{M/2}$ and function f differ from $\det \tilde{F}$ by a factor whose absolute value equals unity almost everywhere on $\mathbb{R}$. The lemma is proved.

PROOF OF THE THEOREM. Introduce function $\tilde{f} := f/\prod_{j=1}^N (k - \lambda_j)$ with a factorization

$$\tilde{f} = \tilde{b} \tilde{f}_e^+ = e^{ik\delta} \tilde{f}_e^-$$

in which $\tilde{b}$ is a BP and $\tilde{f}_e^{\pm}$ are outer functions. It is clear that $\delta > 0$ because for $\delta \leq 0$, $\tilde{f}$ must be a polynomial. According to Lemma 2, function $|\tilde{f}|/(1 + |k|)$ is squarely integrable on $\mathbb{R}$. Theorem 4.1 then implies minimality in $L^2(0, \delta)$ of the exponential family $\{e^{i\mu_n t}\}$, where $\{\mu_n\}_1^{\infty}$ are the zeros of $\tilde{f}$. Now, family $\{e^{i\lambda_n t}\}_{N+1}^{\infty}$ is obviously minimal in $L^2(0, NT)$, since $\tilde{f}(\lambda_n) = 0$ for $n > N$. This completes the proof of the theorem.

Example II.5.3. Generally, minimality in $L^2(0, NT)$ of the whole family of scalar exponentials $\{e^{i\lambda_n t}\}_{n \in \mathbb{N}}$ does not follow from the minimality of the vector family (of elements) $\mathscr{E} = \{\eta_n \, e^{i\lambda_n t}\}_{n \in \mathbb{N}} \subset L^2(0, T; \mathbb{C}^N)$, $\eta_n \in \mathbb{C}^N$. To get a minimal family, one has to discard several exponentials. By Theorem 1, it is enough to throw out N exponentials, but this amount may be excessive.

Let us consider family $\mathscr{E} = \mathscr{E}_1 \cup \mathscr{E}_2$ of vector exponentials of the form

$$\mathscr{E}_1 := \{e^{i(n - \delta_n)t}\binom{1}{0}\}_{n \in \mathbb{Z}}, \qquad \mathscr{E}_2 := \{e^{i(n - \tilde{\delta}_n)t}\binom{0}{1}\}_{n \in \mathbb{Z}}$$

where

$$\delta_{2n} = \text{sign}(n)/3, \qquad \delta_{2n+1} = 0,$$

$$\tilde{\delta}_0 = 1/6; \qquad \tilde{\delta}_{2n} = 0, \qquad n \neq 0,$$

$$\tilde{\delta}_{2n+1} = \text{sign}(2n + 1)/3.$$

Each of scalar families

$$\mathscr{E}_1^{sc} := \{e^{i(n - \delta_n)t}\binom{1}{0}\}_{n \in \mathbb{Z}}, \qquad \mathscr{E}_2^{sc} := \{e^{i(n - \tilde{\delta}_n)t}\binom{0}{1}\}_{n \in \mathbb{Z}}$$

is a basis in $L^2(0, 2\pi)$ space. Actually, let us take Λ-decompositions for the spectra of families $\mathscr{E}_1^{sc}$ and $\mathscr{E}_2^{sc}$ by setting $\alpha_j = 2j + 1/2$, $j \in \mathbb{Z}$ (see Definition 4.5). Then, as one can easily see, the points with the numbers $n = 2j - 1$ and $n = 2j$ enter the same group σ_j, while the sums of $Re \, \delta_n$ and $Re \, \tilde{\delta}_n$ are not greater than one-third by its absolute value. The steps l_j of the decomposition equal 2, and therefore Proposition 4.6 is applicable, which guarantees $\mathscr{E}_1^{sc}$ and $\mathscr{E}_2^{sc}$ to be bases. The basis property of these families implies that vector family $\mathscr{E}$ is a basis: subfamilies $\mathscr{E}_1$ and $\mathscr{E}_2$ are bases in orthogonal subspaces of $L^2(0, T; \mathbb{C}^2)$ spanned over the functions of the form

$$f(t)\binom{1}{0}, \qquad f(t)\binom{0}{1}, \qquad f \in L^2(0, T),$$

respectively.

Consider scalar family $\mathscr{E}^{sc} := \mathscr{E}_1^{sc} \cup \mathscr{E}_2^{sc}$. If element $e^{-i\tilde{\delta}_0 t}$ is thrown out of it, the remaining family takes the form

$$\{e^{i(n/2 + \hat{\delta}_n)t}\}_{n \in \mathbb{Z}},$$

$$\hat{\delta}_{2n} = 0, \qquad \hat{\delta}_{2n+1} = \text{sign}(2n + 1)/6.$$

The latter forms a basis in $L^2(0, 4\pi)$, since the unperturbed harmonics family is a basis, and the points with numbers $n = 2j$ and $n = 2j - 1$ fall in the same group under Λ-decomposition by points $\alpha_j = j - 1/2$. And the conditions of Proposition 4.6 are now fulfilled: representing the sum of shifts as

$$1/6 = |\hat{\delta}_{2j-1}| = dl_j,$$

we have

$$d = 1/6 < 1/4, \qquad l_j = 1.$$

Hence, $\mathscr{E}^{sc}$ is a nonminimal family in $L^2(0, 4\pi)$: element $e^{-i\hat{\delta}_0 t}$ may be expanded in a converging series of other elements of $\mathscr{E}^{sc}$.

5.2. Perturbation of basis families

We noted (Theorem 4.13) that for the scalar family any point λ_n of the basis family spectrum may be "ε-shake" without any loss of the basis property. We now demonstrate the basis property stability for family $\mathscr{E}_T$ of vector exponentials, $\mathscr{E}_T = \{e_n\}_{n \in \mathbb{N}} \subset L^2(0, T; \mathfrak{N})$, $e_n := e^{i\lambda_n t} \cdot \eta_n \in \mathfrak{N}$, under a perturbation of both the spectrum points λ_n and vectors η_n. The assertion concerning the stability when λ_n are perturbed was obtained in collaboration with I. Joó and used by Avdonin, Ivanov, and Joó (1990). First, we need the following definition.

Definition II.5.4. Let family $\Xi = \{\xi_n\}_{n \in \mathbb{N}}$ form an $\mathscr{L}$-basis in Hilbert space $\mathfrak{H}$. The best values of constants c and C in the inequalities

$$c \left\{ \sum_{n \in \mathbb{N}} |c_n|^2 \right\}^{1/2} \leq \left\| \sum_{n \in \mathbb{N}} c_n \xi_n \right\|_{\mathfrak{H}} \leq C \left\{ \sum_{n \in \mathbb{N}} |c_n|^2 \right\}^{1/2}$$

are said to be basis constants of family Ξ and denoted by $q(\Xi)$ and $Q(\Xi)$, respectively (for $\{c_n\} \in \ell^2$ the inequalities are valid with some c, C in view of the Bari theorem).

One easily verifies the expressions of the basis constants via orthogonalizer $\mathscr{V}$ of family Ξ (see Definition I.1.16):

$$q(\Xi) = \|\mathscr{V}^{-1}\|^{-1}, \qquad Q(\Xi) = \|\mathscr{V}\|.$$

Theorem II.5.5. Let family $\mathcal{E}_T$ form an $\mathcal{L}$-basis (Riesz basis) in $L^2(0, T; \mathfrak{N})$, dim $\mathfrak{N} = N < \infty$, and let vectors η_n be almost normed. Then $\varepsilon > 0$ may be found such that any family

$$\tilde{\mathcal{E}}_T = \{\tilde{e}_n\}_{n \in \mathbb{N}} \subset L^2(0, T; \mathfrak{N}), \qquad \tilde{e}_n = \tilde{\eta}_n \, e^{i\tilde{\lambda}_n t},$$

also forms an $\mathcal{L}$-basis (respectively, Riesz basis) in $L^2(0, T; \mathfrak{N})$ as soon as

$$\langle\langle \eta_n - \tilde{\eta}_n \rangle\rangle < \varepsilon \quad \text{and} \quad |\lambda_n - \tilde{\lambda}_n| < \varepsilon, \quad n \in \mathbb{N}.$$

We give the proof for η_n and λ_n perturbations separately:

$$R^{\mathfrak{N}}: \eta_n \, e^{i\lambda_n t} \mapsto \tilde{\eta}_n \, e^{i\lambda_n t}, \qquad R^{\sigma}: \tilde{\eta}_n \, e^{i\lambda_n t} \mapsto \tilde{\eta}_n \, e^{i\tilde{\lambda}_n t}.$$

One shows that operators $R^{\mathfrak{N}}$ and R^{σ} transforming the elements of the unperturbed family into corresponding elements of the perturbed one are close to the identity operator for small ε and are therefore isomorphisms. Under such a separate treatment of perturbations, one should pay additional attention to the uniform estimate of basis constants.

We first consider the perturbation of vectors. Without sacrificing generality, one can assume that Im $\lambda_n \geq c > 0$, $n \in \mathbb{N}$. Let $\mathcal{E}_{\infty}$ denote a family of exponentials $e_n = \eta_n \, e^{i\lambda_n t}$ over a semiaxis (in the space $L^2(0, \infty; \mathfrak{N})$). According to Theorem 3.3(d), family $\mathcal{E}_{\infty}$ forms an $\mathcal{L}$-basis.

On the linear span of family $\mathcal{E}_{\infty}$, define operator $R^{\mathfrak{N}}_{\infty}$:

$$R^{\mathfrak{N}}_{\infty} e_n = \hat{e}_n := \tilde{\eta}_n \, e^{i\lambda_n t} \in L^2(0, \infty; \mathfrak{N}).$$

Set $\gamma_{\mathfrak{N}} := \sup_n \langle\langle \eta_n - \tilde{\eta}_n \rangle\rangle$.

Lemma II.5.6. For a small enough $\gamma_{\mathfrak{N}}$, operator $R^{\mathfrak{N}}_{\infty}$ may be continuously extended to the whole subspace $\bigvee \mathcal{E}_{\infty}$, so that an inequality holds:

$$\|R^{\mathfrak{N}}_{\infty} - I_{\vee \mathcal{E}_{\infty}}\| \leq \kappa \gamma_{\mathfrak{N}}$$

for some $\kappa > 0$ independent on $\gamma_{\mathfrak{N}}$.

PROOF OF THE LEMMA. As noted in Proposition 2.7, the spectrum σ of the $\mathcal{L}$-basis family $\mathcal{E}_{\infty}$ is a unification of not more than $N = \dim \mathfrak{N}$ of Carlesonian sets

$$\sigma = \bigcup_1^M \sigma_j, \qquad \sigma_j = \{\lambda_n\}_{n \in \mathcal{N}_j} \in (C); \quad j = 1, 2, \ldots, M, M \leq N.$$

The $\mathscr{L}$-basis property of $\mathscr{E}_T$ together with the fact that the set $\{\eta_n\}$ is almost normed implies that σ lies in the strip $0 < c \le \operatorname{Im} z \le C$. That is why each of the scalar families $\mathscr{E}_\infty^j = \{e_n^{sc}\}_{n \in \mathcal{N}_j}$, $e_n^{sc} := e^{i\lambda_n t}$ forms an $\mathscr{L}$-basis in $L^2(0, \infty)$. We write q, Q and q_j, Q_j for the basis constants of families $\mathscr{E}_\infty$ and $\mathscr{E}_\infty^{(j)}$, respectively. Let us take an arbitrary finite sequence $c = \{c_n\}$ and for $f := \sum_{n \in \mathbb{N}} c_n e_n^{sc}$ estimate the value $\|(I - R_\infty^{\mathfrak{N}})f\|$ by means of the expansion of vectors $\eta_n - \tilde{\eta}_n$ over the elements of an orthonormal basis $\{\xi_p^0\}$ of space $\mathfrak{N}$:

$$\|(I - R_\infty^{\mathfrak{N}})f\|_{L^2(0, \infty; \mathfrak{N})}^2 = \sum_{p=1}^N \left\| \sum_{n \in \mathbb{N}} c_n \langle \eta_n - \tilde{\eta}_n, \xi_p^0 \rangle e_n^{sc} \right\|_{L^2(0, \infty)}^2$$

$$= \sum_{p=1}^N \left\| \sum_{j=1}^M \sum_{n \in \mathcal{N}_j} c_n \langle \eta_n - \tilde{\eta}_n, \xi_p^0 \rangle e_n^{sc} \right\|_{L^2(0, \infty)}^2. \quad (4)$$

Exploiting an elementary inequality (19) of Section II.2 with

$$z_j = \sum_{n \in \mathcal{N}_j} c_n \langle \eta_n - \tilde{\eta}_n, \xi_p^0 \rangle e_n^{sc}$$

from (4) and the basis constants definition, we derive

$$\|(I - R_\infty^{\mathfrak{N}})f\|_{L^2(0, \infty; \mathfrak{N})}^2 \le M \sum_{p=1}^N \sum_{j=1}^M \left\| \sum_{n \in \mathcal{N}_j} c_n \langle \eta_n - \tilde{\eta}_n, \xi_p^0 \rangle e_n^{sc} \right\|^2$$

$$\le M \sum_{p=1}^N \sum_{j=1}^M Q_j^2 \sum_{n \in \mathcal{N}_j} |c_n \langle \eta_n - \tilde{\eta}_n, \xi_p^0 \rangle|^2$$

$$\le M \max_j Q_j^2 \sum_{p=1}^N \sum_{n \in \mathbb{Z}} |c_n \langle \eta_n - \tilde{\eta}_n, \xi_p^0 \rangle|^2$$

$$\le N \max_j Q_j^2 \sum_{n \in \mathbb{N}} |c_n|^2 \langle\langle \eta_n - \tilde{\eta}_n \rangle\rangle^2$$

$$\le N \max_j Q_j^2 \gamma_{\mathfrak{N}}^2 \sum_{n \in \mathbb{N}} |c_n|^2 \le \left(N \max_j Q_j^2 q^{-2} \right) \gamma_{\mathfrak{N}}^2 \|f\|^2.$$

The lemma is proved.

Let us return to the family $\mathscr{E}_T$. From Theorem 3.3(d) and Lemma I.1.10(a) we conclude that operator

$$P: \bigvee \mathscr{E}_\infty \ni f \mapsto P_{L(0, T^2; \mathfrak{N})} f$$

is an isomorphism on its image. It is easy to see that operator

$$R_T^{\mathfrak{N}} := P R_\infty^{\mathfrak{N}} P^{-1}$$

turns family $\mathscr{E}_T$ into family $\hat{\mathscr{E}}_T = \{\tilde{\eta}_n \, e^{i\lambda_n t}\}_{n \in \mathbb{N}} \in L^2(0, T; \mathfrak{N})$. From Lemma 6 it follows that

$$\|I_{\vee \hat{\mathscr{E}}_T} - R_T^{\mathfrak{N}}\| \le \tilde{\kappa}\gamma_{\mathfrak{N}}, \qquad \tilde{\kappa} := \kappa\|P\| \, \|P^{-1}\|. \qquad (5)$$

Let us choose a small enough $\gamma_{\mathfrak{N}}$ to make $\tilde{\kappa}\gamma_{\mathfrak{N}}$ less than unity. Then inequality (5) provides operator $R_T^{\mathfrak{N}}$ with the property of being an isomorphism on its image, and, for a basis family $\mathscr{E}_T$, the image coincides with the whole space $L^2(0, T; \mathfrak{N})$. Family $\hat{\mathscr{E}}_T$ is thus an $\mathscr{L}$-basis (or a basis for basis $\mathscr{E}_T$). Let us estimate constants $\hat{q}$ and $\hat{Q}$ for family $\hat{\mathscr{E}}_T$. If $\mathscr{V}_T$ stands for the orthogonalizer of family $\mathscr{E}_T$, then the orthogonalizer for $\hat{\mathscr{E}}_T$ is $\mathscr{V}_T(R_T^{\mathfrak{N}})^{-1}$. Estimating $\|R_T^{\mathfrak{N}}\|$ and $\|(R_T^{\mathfrak{N}})^{-1}\|$ with the help of inequality (5), we arrive at

$$\|R_T^{\mathfrak{N}}\| \le 1 + \tilde{\kappa}\gamma_{\mathfrak{N}}, \qquad \|(R_T^{\mathfrak{N}})^{-1}\| \le 1/(1 - \tilde{\kappa}\gamma_{\mathfrak{N}}).$$

From here

$$\hat{q} \ge q/(1 + \tilde{\kappa}\gamma_{\mathfrak{N}}), \qquad \hat{Q} \le Q/(1 - \tilde{\kappa}\gamma_{\mathfrak{N}}).$$

That completes the case of vector η_n perturbation, and we can now deal with the case of λ_n perturbation. On the linear span of $\hat{\mathscr{E}}_T$, we define operator R_T^σ by a formula

$$R_T^\sigma \hat{e}_n = \tilde{e}_n = \tilde{\eta}_n \, e^{i\tilde{\lambda}_n t}.$$

Succeeding the scheme of the previous case of $R_\infty^{\mathfrak{N}}$ operator, we take a finite sequence $\{c_n\}$ and estimate $\|(I_{\vee \hat{\mathscr{E}}_T} - R_T^\sigma)f\|$ for $\delta_n := -\lambda_n + \tilde{\lambda}_n$ and $f := \sum c_n \hat{e}_n$

$$\|f - R_T^\sigma f\|_{L^2(0, \infty; \mathfrak{N})}^2 = \left\| \sum_{n \in \mathbb{N}} c_n \tilde{\eta}_n \, e^{i\lambda_n t}(1 - e^{i\delta_n t}) \right\|_{L^2(0, \infty; \mathfrak{N})}^2$$

$$= \left\| \sum_{n=1}^\infty c_n \tilde{\eta}_n \, e^{i\lambda_n t} \sum_{k=1}^\infty \frac{(i\delta_n)^k}{k!} t^k \right\|$$

$$= \left\| \sum_{k=1}^\infty \frac{t^k}{k!} \sum_{n=1}^\infty c_n \hat{e}_n (i\delta_n)^k \right\|$$

$$\le \sum_{k=1}^\infty \frac{T^k}{k!} \left\| \sum_{n=1}^\infty c_n (i\delta_n)^k \hat{e}_n \right\|$$

$$\le \hat{Q} \sum_{k=1}^\infty \frac{T^k}{k!} \left[\sum_{n \in \mathbb{N}} |c_n \delta_n^k|^2 \right]^{1/2}$$

Setting $\gamma_\sigma := \sup_{n \in \mathbb{N}} |\lambda_n - \tilde{\lambda}_n|$ and again using the definition of the basis constants, we write

$$\|(I_{\vee \hat{\mathscr{E}}_T} - R_T^q)f\|_{L^2(0, \infty; \, \mathfrak{N})} \leq \hat{Q} \sum_{k=1}^{\infty} \frac{(T\gamma_\sigma)^k}{k!} \left[\sum_{n \in \mathbb{N}} |c_n|^2 \right]^{1/2}$$

$$= \hat{Q}(e^{T\gamma_\sigma} - 1) \left(\sum_{n \in \mathbb{N}} |c_n|^2 \right)^{1/2}$$

$$\leq \hat{Q}(e^{T\gamma_\sigma} - 1) \hat{q}^{-1} \|f\|.$$

This inequality allows us to conclude that for a small enough value of γ_σ, operator R_T^q is an isomorphism on its image that, for a basis family $\hat{\mathscr{E}}_T$, coincides with $L^2(0, T; \mathfrak{N})$. The theorem is proved.

Remark II.5.7. The assertion of Theorem 5 about the basis property stability under vector perturbations is of a rather abstract nature. For instance, let family $\{\xi_n\}_{n=1}^{\infty}$ be a finite unification of $\mathscr{L}$-basis (basis) families in Hilbert space $\mathfrak{H}$ and for vector family $\{\eta_n\}_{n=1}^{\infty}$, $\eta_n \in \mathfrak{N}$, dim $\mathfrak{N} \leq \infty$, let family $\{\eta_n \xi_n\}_{n=1}^{\infty}$ form an $\mathscr{L}$-basis (basis) in the space of vector functions $\mathfrak{H}(\mathfrak{N})$. Then $\varepsilon > 0$ may be found such that any family $\{\tilde{\eta}_n \xi_n\}_{n=1}^{\infty}$ also forms an $\mathscr{L}$-basis (basis) as soon as $\langle\langle \eta_n - \tilde{\eta}_n \rangle\rangle < \varepsilon$, $n \in \mathbb{N}$.

So, Theorem 5 remains to be true for dim $\mathfrak{N} = \infty$, under the extra condition $\sigma \in (CN)$, because the finite dimension of $\mathfrak{N}$ is not involved in the proof of stability under the spectrum perturbation.

Remark II.5.8. In a scalar situation, the substitution of an element of the family $\{x_{\lambda_j}\}_{j=1}^{\infty}$ by some other element (outside the family) cannot lead to the loss of the basis property (see Lemma 4.11). This is not so in the vector case. If family $P_\Theta \mathscr{X}_\Pi = \{P_\Theta x_{\lambda_j} \eta_j\}_{j=1}^{\infty}$ forms a basis in K_Θ, then a perturbed family

$$P_\Theta \tilde{\mathscr{X}}_\Pi = \{P_\Theta x_\mu \eta_\mu\} \cup \{P_\Theta x_{\lambda_j} \eta_j\}_{j=2}^{\infty}$$

may be both not complete and not minimal (even when $\mu \notin \{\lambda_j\}_1^{\infty}$). The minimality (completeness) of a perturbed family is equivalent to

$$P_\Theta x_\mu \eta_\mu \notin \bigvee_{j>1} P_\Theta x_{\lambda_j} \eta_j. \tag{6}$$

This relation, in turn, is equivalent to the fact that element $P_\Theta x_\mu \eta_\mu$ is not orthogonal to the subspace

$$K_\Theta \ominus \bigvee_{j>1} P_\Theta x_{\lambda_j} \eta_j,$$

which is one-dimensional and spanned on the element x'_{Θ, λ_1} of family $\mathscr{X}'_{\Theta, \Pi}$ biorthogonal to $P_\Theta \mathscr{X}_\Pi$. Basis family $P_\Theta \mathscr{X}_\Pi$ possesses a GF, so that expressing element x'_{Θ, λ_1} through F (see Lemma 3.5), we arrive at the following criterion of the perturbed family minimality:

$$\left(P_\Theta x_\mu \eta_\mu, \frac{F(k)\eta^1}{k - \lambda_1}\right)_{H^2_+(\mathfrak{N})} \neq 0,$$

where $\eta^1 \in \text{Ker } F(\lambda_1)$, $\eta_1 \neq 0$. Since

$$\left(P_\Theta x_\mu \eta_\mu, \frac{F(k)\eta^1}{k - \lambda_1}\right)_{H^2_+(\mathfrak{N})} = 2\pi i \langle \eta_\mu, F(\mu)\eta^1 \rangle \frac{1}{\bar\mu - \bar\lambda},$$

one can contest the validity of (6) by means of only a rotation of vector η_μ (operator $F(\mu)$ is nondegenerate in $\mathfrak{N}$).

In practice, it is often the case that the investigated family of vector exponentials is asymptotically close to the basis in $L^2(0, T; \mathfrak{N})$ unperturbed family (see, for instance, Sections 2 and 3 of Chapter VII). From Remark 8, we know that in such a case the perturbed family may be both incomplete and nonminimal in $L^2(0, T; \mathfrak{N})$. However, the perturbed family happens to be complete in $L^2(0, T - \varepsilon; \mathfrak{N})$ and forms an $\mathscr{L}$-basis in $L^2(0, T + \varepsilon; \mathfrak{N})$ for any $\varepsilon > 0$. These facts are demonstrated in a more convenient way in Fourier-representation, when simple fractions, rather than exponentials, are studied.

Suppose now that ESF Θ satisfies condition (24) of Section II.3 and that numbers λ_n lie in the strip parallel to the real axis:

$$0 < \inf_{n \in \mathbb{N}} \text{Im } \lambda_n, \qquad \sup_{n \in \mathbb{N}} \text{Im } \lambda_n < \infty. \tag{7}$$

Let us take $\varepsilon > 0$ and set $\Theta_{\pm \varepsilon}(k)$ for functions $e^{\pm ik\varepsilon} \Theta(k)$. For $\varepsilon < \rho$ (ρ is taken from relation (24) in Section II.3), $\Theta_{-\varepsilon}$ is an ESF.

Theorem II.5.9. Let family $P_\Theta \mathscr{X} = \{P_\Theta x_{\lambda_n} \eta_n\}_{n=1}^\infty$, $\eta_n \in \mathfrak{N}$, dim $\mathfrak{N} < \infty$, forms a basis in K_Θ, while family $P_\Theta \tilde{\mathscr{X}} = \{P_\Theta x_{\tilde\lambda_n} \tilde\eta_n\}_{n=1}^\infty$ is linearly independent and asymptotically close to $P_\Theta \mathscr{X}$ in the following sense

$$|\lambda_n - \tilde\lambda_n| \xrightarrow[n \to \infty]{} 0, \qquad \langle\langle \eta_n - \tilde\eta_n \rangle\rangle \xrightarrow[n \to \infty]{} 0. \tag{8}$$

Then

(a) for any $\varepsilon > 0$ family, $P_{\Theta_\varepsilon} \tilde{\mathscr{X}}$ constitutes an $\mathscr{L}$-basis in K_{Θ_ε}; what is more,

when complemented by any finite family

$$P_{\Theta_\varepsilon} \mathcal{X}^0 = \{P_{\Theta_\varepsilon} x_\mu \eta_j^0\}_{j=1}^R, \quad \text{with } \{\mu_j\}_{j=1}^R \cap \{\tilde{\lambda}_n\}_{n=1}^\infty = \varnothing,$$

it forms an $\mathcal{L}$-basis.

(b) For $\varepsilon < \rho$, family $P_{\Theta_{-\varepsilon}} \tilde{\mathcal{X}}$ is complete in $K_{\Theta_{-\varepsilon}}$; what is more, under the removal of any finite set of its elements, the completeness survives.

PROOF. Introduce family

$$P_{\Theta} \hat{\mathcal{X}}^M := P_{\Theta} \mathcal{X}_M \cup P_{\Theta} \tilde{\mathcal{X}}^M,$$

$$\mathcal{X}_M := \{x_{\lambda_n} \eta_n\}_{n=1}^M, \qquad \tilde{\mathcal{X}}^M := \{x_{\tilde{\lambda}_n} \tilde{\eta}_n\}_{n=M+1}^\infty,$$

consisting of M first elements of the unperturbed family and elements numbered $M+1, M+2, \ldots$ of the perturbed one. Condition (8) and Theorem 5 reveal that for a large enough M family, $P_{\Theta} \hat{\mathcal{X}}^M$ is a basis in K_Θ. Later in the proof, we consider this to be true. Basis family $P_{\Theta} \hat{\mathcal{X}}^M$ possesses, by Theorem 3.9(b), a generating function, which we denote by $F_M(k)$. In order to prove assertion (a), let us complement the family by an infinite family preserving the basis property K_{Θ_ε} (for a scalar case, the construction is implemented in Subsection 4.4).

Set

$$f_\varepsilon(k) := e^{ik\varepsilon/2} \sin\{\varepsilon(k - \alpha i)/2\},$$

where $\alpha > \sup \operatorname{Im} \lambda_n$. Function f_ε is obvious to satisfy the estimate

$$|f(k)| \asymp 1, \qquad k \in \mathbb{R},$$

and has a factorization

$$f_\varepsilon = f_+^e b = e^{ik\varepsilon} f_-^e,$$

with $f_\pm^e$ being outer functions, and b being a BP with zeros

$$\sigma(b) = \{2\pi n/\varepsilon + \alpha i\}_{n \in \mathbb{Z}}.$$

Therefore, function $F_\varepsilon := f_\varepsilon F_M$ is a GF for family

$$P_{\Theta_\varepsilon} \hat{\mathcal{X}}_\varepsilon^M := P_{\Theta_\varepsilon} \hat{\mathcal{X}}^M \cup P_{\Theta_\varepsilon} \mathcal{X}_\varepsilon, \qquad \mathcal{X}_\varepsilon := \{x_\lambda \zeta_j^0\}_{j=1, \lambda \in \sigma(b)}^{j=N},$$

$\{\zeta_j^0, j = 1, \ldots, N$, is an orthonormal basis in $\mathfrak{N}$). Using Theorem 3.5, we conclude that family $P_{\Theta_\varepsilon} \hat{\mathcal{X}}_\varepsilon^M$ is minimal. Elements of a biorthogonal family, corresponding to $P_{\Theta_\varepsilon} \mathcal{X}_\varepsilon$, are given by a formula (see formula (26) in Section II.3),

$$x'_{\Theta_\varepsilon, \lambda, j} = \alpha_{\lambda, j} \frac{F_\varepsilon(k)}{k - \lambda} \zeta_j^0, \qquad \lambda \in \sigma(b), j = 1, \ldots, N, \tag{9}$$

($\alpha_{\lambda, j}$ are constants).

We now turn directly to the proof of assertion (a). Consider a family

$$P_{\Theta_\varepsilon}\widetilde{\mathscr{X}} \cup P_{\Theta_\varepsilon}\mathscr{X}^0, \qquad \mathscr{X}^0 = \{x_{\mu_j}\eta_j^0\}_{j=1}^R,$$

which is a unification of $\mathscr{L}$-basis family $P_{\Theta_\varepsilon}\widetilde{\mathscr{X}}^M$ and a finite family $P_{\Theta_\varepsilon}(\widetilde{\mathscr{X}}_M \cup \mathscr{X}^0)$, where family $\widetilde{\mathscr{X}}_M$ has the form $\{x_{\xi_n}\tilde{\eta}_n\}_{n=1}^M$. Since family $P_{\Theta_\varepsilon}\widetilde{\mathscr{X}}^M$ is an $\mathscr{L}$-basis and family $P_{\Theta_\varepsilon}(\widetilde{\mathscr{X}}_M \cup \mathscr{X}^0)$ is the finite one, then either the whole family $P_{\Theta_\varepsilon}\widetilde{\mathscr{X}}^M \cup P_{\Theta_\varepsilon}(\widetilde{\mathscr{X}}_M \cup \mathscr{X}^0)$ forms an $\mathscr{L}$-basis and assertion (a) is proved, or the angle between subspaces $\bigvee P_{\Theta_\varepsilon}\widetilde{\mathscr{X}}^M$ and $\bigvee P_{\Theta_\varepsilon}(\widetilde{\mathscr{X}}_M \cup \mathscr{X}^0)$ is equal to zero. In the latter case, we again use the finiteness of family $P_{\Theta_\varepsilon}(\widetilde{\mathscr{X}}_M \cup \mathscr{X}^0)$. Applying Lemma I.1.3, we conclude that there exists element

$$g \in \bigvee P_{\Theta_\varepsilon}\widetilde{\mathscr{X}}^M \cap \bigvee P_{\Theta_\varepsilon}(\widetilde{\mathscr{X}}_M \cup \mathscr{X}^0), \qquad g \neq 0.$$

Since $P_{\Theta_\varepsilon}\widetilde{\mathscr{X}}^M \in (LB)$, this means that element g being a linear combination of elements of $P_{\Theta_\varepsilon}(\widetilde{\mathscr{X}}_M \cup \mathscr{X}^0)$,

$$g = \sum_1^R c_n P_{\Theta_\varepsilon}\eta_n^0/(k - \bar{\mu}_n) + \sum_1^M \tilde{c}_n P_{\Theta_\varepsilon}\tilde{\eta}_n/(k - \bar{\tilde{\lambda}}_n),$$

can be decomposed in convergent series in elements of the family $P_{\Theta_\varepsilon}\widetilde{\mathscr{X}}^M$:

$$g = \sum_{M+1}^\infty \tilde{c}_n P_{\Theta_\varepsilon}\tilde{\eta}_n/(k - \lambda_n).$$

Multiplying these equalities by elements (9) orthogonal to $P_{\Theta_\varepsilon}\widetilde{\mathscr{X}}^M$, we obtain

$$\sum_1^R c_n\langle \eta_n^0, F_\varepsilon(\mu_n)\xi_j^0\rangle/(\mu_n - \lambda) + \sum_1^M \tilde{c}_n\langle \tilde{\eta}_n, F_\varepsilon(\tilde{\lambda}_n)\xi_j^0\rangle/(\tilde{\lambda}_n - \lambda) = 0, \quad (10)$$

$$\lambda \in \sigma(b), \qquad j = 1, \ldots, N.$$

Introducing rational vector function

$$\mathscr{R}(k) := \sum_{n=1}^R h_n/(\mu_n - k) + \sum_{n=1}^M \tilde{h}_n/(\tilde{\lambda}_n - k),$$

$$h_n := c_n F_\varepsilon^*(\mu_n)\eta_n^0, \qquad \tilde{h}_n := \tilde{c}_n F_\varepsilon^*(\tilde{\lambda}_n)\tilde{\eta}_n,$$

we find from (10) that $\mathscr{R}(\lambda) = 0$ for $\lambda \in \sigma(b)$. A rational function has an infinite number of roots and thus is identical to zero; hence, $h_n = 0$, $n = 1, \ldots, R$, and $\tilde{h}_n = 0$, $n = 1, \ldots, M$. If also $c_n \neq 0$, then $\eta_n \in \text{Ker } F_\varepsilon^*(\mu_n)$. This means that element $\eta_n^0 x_{\mu_n}$ coincides with some of the elements of family $P_{\Theta_\varepsilon}\widetilde{\mathscr{X}}^M$, which is impossible by the assumption of the theorem. So $c_n = 0$. By analogy, $\tilde{c}_n = 0$, $n = 1, \ldots, M$, and we have $g = 0$. Assertion (a) is justified.

With regard to (b), let us demonstrate family $P_{\Theta_{-\varepsilon}}\widetilde{\mathscr{X}}^M$ to be complete in $K_{\Theta_{-\varepsilon}}$. Supposing the opposite, we take a nonzero function $g \in K_{\Theta_{-\varepsilon}}$ orthogonal to $\bigvee P_{\Theta_{-\varepsilon}}\widetilde{\mathscr{X}}^M$. Now, $g \perp P_{\Theta}\widetilde{\mathscr{X}}^M$. Indeed, for any $h \in P_{\Theta}\widetilde{\mathscr{X}}^M$ we have

$$(g, h) = (P_{\Theta_{-\varepsilon}}g, h) = (g, P_{\Theta_{-\varepsilon}}h) = 0$$

since $P_{\Theta_{-\varepsilon}}h \in \bigvee P_{\Theta_{-\varepsilon}}\widetilde{\mathscr{X}}^M$.

Recall that family $P_{\Theta}\widehat{\mathscr{X}}^M = P_{\Theta}\mathscr{X}_M \cup P_{\Theta}\widetilde{\mathscr{X}}^M$ forms a basis in K_{Θ}, and that F_M is a GF of this family. The biorthogonal to family $P_{\Theta}\widehat{\mathscr{X}}^M$ is of the form (see Lemma 3.5)

$$\left\{\frac{F_M(k)\eta^n}{k - \lambda_n}\right\}_{n=1}^M \cup \left\{\frac{F_M(k)}{k - \tilde{\lambda}_n}\tilde{\eta}^n\right\}_{n=M+1}^\infty,$$

where $\eta^n \in \operatorname{Ker} F_M(\lambda_n)$, $\tilde{\eta}^n \in \operatorname{Ker} F_M(\tilde{\lambda}_n)$. Since $g \perp P_{\Theta}\widetilde{\mathscr{X}}^M$, g belongs to the linear span of the first M elements of the biorthogonal family. Therefore,

$$g = \sum_{n=1}^M F_M(k)\eta^n/(k - \lambda_n) = F_M\tilde{g}, \tag{11}$$

where we set

$$\tilde{g} = \sum_{n=1}^M \eta^n/(k - \lambda_n).$$

By the force of (11) and factorization $F_M = \Theta F_e^-$, we have

$$\Theta F_e^- \tilde{g} \in K_{\Theta_{-\varepsilon}}. \tag{12}$$

Since (see formula (14) in Section II.3)

$$\Theta^* e^{ik\varepsilon} K_{\Theta_{-\varepsilon}} = H_-^2(\mathfrak{N}) \ominus \Theta^* e^{ik\varepsilon} H_-^2(\mathfrak{N}),$$

(12) implies an inclusion

$$e^{ik\varepsilon} F_e^- \tilde{g} \in H_-^2(\mathfrak{N}). \tag{13}$$

But we can show that the latter is impossible for $\tilde{g} \neq 0$. We rely here on the fact that outer operator functions and rational function $\tilde{g}$ cannot fall exponentially with $\operatorname{Im} k \to -\infty$ while $e^{ik\varepsilon}$ grows exponentially in $\mathbb{C}_-$.

Lemma II.5.10. *Let F_e^- be an outer in $\mathbb{C}_-$ operator function. Then for any $x \in \mathbb{R}$ and $\delta > 0$, there exist $q > 0$ and constant C_x such that for $y > q$ an estimate holds*

$$\langle\!\langle\{F_e^-(x - iy)\}^{-1}\rangle\!\rangle \leq C_x e^{\delta y}.$$

PROOF OF THE LEMMA. Element ij of matrix $\{F_e^-(x - iy)\}^{-1}$ is a ratio of the algebraic complement φ_{ij} of ij entry to scalar outer function $\det F_e^-$.

Since an outer operator function is, by definition, a strong one, function $(F_e^-(k))_{ij}/(k - i)$ lies in H_-^2. Hence, for $\operatorname{Im} k < -\kappa < 0$, this function is bounded (see formula (1) in Section II.1), so that an inequality

$$|(F_e^-(k))_{ij}| \le \operatorname{Const}(1 + |k|)$$

is valid. For algebraic complements from here, we conclude that

$$|\varphi_{ij}(k)| \le \operatorname{Const}(1 + |k|)^{N-1}.$$

To complete the proof of the lemma it remains to look at restriction (4) in Section II.1 on the growth of the scalar outer function $1/\det F_e^-$.

PROOF OF THE THEOREM. For large $|k|$ there is an obvious estimate

$$\langle\langle \tilde{g}(k) \rangle\rangle \succ |k|^{-M}.$$

So for large y, one has

$$\langle\langle e^{i(x-iy)\varepsilon} F_e^-(x - iy) \tilde{g}(x - iy) \rangle\rangle$$

$$\ge e^{y\varepsilon} (\langle\langle \{F_e^-(x - iy)\}^{-1} \rangle\rangle)^{-1} \langle\langle \tilde{g}(x - iy) \rangle\rangle$$

$$\succ e^{y\varepsilon}(x^2 + y^2)^{-M/2} (\langle\langle \{F_e^-(x - iy)\}^{-1} \rangle\rangle)^{-1}.$$

If we now use Lemma 10 with $\delta < \varepsilon$, we end up with the unbounded function $e^{ik\varepsilon} F_e^- \tilde{g}$ for large $\operatorname{Im} k$. On the other hand, from (13) and relation (1) in Section I.1 it follows that the same function is bounded for $\operatorname{Im} k < -\kappa$. This contradiction proves assertion (b). The theorem is proved completely.

5.3. Exponential bases in Sobolev spaces

Any family of scalar exponentials that forms a basis in $L^2(0, T)$ is incomplete in Sobolev space $H^s(0, T)$, $s \in \mathbb{N}$. This assertion follows from Proposition 12 given below. We now illustrate it by the following example.

Example II.5.11. A family of harmonics $\{\exp(int)\}_{n \in \mathbb{Z}}$ joined by element $\sinh(t)$ forms a complete orthogonal family in $H^1(-\pi, \pi)$. The fact that this family is orthogonal is checked directly; let us demonstrate that the orthogonal complement to the family of harmonics is one-dimensional.

Let g be orthogonal in $H^1(-\pi, \pi)$ to e^{int} for each $n \in \mathbb{Z}$. Then

$$0 = \int_{-\pi}^{\pi} ing'(t) e^{int} dt + \int_{-\pi}^{\pi} g(t) e^{int} dt$$

$$= \int_{-\pi}^{\pi} + n^2 e^{int} g(t) dt + ing(t) e^{int}|_{-\pi}^{\pi} + \int_{-\pi}^{\pi} e^{int} g(t) dt$$

$$= in(-1)^n[g(\pi) - g(-\pi)] + \int_{-\pi}^{\pi} (1 + n^2)g(t) e^{int} dt.$$

Writing g_n for the Fourier coefficients of g in space L^2, we find

$$0 = in(-1)^n[g(\pi) - g(-\pi)] + g_n(1 + n^2), \qquad n \in \mathbb{Z},$$

so that $g_n = c(-1)^n n/(1 + n^2)$. Thus, the family deficiency is one-dimensional. Hence, by adding function $\sinh(t)$ to it, one makes it both orthogonal and complete in $H^1(-\pi, \pi)$.

We now formulate D. L. Russell's result about exponential bases in $H^s(0, T)$. For the basis properties in Sobolev spaces with noninteger exponents, we refer to Narukawa and Suzuki (1986).

Proposition II.5.12 (Russell 1982). *Let family* $\{e^{i\lambda t}\}_{\lambda \in \sigma}$ *form a Riesz basis in* $L^2(0, T)$ *and points* $\mu_1, \mu_2, \ldots, \mu_s$ *be different and not lying in* σ. *Then family*

$$\{e^{i\lambda t}/(1 + |\lambda|^s)\}_{\lambda \in \sigma} \cup \{e^{i\mu_j t}\}_{j=1}^s$$

forms a basis in $H^s(0, T)$ *space.*

We need a generalization of this statement for the subspaces formed by groups of exponentials.
reason for the differences in particle size of the drug-polymer
Theorem II.5.13. *Let* $\mathfrak{H}^0$ *be a family of subspaces*

$$\mathfrak{H}^0 := \{\mathfrak{H}_n\}_{n=1}^\infty, \quad \text{where } \mathfrak{H}_n := \bigvee_{m=1}^{N_n} e^{i\lambda_{mn} t},$$

and let $\mathfrak{H}_0$ *be an s-dimensional family*

$$\mathfrak{H}_0 := \bigvee_1^s e^{i\mu_j t} \quad \text{with } \{\mu_j\}_{j=1}^s \cap \{\lambda_{mn}\}_{n=1,m=1}^{\infty,N_n} = \varnothing.$$

If $\mathfrak{H}^0$ *forms a Riesz basis in* $L^2(0, T)$, *then family* $\{\mathfrak{H}_n\}_1^\infty \cup \mathfrak{H}_0$ *is a Riesz basis in* $H^s(0, T)$.

PROOF. The proof consists of three steps:

(i) The family $\mathfrak{H}^0$ forms a Riesz basis in the closure of its span in $H^s(0, T)$ (say, $\mathscr{L}$-basis). It is Lemma 14.
(ii) The subspace $\mathfrak{H}_0$ has no nonnull elements in common with subspace $\mathrm{Cl}_{H^s(0,T)} \mathrm{Lin}\, \mathfrak{H}^0$ (Lemma 15).
(iii) The family $\mathfrak{H}^0 \cup \mathfrak{H}_0$ forms a Riesz basis in the Sobolev space $H^s(0, T)$.

We introduce the operator A

$$Af = \prod_1^s \left(\frac{d}{dt} - i\mu_j \right) f$$

with $D(A) = H^s(0, T)$ and the operator $\mathscr{A} := A|_{0H^s}$ where

$$_0H^s := \{ f \in H^s(0, T) | f(0) = f'(0) = \cdots = f^{(s-1)}(0) = 0 \}.$$

Notice that $\mathrm{Ker}\, A = \mathfrak{H}_0$ and $\mathrm{Ker}\, \mathscr{A} = \{0\}$.

Lemma II.5.14. The family $\mathfrak{H}^0$ forms an $\mathscr{L}$-basis in $H^s(0, T)$.

PROOF OF LEMMA 14. In $H^s(0, T)$ there exists the equivalent norm

$$\|f\|^2_{H^s} \asymp \|f\|^2_{L^2} + \|Af\|^2_{L^2}, \qquad f \in H^s(0, T).$$

Let f be a finite sum of elements h_n, $h_n \in \mathfrak{H}_n$, $n \in \mathbb{N}$. We have

$$\left\| \sum_n h_n \right\|^2_{H^s} \asymp \left\| \sum_n h_n \right\|^2_{L^2} + \left\| \sum_n Ah_n \right\|^2_{L^2} \asymp \sum_n \|h_n\|^2_{L^2} + \sum_n \|Ah_n\|^2_{L^2} \asymp \sum_n \|h_n\|^2_{H^s}$$

Indeed, the operator A takes subspace $\mathfrak{H}_n$ to itself, and family $\mathfrak{H}^0$ is a Riesz basis in $L^2(0, T)$. Thus the assertion of Lemma 14 follows from the Bari theorem (Proposition I.1.17(b)).

Lemma II.5.15.

$$\mathfrak{H}_0 \cap \mathrm{Cl}_{H^s} \mathrm{Lin}\, \mathfrak{H}^0 = \{0\}. \tag{14}$$

PROOF OF LEMMA 15. Let $h_0 \in \mathfrak{H}_0$ and $h_0 \in \mathrm{Cl}_{H^s} \mathrm{Lin}\, \mathfrak{H}^0$. By virtue of Lemma 14,

$$h_0 = \sum_n h_n, \qquad h_n \in \mathfrak{H}_n, \tag{15}$$

where the series converges in $H^s(0, T)$. Acting on both sides of (15) by the operator A, we obtain

$$0 = Ah_0 = \sum_n Ah_n$$

in L^2. From the fact $Ah_n \in \mathfrak{H}_n$ and basis property, we have $Ah_n = 0$. Then for any $n \in \mathbb{N}$, $h_n = 0$, since $\mu \notin \sigma := \{\lambda_{mn}\}_{n=1, m=1}^{\infty, N_n}$. Lemma 2 is proved.

Lemma II.5.16. The family $\{\mathscr{A}^{-1}\mathfrak{H}_n\}_1^\infty$ forms an $\mathscr{L}$-basis in $H^s(0, T)$.

PROOF OF LEMMA 16. It is sufficient to prove that operator $\mathscr{A}$ is an isomorphism $_0H^s$ and L^2. The boundedness of $\mathscr{A}^{-1}$ follows from the expression

$$f(t) = \int_0^t dt_1 \, e^{i\mu_1(t-t_1)} \int_0^{t_1} dt_2 \, e^{i\mu_2(t_1-t_2)} \cdots \int_0^{t_{s-1}} dt_s \, e^{i\mu_s(t_{s-1}-t_s)} g(t_s)$$

$$= e^{i\mu_1 t} \int_0^t dt_1 \, e^{i(-\mu_1+\mu_2)t_1} \int_0^{t_1} dt_2 \, e^{i(-\mu_2+\mu_3)t_2} \cdots \int_0^{t_{s-1}} dt_s \, e^{-i\mu_s t_s} g(t_s)$$

$$(16)$$

for the solution f of the problem

$$\mathscr{A}f = g \in L^2(0, T).$$

Lemma 16 is proved.

Lemma II.5.17. The family of subspaces

$$\mathfrak{H}_0 \cup \{\mathscr{A}^{-1}\mathfrak{H}_n\}_1^\infty$$

forms a Riesz basis in $H^s(0, T)$.

The fact that

$$\mathfrak{H}_0 \cap Cl_{H^s} \, Lin\{\mathscr{A}^{-1}\mathfrak{H}^0\} = \{0\}$$

may be proved completely analogously to (14). The lemma now follows from the equality

$$dim[H^s \ominus {}_0H^s] = s = dim \, \mathfrak{H}_0$$

and Lemma 16.

We can now finish the proof of Theorem 13. The subspace $\mathscr{A}^{-1}\mathfrak{H}_n$ is situated in $\mathfrak{H}_0 + \mathfrak{H}_n$, which is easily obtained from (16) for $g(t) = e^{i\lambda_{mn}t}$. So the subspace family $\mathfrak{H}_0 \cup \mathfrak{H}^0$ is complete in H^s in view of Lemma 17. Theorem 13 follows from Lemmas 14 and 15.

Remark II.5.18. Russell (1982) has also shown that a basis from exponentials in $L^2(0, T)$ has an excess in $H^{-s}(0, T)$ equal to s. Note that Proposition 12 is specific for exponential functions. For instance, a polynomial family, which forms a basis in the $L^2(0, T)$, is complete in H^s for any integer $s > 0$, but the basis property is lost there.

Remark II.5.19. Proposition 12, as well as Theorem 13, does not allow a direct generalization for vector exponential families. The situation here is similar to the replacement of exponentials in $L^2(0, T; \mathfrak{N})$. For scalars, it is possible to substitute any exponential for any other one (Lemma 4.11) conserving the basis property; one only needs to check on the linear independence of the family. At the same time, with vectors there exist, for any μ, exceptional directions $\eta(\mu)$ such that the replacement of $\eta_\lambda e^{i\lambda t}$ by $\eta(\mu) e^{i\mu t}$ leads to the loss of both minimality and basis properties (see Remark 8).

5.4. Relationship between minimality of exponential families of parabolic and hyperbolic types

Let $\mathbb{K}$ denote the set $\mathbb{Z}\backslash\{0\}$, let numbers ω_n, $n \in \mathbb{K}$, be real, and let $\omega_n = -\omega_n$. In addition, let $\lambda_n := \omega_n^2$, $n \in \mathbb{N}$. Consider two exponential families

$$\mathscr{E}_{\text{hyp}} = \{\eta_n e^{i\omega_n t}\}_{n \in \mathbb{K}}, \qquad \mathscr{E}_{\text{par}} = \{\eta_n e^{-\lambda_n t}\}_{n \in \mathbb{N}},$$

$$\eta_n = \eta_{-n}, \qquad \eta_n \in \mathfrak{N}, \qquad \dim \mathfrak{N} \leq \infty.$$

The first family we call the hyperbolic type, the second one the parabolic type. The names derive from the fact that the Fourier method for hyperbolic and parabolic problems gives rise to the $\mathscr{E}_{\text{hyp}}$ and $\mathscr{E}_{\text{par}}$ families, respectively (see Chapter III).

We prove here minimality of family $\mathscr{E}_{\text{par}}$ under the condition of minimality of $\mathscr{E}_{\text{hyp}}$. The space $\mathfrak{N}$ may be infinite dimensional.

The proof consists of an extension to an abstract situation of D. L. Russell's (1973) proof of a similar assertion valid for particular exponential families arising in the boundary control problem.

Theorem II.5.20. Let family $\mathscr{E}_{\text{hyp}} = \{e_n\}_{n \in \mathbb{K}}$ be minimal in space $L^2(0, T; \mathfrak{N})$, and let $\mathscr{E}'_{\text{hyp}} := \{e'_n\}_{n \in \mathbb{K}}$ be a biorthogonal family. Then family $\mathscr{E}_{\text{par}}$ is minimal in $L^2(0, T; \mathfrak{N})$ for any $\tau > 0$ and the norms of elements $\tilde{e}'_n$ of the biorthogonal *to $\mathscr{E}_{par}$ family satisfy an estimate*

$$\|e'_n\|_{L^2(0,\tau;\,\mathfrak{N})} \le C(\tau)\|e'_n\|_{L^2(0,\tau;\,\mathfrak{N})}\,e^{\beta\sqrt{|\lambda_n|}},$$

where $C(\tau)$ and β are positive constants.

PROOF. Introduce for $m \in \mathbb{N}$ entire operator functions

$$\hat{G}_m(z) := \int_0^T e^{-izt}\,e'_m(t)\,dt, \qquad z \in \mathbb{C},$$

and set $\tilde{G}_m(z) := \hat{G}_m(z) + \hat{G}_m(-z)$ ($\tilde{G}_m$ is the doubled even part of $\hat{G}_m$). The estimate is obvious:

$$\langle\langle\tilde{G}_m(z)\rangle\rangle \le \sqrt{T}\alpha_m\,e^{|z|T}, \qquad \alpha_m := \|e'_m\|_{L^2(0,T;\,\mathfrak{N})}. \tag{17}$$

Let us check the equality

$$\langle\eta_n, \tilde{G}_m(\omega_n)\rangle = \delta_n^m, \qquad m, n \in \mathbb{N}. \tag{18}$$

Indeed, from the definition of $\hat{G}_m$ we have

$$\langle\eta_n, \hat{G}_m(\omega_n)\rangle = \int_0^T \langle\eta_n, e^{-i\omega_n t}\,e'_m\rangle\,dt = (e_n, e'_m)_{L^2(0,T;\,\mathfrak{N})} = \delta_n^m,$$

$$\langle\eta_n, \hat{G}_m(-\omega_n)\rangle = \int_0^T \langle\eta_n, e^{i\omega_n t}\,e'_m\rangle\,dt = (e_{-n}, e'_m)_{L^2(0,T;\,\mathfrak{N})} = 0.$$

By adding these equalities, we get (18).

We have to relate the minimality of two exponential families with spectra $\{\omega_n\}$ and $\{i\omega_n^2\}$. It is therefore natural to pass from functions of variable z to functions of z^2:

$$\Phi_m(z^2) := \tilde{G}_m(z), \qquad m \in \mathbb{N}.$$

Since $\tilde{G}_m$ is an even function, it may be represented in the form of a series in even powers of z, and Φ_m are entire functions.

Introduce functions $Q_m(z) := \Phi_m(-iz)$. Then $Q_m(iz^2) = \Phi_m(z^2) = \tilde{G}_m(z)$ and $Q_m(k) = \tilde{G}_m(\sqrt{k/i})$. ($Q_m$ does not depend on the choice of the square root branch.)

Let us rewrite relations (17) and (18) for functions Q_m:

$$\langle\langle Q_m(k)\rangle\rangle \le \sqrt{T}\exp(T\sqrt{|k|})\alpha_m, \tag{19}$$

$$\langle\eta_n, Q_m(i\lambda_n)\rangle = \delta_n^m, \tag{20}$$

Functions Q_m generally may be increasing on $\mathbb{R}$. To make the inverse Fourier transform possible, we multiply Q_m by function $E(z)$, decreasing on the real axis whose existence and properties are described by the following statement.

Proposition II.5.21 (Fattorini and Russell 1971). *For every $\tau > 0$ there may be found a function $E(z)$ of the exponential type such that its zeros are real and differ from zero; E is real on the real axis and*

(i) $|E(s) \exp(\sqrt{|s|}\, T)| \leq c_1(\tau)/(1 + |s|)$, $s \in \mathbb{R}$,
(ii) $|E(-is)| \leq c_2(\tau) \exp(\tau s)$, $s \geq 0$,
(iii) $1 \geq |E(is)| \geq c_3(\tau) \exp(-\beta\sqrt{s})$, $s \geq 0$, $\beta \geq 0$, $c_3(\tau) > 0$.

Let us take arbitrary $\tau > 0$ and, using $E(z)$ (corresponding to this value of τ), set

$$G_m(z) := E(z)Q_m(z),$$

$$\tilde{e}'_m(t) := \frac{1}{2\pi E(i\lambda_m)} \int_{\mathbb{R}} e^{ist}\, G_m(s)\, ds, \qquad m \in \mathbb{N}. \tag{21}$$

Let us first verify vector functions $\tilde{e}'_m$ to be properly defined. From inequality (19) and the first property of function E, we derive for $s \in \mathbb{R}$ the estimate

$$\langle\!\langle G_m(z) \rangle\!\rangle = |E(s)| \langle\!\langle Q_m(s) \rangle\!\rangle \leq \sqrt{T}\, \alpha_m |E(s)| \exp(\sqrt{s}\,T)$$

$$\leq \sqrt{T}\, \alpha_m c_1(\tau)/(1 + |s|). \tag{22}$$

Therefore, $G_m(s) \in L^2(\mathbb{R}, \mathfrak{N})$, and the inverse Fourier transform is correct.

Lemma II.5.22. Functions $\tilde{e}'_m$ have the properties

(a) $\operatorname{supp} e'_m \subset [0, \tau]$,
(b) $\|\tilde{e}'_m\|_{L^2(0, \tau;\, \mathfrak{N})} \leq c(\tau)\alpha_m \exp(\beta\sqrt{\lambda_m})$.

PROOF OF THE LEMMA. Estimates (19) of the norm growth of operator functions Q_m testify that G_m are functions of the exponential type zero. Then from estimates (i)–(iii) of Proposition 21 it follows that for any ε, $0 < \varepsilon < \tau$,

$$e^{iz\varepsilon}\, G_m(z) \text{ belongs to } H^2_+ \text{ and}$$

$$e^{i(-\tau+\varepsilon)z}\, G_m(z) \text{ belongs to } H^2_-,$$

and thus $e^{ize} G_m(z) \in K_\tau$. Now the Paley–Wiener theorem implies that supp $e'_m \subset [-\varepsilon, \tau - \varepsilon]$. Since ε is arbitrarily small, the first statement of our lemma is valid.

Formulas (21), (22), and the third property of E function provide the estimates

$$\|\tilde{e}'_m\|_{L^2(0,\tau;\,\mathfrak{N})} = \frac{\|E^{-1}(i\lambda_m)G_m\|_{L^2(\mathbb{R},\,\mathfrak{N})}}{\sqrt{2\pi}\,|E(i\lambda_m)|} \le \frac{\sqrt{T}\,c_1(\tau)\alpha_m}{\sqrt{\pi}\,|E^{-1}(i\lambda_m)|}$$

$$\le \frac{\sqrt{T}\,c_1(\tau)\alpha_m}{\sqrt{\pi}\,c_3(\tau)}\,\exp(\beta\sqrt{\lambda_m}).$$

The lemma is proved.

Let us now check whether $\{\tilde{e}'_m\}$ is a biorthogonal in space $L^2(0,\tau;\mathfrak{N})$ to family $\mathcal{E}_{\mathrm{par}}$. Expressing G_m via $\tilde{e}'_m$ and using Lemma 22(a), we get

$$G_m(s) = E(i\lambda_m) \int_{\mathbb{R}} \tilde{e}'_m(t)\, e^{-ist}\, dt = E(i\lambda_m) \int_0^\tau \tilde{e}'_m(t)\, e^{-ist}\, dt.$$

So for elements $\tilde{e}_n$ of family $\mathcal{E}_{\mathrm{par}}$, we find

$$(\tilde{e}'_m, \tilde{e}_n)_{L^2(0,\tau;\,\mathfrak{N})} = \int_0^\tau \langle \tilde{e}'_m(t), e^{-\lambda_n t}\eta_n \rangle\, dt = \left\langle \int_0^\tau e^{-\lambda_n t}\tilde{e}'_m(t)\, dt, \eta_n \right\rangle$$

$$= \left\langle \frac{G_m(i\lambda_n)}{E(i\lambda)}, \eta_n \right\rangle = \langle G_m(i\lambda_n), \eta_n \rangle.$$

From (20) we conclude that

$$(\tilde{e}'_m, \tilde{e}_n) = \delta_n^m;$$

that is, family $\{\tilde{e}'_m\}$ is biorthogonal to $\{\tilde{e}_m\}$. The estimate for the elements of this family is obtained in Lemma 22. The theorem is proved.

Remark II.5.23. A small modification of the proof enables one to obtain the same result when several of the numbers ω_n are imaginary.

6. W-linear independence of exponential families

In this section we examine the conditions providing the uniqueness of a weak sum of series $\sum a_n e^{\mu_n t}$ in $L^2(0, T)$ or $L^2(0, \infty)$. We also prove the "excessiveness" of a particular exponential family arising in the problem of boundary control of rectangle membrane vibrations.

6.1. Dirichlet series

Formal series of the form $\sum_{n=1}^{\infty} a_n e^{-\lambda_n t}$ considered in $\mathbb{C}$ are called Dirichlet series, and for their theory we may refer, for instance, to Leont'ev (1976). We treat them for positive λ_n and prove the uniqueness of the weak series sum. In view of their applications to the DPS control problem, it is sufficient to consider λ_n growing faster than $\log n$ with n.

Proposition II.6.1 (Leont'ev 1976). *Let* $0 < \lambda_1 < \lambda_2 < \cdots$ *and*

$$\lim_{n \to \infty} \lambda_n / \log n = \infty. \tag{1}$$

If series $f(z) = \sum_{n=1}^{\infty} a_n e^{-\lambda_n z}$ *converges for* $z = t_0 \in \mathbb{R}$, *then it converges absolutely and uniformly in the half-plane* $\operatorname{Re} z \geq t_0 + \varepsilon$ *for any* $\varepsilon > 0$, *where it represents an analytical function. In this half-plane, function* $f(z)$ *is bounded and decreases exponentially when* $\operatorname{Re} z \to +\infty$:

$$|f(z)| \prec e^{-\delta \operatorname{Re} z}, \qquad \delta > 0. \tag{2}$$

PROOF. Convergence of the series at t_0 implies that its general term is bounded:

$$|a_n| e^{-\lambda_n t_0} \leq C.$$

Let us take $\varepsilon > 0$ and let $\operatorname{Re} z \geq t_0 + \varepsilon$. Then

$$\sum_{n=1}^{\infty} |a_n e^{-\lambda_n z}| = \sum_{n=1}^{\infty} |a_n| e^{-\lambda_n t_0} e^{-\lambda_n (\operatorname{Re} z - t_0)} \leq C \sum_{n=1}^{\infty} e^{-\varepsilon \lambda_n}.$$

From (1) it is seen that for large enough n, $\varepsilon \lambda_n > 2 \log n$. So

$$\sum_{n=N}^{\infty} |a_n e^{-\lambda_n z}| \leq C \sum_{n=N}^{\infty} e^{-2 \log n} = C \sum_{n=N}^{\infty} 1/n^2 < \infty,$$

and the series of analytical functions converges uniformly. Hence, it represents an analytical function.

Let us establish exponential decay of f. Represent f in the form

$$f(z) = e^{-\lambda_1 z} \sum_{n=1}^{\infty} a_n e^{-(\lambda_n - \lambda_1) z}.$$

The Dirichlet series in this formula evidently satisfies the conditions of the proposition. As proved, its sum is bounded for $\operatorname{Re} z \geq t_0 + \varepsilon$. Therefore, f is exponentially decreasing: estimate (2) is valid for $0 < \delta \leq \lambda_1$.

Corollary II.6.2. Let sequence $\{\lambda_n\}$ obey the condition of Proposition 1, ε be an arbitrary positive number, and series $\sum_{n=1}^{\infty} a_n e^{-\lambda_n t}$ converge to zero on interval $(t_0, t_0 + \varepsilon)$. Then all the coefficients a_n equal zero.

PROOF. From Proposition 1, Dirichlet series $f(t) = \sum_{n=1}^{\infty} a_n e^{-\lambda_n t}$ is analytically continued to the half-plane Re $z > t_0$. Hence, $f(z) \equiv 0$ for Re $z > t_0$.

Suppose that not all the coefficients a_n are zeros, and let a_j be a nonzero coefficient with the smallest number. Write f in the form

$$f(t) = a_j e^{-\lambda_j t}\left(1 + \sum_{n=j+1}^{\infty} a_n/a_j \, e^{-(\lambda_n - \lambda_j)t}\right).$$

Dirichlet series

$$\sum_{n=j+1}^{\infty} a_n/a_j \, e^{-(\lambda_n - \lambda_j)t}$$

satisfies conditions of Lemma 1 and, consequently, exponentially decreases with $t \to \infty$. Therefore, equality $f(t) \equiv 0$, $t > t_0$, is impossible for $a_j \neq 0$.

6.2. Parabolic family on an interval

Theorem II.6.3. Let $0 < \lambda_1 < \lambda_2 < \cdots$ and condition (1) be valid. If series $f(t) = \sum_{n=1}^{\infty} a_n e^{-\lambda_n t}$ converges weakly to zero in $L^2(0, T)$, $T > 0$, then all the coefficients a_n are zeros.

PROOF. If partial sums of Dirichlet series $\sum_{n=1}^{\infty} a_n e^{-\lambda_n t}$ weakly converge to zero in $L^2(0, T)$, then they are bounded in the norm. Hence, the norm of any term of the series is bounded uniformly with respect to n. Since for $\lambda \geq \lambda_0 > 0$

$$\|e^{-\lambda t}\|_{L^2(0, T)}^2 = \frac{(1 - e^{-2\lambda T})}{2\lambda} \asymp \lambda^{-1},$$

then

$$|a_n|/\sqrt{\lambda_n} \prec 1, \qquad n \in \mathbb{N}. \tag{3}$$

Let us demonstrate that in such a case Dirichlet series converges absolutely for $t > 0$ and uniformly for $t \geq t_0 > 0$. From estimate (3) we have

$$\sum_{n=1}^{\infty} |a_n| e^{-\lambda_n t} \prec \sum_{n=1}^{\infty} \sqrt{\lambda_n} e^{-\lambda_n t} = \sum \sqrt{\lambda_n} \exp(-\lambda_n t) \, e^{-\lambda_n t/2}.$$

Function $\lambda e^{-\lambda t}$ is bounded for $t > t_0 > 0$ and $\lambda \geq 0$. So this inequality implies

$$\sum_{n=1}^{\infty} |a_n| \, e^{-\lambda_n t} \prec \sum_{n=1}^{\infty} e^{-\lambda_n t/2}.$$

If $t \geq t_0 > 0$, then for $n \geq N(t_0)$ estimates $\lambda_n t/2 \geq 2 \log n$ follow from condition (1). Then

$$\sum_{n=1}^{\infty} |a_n| \, e^{-\lambda_n t} \prec \sum_{n=1}^{\infty} e^{-2 \log n} = \sum_{n=1}^{\infty} 1/n^2 < \infty.$$

Uniformly convergence of the Dirichlet series implies convergence in $L^2(t_0, T)$. But then the weak limit of partial sums of the series equals the limit in the norm, and the Dirichlet series turns to zero for $t \in [t_0, T]$. The assertion of the theorem follows from the uniqueness of the Dirichlet series with real λ_n (Corollary 2).

6.3. Hyperbolic family on the semiaxis

Theorem II.6.4. Let sequence $\{\omega_n\}_{n \in \mathbb{K}}$, $\mathbb{K} = \mathbb{Z} \backslash \{0\}$, be strictly increasing and $|\omega_n| \succ |n|^\alpha$, $\alpha > 0$. Then for any $R > 0$ and any $\varepsilon > 0$, condition $|a_n| \prec |n|^R$ and weak convergence to zero in $L^2(0, \infty)$ of partial sums of series

$$\sum_{n \in \mathbb{K}} a_n \, e^{i(\omega_n + i\varepsilon)t}$$

imply that all coefficients a_n equal zero.

PROOF. Let v_n denote numbers $\omega_n + i\varepsilon$, and with the help of the inverse Fourier transform, turn from exponentials to simple fractions in the Hardy space H_+^2.

Weak convergence of the exponential series to zero turns into the equality

$$\sum_{n \in \mathbb{K}} a_n f(v_n) = 0,$$

which holds for any function f from H_+^2. Suppose $a_j \neq 0$. Consider a family of sample functions

$$f_N(k) := (2i\varepsilon)^N / (k - \bar{v}_j)^N, \qquad j \text{ is fixed}, \ N = 1, 2, \ldots,$$

so that $f_N(v_j) = 1$ and for $n \neq j$

$$f_N(v_n) = \frac{(2i\varepsilon)^N}{[(\omega_n - \omega_j) + 2i\varepsilon]^N} = \left[1 + \frac{\omega_n - \omega_j}{2i\varepsilon}\right]^{-N}.$$

It is clear that $f_N(v_n) \xrightarrow[N \to \infty]{} 0$ and that an estimate follows from the condition on ω_n

$$|f_N(v_n)| \prec n^{-\alpha N}.$$

Since a_n are of not more than power growth, for large enough N, series $\sum a_n f_N(v_n)$ converges to zero uniformly in N. All the terms of the series, except the jth one, tend to zero with $n \to \infty$; therefore $a_j = 0$. The theorem is proved.

6.4. Hyperbolic family on an interval

Theorem II.6.5. Let $\{\omega_n\}_{n \in \mathbb{K}}$ be a strictly monotonic sequence, sgn $\omega_n =$ sgn(n), and

$$\lim_{n \to \pm\infty} \frac{\omega_n}{\log|n|} = \pm\infty.$$

Then for any $T > 0$ and $\varepsilon > 0$ condition, $\{a_n\} \in \ell^1$ and weak convergence to zero in space $L^2(0, T)$ of partial sums of

$$\sum_{n \in \mathbb{Z}} a_n e^{-|\omega_n|\varepsilon} e^{i\omega_n t}$$

imply that all a_n are zeros.

PROOF. Introduce function

$$F(z) = \sum_{n \in \mathbb{K}} a_n e^{-|\omega_n|\varepsilon} e^{i\omega_n z}.$$

For $|\text{Im } z| \leq \varepsilon$, exponentials $e^{-|\omega_n|\varepsilon} e^{i\omega_n z}$ are bounded by unity in the absolute value and therefore the series converges uniformly; that is, it represents an analytical in the strip $|\text{Im } z| < \varepsilon$ function. From weak convergence to zero in $L^2(0, T)$ one obtains then that $F(z) \equiv 0$ for $|\text{Im } z| < \varepsilon$. Consider functions

$$F_+(z) = \sum_{n=1}^{\infty} a_n e^{-|\omega_n|\varepsilon} e^{i\omega_n z}, \tag{4}$$

$$F_-(z) = -\sum_{n=-\infty}^{1} a_n e^{-|\omega_n|\varepsilon} e^{i\omega_n z} \tag{5}$$

The first of them is analytical in the half-plane $\text{Im } z > -\varepsilon$, and the second one in the half-plane Im $z < \varepsilon$. Since

$$0 = F(z) = F_+(z) - F_-(z),$$

they coincide on their common domain. Hence, F_+ and F_- are analytically extended to the whole plane and coincide everywhere, $F_+(z) = F_-(z)$.

By Proposition 1, function F_+ is bounded at Im $z \geq -\varepsilon$ and

$$F_+(z) \xrightarrow[\text{Im } z \to +\infty]{} 0.$$

Similarly, $F_-(z)$ is bounded at Im $z \leq \varepsilon$. According to the Liouville theorem, $F_\pm(z) = 0$. Set $z = i(x - \varepsilon)$ in (4). Then

$$\sum_{n=1}^{\infty} a_n \exp(-\omega_n x) = 0.$$

Using Corollary 2, we find that $a_n = 0$, $n > 0$. Analogously, $a_n = 0$ for $n < 0$. The theorem is proved.

6.5. *ω-linear independence in $L^2(0, T)$*

Theorem II.6.6. Let family $\mathscr{E} = \{e^{i\lambda t}\}_{\lambda \in \Lambda}$ possess a subfamily $\mathscr{E}_0 = \{e^{i\lambda t}\}_{\lambda \in \Lambda_0}$ being a Riesz basis in $L^2(0, T)$, with set $\Lambda \backslash \Lambda_0$ containing not less than $s + 1$ points, $s \geq 0$. Then there exists a nonzero sequence $\{a_\lambda\}_{\lambda \in \Lambda}$ such that series $\sum_{\lambda \in \Lambda} a_\lambda \exp(i\lambda t)$ converges to zero in $L^2(0, T)$ and $\sum_{\lambda \in \Lambda} |a_\lambda|^2 |\lambda|^{2s} < \infty$.

PROOF. Choose subset Λ_1 of set Λ in a way that contains Λ_0 and s extra points. Then by Proposition 5.12 family

$$\{\lambda^{-s} e^{i\lambda t}\}_{\lambda \in \Lambda_1}$$

is a Riesz basis in $H^s(0, T)$ (for $\lambda = 0$ we take unity instead of λ^s in the denominator).

Expand function $e^{i\lambda_0 t}$, $\lambda_0 \in \Lambda \backslash \Lambda_1$, over this basis:

$$e^{i\lambda_0 t} = \sum_{\lambda \in \Lambda_1} b_\lambda e^{i\lambda t} / \lambda^s;$$

then $\sum_{\lambda \in \Lambda_1} |b_\lambda|^2 < \infty$. Setting $a_\lambda := b_\lambda / \lambda^s$ for $\lambda \in \Lambda_1$, $a_{\lambda_0} := -1$, and $a_\lambda := 0$ for other cases, we obtain the assertion of the theorem.

Remark II.6.7. From the theorem, the larger the "excess" of family $\mathscr{E}$, the "smaller" the coefficients a_λ that may be taken in the series $\sum a_\lambda \exp(i\lambda t)$ converging to zero. But by Theorem 5, they cannot be exponentially small; otherwise the W-linear independent family will appear.

Remark II.6.8. When studying the controllability of distributed parameter systems, one may face families in which, along with the exponentials, the term $\{t\}$ is present (see Section III.2). It is not difficult to show that Theorems 3 and 5 concerning the uniqueness of a weak sum on an interval remain true with the addition to the series of term $\tilde{a}_0 t$.

6.6. Excessiveness of the exponential family arising in the problem of controlling rectangular membrane oscillation

Let us apply Theorem 5.13 about a subspace basis in H^s to the investigation of a particular family. Set

$$\lambda_{mn} = \text{sgn}(n)\sqrt{\alpha^2 m^2 + \beta^2(|n| - 1/2)^2}; \qquad \alpha, \beta > 0, \ m \in \mathbb{N}, \ n \in \mathbb{Z}.$$

Theorem II.6.9. For any $R > 0$, $T > 0$, there exists sequence $\{c_{mn}\}_{m \in \mathbb{N}, n \in \mathbb{K}}$ such that

(a) series $\sum_{m,n} c_{mn} \exp(i\lambda_{mn}t)$ converges to zero in $L^2(0, T)$,
(b) $\sum_{m,n} |c_{mn}| \lambda_{mn}^{2R} < \infty$,
(c) for some m_0 and $m \geq m_0$, $c_{mn} = 0$.

Here is the plan of the proof. Family $\{\lambda_{mn}\}$ is naturally decomposed in series, namely, sequences $\{\lambda_{mn}\}_{n \in \mathbb{K}}$ (with fixed m). Consider M series corresponding to $m = 1, 2, \ldots, M$. At $n \to +\infty$ the series points cluster around $\beta(|n| - 1/2)$. The sequence of exponentials corresponding to any of the series forms a basis in $L^2(0, T_0)$, $T_0 := 2\pi/\beta$. However, the exponential sequence of even two series is not uniformly minimal in $L^2(0, T)$ for every T. Therefore, one has to examine linear spans of groups of exponentials. It so happens that they constitute a basis in $L^2(0, MT_0)$. By taking a qualified (see (7) below) nearness of the spectrum points inside the groups into account, it becomes possible to proceed from the series in exponential groups to the series of individual exponentials. Using the infinite excess of $\{\lambda_{mn}\}$ and Theorem 6, one is then able to obtain assertions of the theorem.

Let us assume there are no coinciding values among $\{\lambda_m\}$ (it takes place when $\alpha^2/\beta^2 \notin \mathbb{Q}$; the theorem is trivial in the opposite case). Specify R, T and choose integer M such that $T \leq MT_0$. Evidently, it is enough to check the assertion of the theorem for $T = MT_0$. Further, consider, instead of $\{\lambda_{mn}\}$, points shifted to the upper half-plane by the value $i/2$: $\nu_{mn} := \lambda_{mn} + i/2$. If the theorem is proved for $\{\nu_{mn}\}$, it is also valid for $\{\lambda_{mn}\}$ because

$$\sum c_{mn} e^{i\nu_{mn}t} = e^{-t/2} \sum c_{mn} e^{i\lambda_{mn}t}$$

and the series converge in $L^2(0, T)$ simultaneously.

Let σ_n, $n \in \mathbb{K}$, denote the set $\{v_{mn}\}_{m=1}^M$, and let σ_0 denote the set $\{\mu_j\}_{j=1}^s$, where $\mu_j := v_{M+1,j}$ are the "first" s points of the series with number $M + 1$. Set $\sigma = \bigcup_{n \in \mathbb{Z}} \sigma_n$ and suppose s to be large enough; for instance, $s > R + 2M$. Introduce subspaces $\mathfrak{H}_n := \bigvee_{v \in \sigma_n} \exp(ivt)$, $n \in \mathbb{Z}$. Write δ_n for the Carleson constant of set σ_n.

Lemma II.6.10.

(a) *Family of subspaces $\{\mathfrak{H}_n\}_{n \in \mathbb{K}}$ forms a Riesz basis in the closure of its linear span in $L^2(0, \infty)$.*

(b) *Orthoprojector P_T from $\mathrm{Cl}_{L^2(0, \infty)} \mathrm{Lin}\{\mathfrak{H}_n\}_{n \in \mathbb{K}}$ on $L^2(0, T)$ is an isomorphism.*

(c) *$c > 0$ may be found such that for any $n \in \mathbb{K}$ an estimate*

$$\delta_n \geq c|v|^{-M}$$

holds for all $v \in \sigma_n$.

We first prove the theorem and then demonstrate the lemma.

From (a) and (b), it follows that $\{\mathfrak{H}_n\}_{n \in \mathbb{K}}$ is a Riesz basis in $L^2(0, T)$. Then by Theorem 5.13, family $\{\mathfrak{H}_n\}_{n \in \mathbb{Z}}$ is a Riesz basis in $H^s(0, T)$. Expand element $\exp(iv_0 t)$, $v_0 := v_{M+2,1}$, not entering $\{\mathfrak{H}_n\}_{n \in \mathbb{Z}}$, on this basis:

$$e^{iv_0 t} = \sum_{n \in \mathbb{Z}} h_n, \qquad h_n = \sum_{v \in \sigma_n} b_v e^{ivt} \in \mathfrak{H}_n.$$

Series $\sum_{n \in \mathbb{Z}} h_n$ converges in space $H^s(0, T)$ and, according to a Riesz basis property, we have an estimate

$$\sum_{n \in \mathbb{Z}} \left\| \left(\frac{d}{dt}\right)^s h_n \right\|_{L^2(0, T)}^2 < \infty.$$

So assertion (b) of the lemma implies

$$\infty > \sum_{n \in \mathbb{Z}} \left\| \sum_{v \in \sigma_n} b_v (iv)^s e^{ivt} \right\|_{L^2(0, T)}^2 \succ \sum_{n \in \mathbb{Z}} \left\| \sum_{v \in \sigma_n} b_v (iv)^s e^{ivt} \right\|_{L^2(0, \infty)}^2.$$

If we apply estimate (21) of Section II.1 to each of the inner sums, with the assertion (c) of the lemma we arrive at

$$\infty > \sum_{n \in \mathbb{Z}} \frac{\delta_n^2}{32(1 + 2 \log 1/\delta_n)} \sum_{v \in \sigma_n} |b_v|^2 |v|^{2s} \succ \sum_{n \in \mathbb{Z}} \sum_{v \in \sigma_n} |b_v|^2 |v|^{2s - 3M}$$

$$\geq \sum_{n \in \mathbb{Z}} \sum_{v \in \sigma_n} |b_v|^2 |v|^{2R}.$$

Thus, series $e^{iv_0t} - \sum h_n$ is shown to satisfy the conditions of the theorem.

Let us now check the statements of the lemma.

(a) Denote $S_j := \{v_{jn}\}_{n \in \mathbb{K};\, j=1,\ldots,M}$. Numbers v_{jn} lie on the straight line $\operatorname{Im} k = 1/2$ and are separable for a given j. Therefore, each set S_j is Carlesonian. Since at $n \to \infty$, we have $\operatorname{diam} \sigma_n \to 0$ and then $\{\mathfrak{H}_n\}_{n \in \mathbb{K}} \in (LB)$ by Proposition 2.11.

(b) To prove the basis property for family $\{\mathfrak{H}_n\}_{n \in \mathbb{K}}$ in $L^2(0, T)$, we construct a GF of the family (more precisely, GF of family $\{P_T(k - \bar{v}_{mn})^{-1}\}_{n \in \mathbb{K},\, m=1,\ldots,M}$).

Following Fattorini (1979), set

$$F_\mu(z) := \cos \sqrt{\pi^2 z^2/\beta^2 - \mu}.$$

This function is of the exponential type, and its indicator diagram is easily seen to coincide with segment $[-iT_0/2, iT_0/2]$.

For $\mu = \mu_m := m^2 \pi^2 \alpha^2/\beta^2$, the zeros of function $F_{\mu_m}(z - i/2)$ are points v_{mn}. From estimates

$$|F_\mu(z)| \asymp \cos\left(\frac{\pi z}{\beta} + \mathcal{O}(1/z)\right) \asymp 1, \qquad \text{for } \operatorname{Im} z = -1/2, \qquad (6)$$

it follows that on the real axis $F_{\mu_m}(x - i/2)$ and $F_{\mu_m}^{-1}(x - i/2)$ are bounded. Now introduce function

$$F(z) := e^{izMT_0/2} \prod_{m=1}^{M} F_{\mu_m}(z - i/2).$$

From the foregoing it is clear that F_M is a GF for family

$$\{P_T(k - \bar{v}_{mn})\}_{m=1,\, n \in \mathbb{K}}^{M}.$$

Formula (6) and Corollary 3.22 thus imply assertion (b).

(c) From the explicit form of λ_{mn} for $r \neq m$ we derive the following estimate:

$$|\lambda_{mn} - \lambda_{rn}| = \left|\frac{\lambda_{mn}^2 - \lambda_{rn}^2}{\lambda_{mn} + \lambda_{rn}}\right| = \left|\frac{\alpha^2(m^2 - r^2)}{\lambda_{mn} + \lambda_{rn}}\right| > \frac{1}{|n|}. \qquad (7)$$

This, by means of the identity

$$\left|\frac{z_1 - z_2}{z_1 - \bar{z}_2}\right|^2 = \frac{|\operatorname{Re} z_1 - \operatorname{Re} z_2|^2}{1 + |\operatorname{Re} z_1 - \operatorname{Re} z_2|^2}, \qquad \text{for } \operatorname{Im} z_1 = \operatorname{Im} z_2 = 1/2,$$

provides

$$\left| \frac{v_{mn} - v_{rn}}{v_{mn} - \bar{v}_{rn}} \right|^2 < \frac{1}{n^2}, \qquad m \neq r.$$

Hence

$$\delta_n := \inf_{r=1,\ldots,M} \prod_{m=1,\, m \neq r}^{m=M} \left| \frac{v_{mn} - v_{rn}}{v_{mn} - \bar{v}_{rn}} \right| > |n|^{-M+1} \geq |n|^{-M}.$$

For points v_{mn} of one group σ_n, the estimates $|v_{mn}| \asymp |n|$, $n \in \mathbb{Z}$, $m = 1, \ldots, M$ is evident. Therefore, assertion (c) of the lemma is proved, and the proof of Theorem 9 is also completed.

III

Fourier method in operator equations and controllability types

1. Evolution equations of the first order in time

1.1. Let V and H be Hilbert spaces, V being dense and continuously embedded in H. Identify H with the space dual to it and let V' denote the space dual to V. Then H may be identified with some dense subspace in V' in such a way that the embedding $H \subset V'$ is continuous (Lions and Magenes 1968: chap. 1). Let $a[\varphi, \psi]$ be a continuous symmetric bilinear form on V,

$$a_\alpha[\varphi, \psi] = a[\varphi, \psi] + \alpha(\varphi, \psi)_H, \qquad \alpha \geq 0.$$

Suppose that for some $\gamma > 0$ an estimate

$$a_\alpha[\varphi, \varphi] \geq \gamma \|\varphi\|_V^2 \tag{1}$$

holds. Then a self-adjoint semibounded from below operator A uniquely corresponds to the form a (see, for example, Birman and Solomyak 1980: chap. 10)

$$\mathcal{D}(A) \subset V, \qquad (A\varphi, \psi)_H = a[\varphi, \psi]; \qquad \varphi \in \mathcal{D}(A), \psi \in V.$$

In turn, the form a_α corresponds to self-adjoint positive definite operator $A_\alpha = A + \alpha I \geq \nu I, \nu > 0, \mathcal{D}(A_\alpha) = \mathcal{D}(A)$. The norm generated by form a_α is equivalent by the force of (1) to the norm of space V. Therefore (Birman and Solomyak 1980: chap. 10),

$$\mathcal{D}(A^{1/2}) = V,$$

$$(A_\alpha^{1/2}\varphi, A_\alpha^{1/2}\psi)_H = a_\alpha[\varphi, \psi]; \qquad \varphi, \psi \in V. \tag{2}$$

146

We assume that operator A has a set of eigenvalues $\{\lambda_n\}$, $n \in \mathbb{N}$, and eigenfunctions $\{\varphi_n\}$ that form an orthonormal basis in space H. This assumption is valid in the following chapters, where the theory developed in Chapters I–III is applied to control problems for equations of mathematical physics.

One can associate the following spaces used constantly below with operators A and A_α: spaces ℓ_r^2, $r \in \mathbb{R}$, of sequences $c = \{c_n\}$, $n \in \mathbb{N}$, with the norm

$$\|c\|_r = \left[\sum_{n=1}^{\infty} |c_n|^2 (\lambda_n + \alpha)^r \right]^{1/2},$$

and corresponding to them spaces W_r,

$$W_r = \left\{ f \mid f = \sum_{n=1}^{\infty} c_n \varphi_n, \ \|f\|_{W_r} := \|\{c_n\}\|_r < \infty \right\}$$

(the latter are understood as the completion of finite sums of this type in the norm $\|\cdot\|_{W_r}$). Spaces ℓ_r^2 and W_r become Hilbert spaces after the standard scalar products are introduced.

For $r > 0$, spaces W_r coincide, by the spectral theorem, with the domains of powers of operator A_α, namely, $W_r = \mathcal{D}(A_\alpha^{r/2})$. It is evident that $W_0 = H$. We have just identified this space with the dual one. We shall write W_r' for the space dual to W_r and reserve the notation $\langle f, \varphi \rangle_*$ for the value of functional $f \in W_r'$ on element $\varphi \in W_r$. It is easy to verify that $W_r' = W_{-r}$. Note also that (2) implies $W_1 = V$.

Later in the chapter, we shall work with operators in the entire scale of spaces W_r. We use the same letter A (no ambiguity will appear) to denote a bounded operator acting from W_{r+2} to W_r according to the rule: $A(\sum_{n=1}^{\infty} c_n \varphi_n) = \sum_{n=1}^{\infty} \lambda_n c_n \varphi_n$. A similar extension is considered for operator A_α as well:

$$A_\alpha \left(\sum_{n=1}^{\infty} c_n \varphi_n \right) = \sum_{n=1}^{\infty} (\lambda_n + \alpha) c_n \varphi_n.$$

Obviously, $\|A_\alpha f\|_{W_r}^2 = \|f\|_{W_{r+2}}^2$.

1.2. Consider a differential equation

$$\frac{dx(t)}{dt} + Ax(t) = f(t), \qquad 0 < t < \infty, \tag{3}$$

with the initial condition

$$x(0) = x_0. \tag{4}$$

First of all, we have to state how one should understand the solution to the problem (1), (2), as well as to the other initial and initial boundary-value problems treated below. To do this, let us introduce spaces $L^2(0, T; W_r)$ of measurable functions $g: (0, T) \mapsto W_r$ such that

$$\|g\|_{L^2(0, T; W_r)}^2 := \int_0^T \|g(t)\|_{W_r}^2 \, dt < \infty,$$

and also spaces $C(0, T; W_r)$ of continuous functions on $[0, T]$ with the values in W_r. Let $\mathscr{D}(0, T)$ be a space of basic functions, that is, of infinitely differentiable and finite scalar functions on $(0, T)$. For a function $g \in L^2(0, T; W_r)$ we define its generalized derivative as an element of $\mathscr{L}(\mathscr{D}(0, T); W_r)$ as follows.

Function $g \in L^2(0, T; W_r)$ may be represented in the form

$$g(t) = \sum_{n=1}^{\infty} g_n(t)\varphi_n, \qquad g_n \in L^2(0, T), \qquad \sum_{n=1}^{\infty} \|g_n\|_{L^2(0, T)}^2 (\lambda_n + \alpha)^r < \infty.$$

Set

$$\frac{dg}{dt} = \sum_{n=1}^{\infty} \frac{dg_n}{dt}\varphi_n, \qquad \frac{dg}{dt}(\psi) = \sum_{n=1}^{\infty} \frac{dg_n}{dt}(\psi)\varphi_n,$$

where $\psi \in \mathscr{D}(0, T)$, dg_n/dt is the generalized derivative of g_n. Since

$$\frac{dg_n}{dt}(\psi) = -g_n\left(\frac{d\psi}{dt}\right) = -\int_0^T g_n(t)\psi'(t) \, dt,$$

then

$$\left|\frac{dg_n}{dt}(\psi)\right|^2 \leq \|g_n\|_{L^2(0, T)}^2 \|\psi'\|_{L^2(0, T)}^2.$$

From this one easily sees that dg/dt is actually a continuous linear mapping from $\mathscr{D}(0, T)$ to W_r and that it satisfies the equality

$$\frac{dg}{dt}(\psi) = -g\left(\frac{d\psi}{dt}\right).$$

We shall be interested in a continuous (with respect to t) solution of equation (3) in which the values belong to one of the spaces W_r.

Let $f \in L^2(0, T; W_{r-1})$, $x_0 \in W_r$. Function x from the space $C(0, T; W_r)$ is said to be a solution of equation (3) with initial condition (4) if the sum $(dx/dt) + Ax$ belongs to the space $L^2(0, T; W_{r-1})$, (3) is valid as the equality of the elements of this last space, and condition (4) is understood as the equality of the elements of space W_r.

Theorem III.1.1. Let $f \in L^2(0, T; W_{r-1})$, $x_0 \in W_r$. Then there exists a unique solution of problem (3), (4), and mapping $\{f, x_0\} \mapsto x$ of space $L^2(0, T; W_{r-1}) \times W_r$ to $C(0, T; W_r)$ is continuous.

PROOF. Let us represent functions f and x_0 as

$$f(t) = \sum_{n=1}^{\infty} f_n(t) \varphi_n, \qquad x_0 = \sum_{n=1}^{\infty} x_n^0 \varphi_n,$$

$$\sum_{n=1}^{\infty} \|f_n\|_{L^2(0, T)}^2 (\lambda_n + \alpha)^{r-1} < \infty, \qquad \sum_{n=1}^{\infty} |x_n^0|^2 (\lambda_n + \alpha)^r < \infty,$$

while we write the sought-for function x in the form

$$x(t) = \sum_{n=1}^{\infty} x_n(t) \varphi_n. \tag{5}$$

Inserting these expansions into (3), (4) and equaling the coefficients at φ_n, we get

$$\dot{x}_n(t) + \lambda_n x_n(t) = f_n(t), \qquad x_n(0) = x_n^0, \qquad n \in \mathbb{N}. \tag{6}$$

From this,

$$x_n(t) = x_n^0 e^{-\lambda_n t} + \int_0^t e^{-\lambda_n(t-\tau)} f_n(\tau) \, d\tau, \qquad n \in \mathbb{N}. \tag{7}$$

Let us check that the function x constructed by these coefficients is really the desired solution to the problem (3), (4). From (7) we derive

$$|x_n(t)| \le |x_n^0| e^{-\lambda_n t} + \|e^{-\lambda_n(t-\tau)}\|_{L^2(0, t)} \|f_n\|_{L^2(0, t)},$$

and then using the estimate $(1 - e^{-\lambda_n t})/\lambda_n \prec (\lambda_n + \alpha)^{-1}$ we obtain

$$|x_n(t)|^2 \prec |x_n^0|^2 + \|f_n\|_{L^2(0, t)}^2 (\lambda_n + \alpha)^{-1}, \qquad n \in \mathbb{N}, \ t \in [0, T]. \tag{8}$$

Multiplying the latter relation by $(\lambda_n + \alpha)^r$ and summing up over n, we obtain

$$\|x(t)\|_{W_r}^2 \prec \|x_0\|_{W_r}^2 + \|f\|_{L^2(0, t; W_{r-1})}^2, \qquad t \in [0, T]. \tag{9}$$

Thus we have demonstrated that for all $t \in [0, T]$, function x constructed by formula (5) with coefficients (7) obeys inclusion $x(t) \in W_r$. Moreover, series (5) converges in this space uniformly in t, $t \in [0, T]$. It follows from estimates (8), (9) and the Weierstrass theorem. Therefore, function x is continuous in t in the norm of W_r.

Equality (3), as equality of elements of $L^2(0, T; W_{r-1})$, and equality (4) in W_r follow directly from (6).

To prove uniqueness, let us note that if x is the solution of problem (3), (4), it is represented by definition in the form (5). Then equality (3), understood as such of the elements of $L^2(0, T; W_{r-1})$, together with (4), leads to (7).

Estimate (9) implies

$$\|x\|^2_{C(0, T; W_r)} \prec \|x_0\|^2_{W_r} + \|f\|^2_{L^2(0, T; W_{r-1})}. \tag{10}$$

Theorem 1 is proved.

Note that the theorem is precise in the following sense. If $\lambda_k \to \infty$, then for any $p > r$, one is able to find $f \in L^2(0, T; W_{r-1})$ such that $x(T) \notin W_p$. This may be easily obtained by setting $f_n(t) = \gamma_n e^{-\lambda_n(T-t)}$ in (7) with appropriate γ_n.

Later, we will also need a solution of problem (3), (4) in space $X_r(0, T)$, $r \in \mathbb{R}$, where

$$X_r(0, T) := \left\{ g \mid g \in L^2(0, T; W_{r+1}), \frac{dg}{dt} \in L^2(0, T; W_{r-1}) \right\},$$

$$\|g\|^2_{X_r(0, T)} = \|g\|^2_{L^2(0, T; W_{r+1})} + \left\| \frac{dg}{dt} \right\|^2_{L^2(0, T; W_{r-1})}.$$

It is known (Lions and Magenes 1968: chap. 1) that when any function from the space $X_r(0, T)$ is changed, if needed, on some set of the zero measure, it is a continuous function on $[0, T]$ with the values in W_r. This fact may be written in the form of an inclusion

$$X_r(0, T) \subset C(0, T; W_r).$$

Theorem III.1.2. Under the conditions of Theorem 1, the solution to the problem (3), (4) constructed there belongs to space $X_r(0, T)$, and the mapping $\{f, x_0\} \mapsto x$ of space $L^2(0, T; W_{r-1}) \times W_r$ to $X_r(0, T)$ is continuous.

PROOF. Let us show that the solution constructed in Theorem 1 belongs to space $X_r(0, T)$. Suppose first that functions x_n are real. Multiply equality (6) by $x_n(t)(\lambda_n + \alpha)^r$ and integrate the result in t from 0 to T. Then

$$\tfrac{1}{2}(\lambda_n + \alpha)^r[|x_n(T)|^2 - |x_n(0)|^2] + \lambda_n(\lambda_n + \alpha)^r \int_0^T |x_n(t)|^2 \, dt$$

$$= \int_0^T (\lambda_n + \alpha)^{(r-1)/2} f_n(t)(\lambda_n + \alpha)^{(r+1)/2} x_n(t) \, dt.$$

Adding $\alpha(\lambda_n + \alpha)^r \int_0^T |x_n(t)|^2\, dt$ to both sides of the equality and using Cauchy inequality with $\varepsilon > 0$, we have

$$\tfrac{1}{2}(\lambda_n + \alpha)^r [|x_n(T)|^2 - |x_n(0)|^2] + (\lambda_n + \alpha)^{r+1} \|x_n\|^2_{L^2(0,\,T)}$$

$$\leq \alpha(\lambda_n + \alpha)^r \|x_n\|^2_{L^2(0,\,T)} + \frac{1}{2\varepsilon}(\lambda_n + \alpha)^{r-1} \|f_n\|^2_{L^2(0,\,T)}$$

$$+ \frac{\varepsilon}{2}(\lambda_n + \alpha)^{r+1} \|x_n\|^2_{L^2(0,\,T)}, \qquad n \in \mathbb{N}.$$

Performing the summation of these inequalities in n and transferring the terms containing $x_n(T)$ and $x_n(0)$ to the right-hand side, we obtain

$$\|x\|^2_{L^2(0,\,T;\,W_{r+1})} \prec \|f\|^2_{L^2(0,\,T;\,W_{r-1})} + \|x\|^2_{C(0,\,T;\,W_r)}. \tag{11}$$

In the complex case of the derivation of estimate (11), one should multiply (6) by $\overline{x_n(t)}(\lambda_n + \alpha)^r$, then multiply the conjugate to (6) equalities by $x_n(t)(\lambda_n + \alpha)^r$, and put together the resulting expressions. The other calculations remain unperturbed.

Now, using Theorem 1 from estimates (10), (11), we conclude that $x \in L^2(0,\,T;\,W_{r+1})$. Equalities (6) then imply $dx/dt \in L^2(0,\,T;\,W_{r-1})$. Finally, we arrive at the estimate

$$\|x\|^2_{X_r(0,\,T)} \prec \|x_0\|^2_{W_r} + \|f\|^2_{L^2(0,\,T;\,W_{r-1})}. \tag{12}$$

Theorem 2 is proved.

Remark III.1.3. Theorems 1 and 2 may be extracted from the theory of nonhomogeneous boundary-value problems developed in Lions and Magenes (1968: chap. 3). In our case (under the assumptions about operator A), there is no need to apply the general theory, because it is rather easier to present a direct proof by means of the Fourier method. Equalities (7) and estimates (10) and (11) produced during the demonstration will be exploited later in the study of control problems. This remark also pertains to Theorem 2.1 proved in the next section.

1.3. Let U be a Hilbert space, $\mathcal{U} = L^2(0,\,T;\,U)$, and B be a linear bounded operator acting from U to W_{r-1}, $r \in \mathbb{R}$. Consider a control system

$$\frac{dx(t)}{dt} + Ax(t) = Bu(t), \qquad 0 < t < T,\, u \in \mathcal{U}, \tag{13}$$

with the initial condition

$$x(0) = x_0, \qquad x_0 \in W_r. \tag{14}$$

From Theorem 1 it follows that for a given control $u \in \mathcal{U}$, there exists a unique function $x(\cdot, u, x_0) \in C(0, T; W_r)$ satisfying relations (13), (14). This enables one to define correctly a reachability set $R(T, x_0)$ of system (13) in time T from the state x_0 as the set of the final points of phase trajectories under all possible controls:

$$R(T, x_0) = \{x(T, u, x_0) \in W_r \mid u \in \mathcal{U}\}.$$

Relations (7) and (9) imply operators

$$S(T) \in \mathcal{L}(W_r, W_r) \quad \text{and} \quad K(T) \in \mathcal{L}(\mathcal{U}, W_r)$$

to exist such that

$$x(T, u, x_0) = x(T, 0, x_0) + x(T, u, 0) = S(T)x_0 + K(T)u, \qquad (15)$$

namely,

$$S(T)x_0 = \sum_{n=1}^{\infty} (x_n^0 \, e^{-\lambda_n T})\varphi_n \qquad (16)$$

$$K(T) = \sum_{n=1}^{\infty} \left[\int_0^T e^{-\lambda_n(T-t)} (Bu(t))_n \, dt \right]\varphi_n. \qquad (17)$$

Here, functions $(Bu(t))_n$, $n \in \mathbb{N}$, are the "coordinates" of the element $Bu(t) \in W_{r-1}$:

$$Bu(t) = \sum_{n=1}^{\infty} (Bu(t))_n \varphi_n. \qquad (18)$$

We are interested mainly in the set $R(T) := R(T, 0)$ which is the image of operator $K(T)$. Set $R(T, x_0)$ is obtained out of $R(T)$ by a shift $S(T)x_0$.

As already mentioned, spaces W_r and W_{-r} ($r \in \mathbb{R}$) are dual, and the elements of one may be treated as the functionals on elements of the other. We have decided to denote the value of functional $f \in W_r$ on element $\psi \in W_{-r}$ as $\langle f, \psi \rangle_*$. If

$$f = \sum_{n=1}^{\infty} c_n \varphi_n, \quad \psi = \sum_{n=1}^{\infty} b_n \varphi_n, \quad \text{then} \quad \langle f, \psi \rangle_* = \sum_{n=1}^{\infty} c_n \bar{b}_n.$$

Using operator $B \in \mathcal{L}(U, W_{r-1})$, let us define operator $B^* \in \mathcal{L}(W_{1-r}, U)$ by an equality

$$\langle Bv, \psi \rangle_* = (v, B^*\psi)_U, \qquad v \in U, \psi \in W_{1-r}, \qquad (19)$$

in which $(\cdot, \cdot)_U$ denotes the scalar product in space U. Now, functions $(Bu(t))_n$ (see (18)) may be represented as

$$(Bu(t))_n = \langle Bu(t), \varphi_n \rangle_* = (u(t), B^*\varphi_n)_U. \qquad (20)$$

Set

$$x_n(T, u, 0) = \langle x(T, u, 0), \varphi_n \rangle_*, \qquad x(T, u, 0) = \sum_{n=1}^{\infty} x_n(T, u, 0)\varphi_n.$$

Since $\|x(T, u, 0)\|_{W_r} = \|\{x_n(T, u, 0\}\|_r$, reachability set $R(T) \subset W_r$ is isometric to the set $\hat{R}(T)$ of sequences $\{x_n(T, u, 0)\}$ scanned in space ℓ_r^2 while control u is running through the whole space $\mathscr{U}$.

Formulas (17) and (20) yield

$$x_n(T, u, 0) = \int_0^T (u(t), e^{-\lambda_n(T-t)} B^* \varphi_n)_U, \qquad n \in \mathbb{N}. \tag{21}$$

The right-hand sides of relations (21) may be rewritten in the form of scalar products in space $\mathscr{U} = L^2(0, T; U)$:

$$x_n(T, u, 0) = (u, e^{-\lambda_n(T-t)} B^* \varphi_n)_{\mathscr{U}}, \qquad n \in \mathbb{N}. \tag{22}$$

Let us introduce a family of functions $\mathscr{E} \subset \mathscr{U}$:

$$\mathscr{E} = \{e_n\}, \, n \in \mathbb{N}, \qquad e_n(t) = e^{-\lambda_n t} B^* \varphi_n. \tag{23}$$

Sometimes, for brevity, we call family $\mathscr{E}$ and the other families of this kind vector exponential families, on the grounds that each function e_n consists of two cofactors, one of them being a scalar exponential function from the space $L^2(0, T)$, while the other is a vector from the space U. Note that space U connected with specific control problems may be either finite dimensional or infinite dimensional. In this and the following chapters we establish the relationship between the "quality" of the system's controllability and the properties of the corresponding vector exponential family. Furthermore, we study these properties and present our conclusions regarding the controllability of systems described by parabolic and hyperbolic equations under various kinds of control (distributed, boundary, and pointwise).

By changing a variable $t' = T - t$ in formula (21) (or in (22)) we now complete the reduction of the problem of reachability set description for system (13) to the problem of moments with respect to family $\mathscr{E}$. We can state the result of our arguments in the form of a theorem.

Theorem III.1.4. Reachability set $R(T) \subset W_r$ of system (13) is isometric to set $\hat{R}(T) \subset \ell_r^2$ where $\hat{R}(T)$ coincides with the set of sequences $c = \{c_n\}$, $n \in \mathbb{N}$, for which the problem of moments

$$c_n = (u, e_n)_{\mathscr{U}}, \qquad n \in \mathbb{N}, \tag{24}$$

has the solution $u \in \mathscr{U}$.

When solving the problem of moments, it is convenient to transfer from space ℓ_r^2 to space ℓ^2. Let $\{x_n\} \in \ell_r^2$. Set $\tilde{x}_n = x_n(\lambda_n + \alpha)^{r/2}$, $\tilde{e}_n = e_n(\lambda_n + \alpha)^{r/2}$,

$$\tilde{R}(T) = \{\{\tilde{x}_n\} \in \ell^2 \mid \{x_n\} \in \hat{R}(T)\}.$$

Corollary III.1.5. *Set* $\tilde{R}(T) \subset \ell^2$ *coincides with the set of sequences* $\{\tilde{c}_n\}$ *for which the problem of moments*

$$\tilde{c}_n = (u, \tilde{e}_n)_{\mathcal{U}}, \qquad n \in \mathbb{N}, \tag{25}$$

is solvable in space $\mathcal{U}$.

2. Evolution equations of the second order in time

2.1. Let operator A be defined as in Section 1. Consider a differential equation

$$\frac{d^2 y(t)}{dt^2} + Ay(t) = f(t), \qquad 0 < t < T, \tag{1}$$

with initial conditions

$$y(0) = y_0, \qquad \dot{y}(0) = y_1. \tag{2}$$

Expression $d^2 y/dt^2$ is meant in the sense of the distributions (generalized functions) with the values in the scale of spaces W_r.

We can define the solution of the problem (1), (2) in the same way as for problem (3), (4) of Section III.1. Let $f \in L^2(0, T; W_r)$, $y_0 \in W_{r+1}$, $y_1 \in W_r$. Function y is said to be a solution to equation (1) with initial conditions (2) if

(i) $y \in C(0, T; W_{r+1})$, $\dot{y} \in C(0, T; W_r)$;
(ii) the sum $\ddot{y} + Ay$ belongs to $L^2(0, T; W_r)$ and (1) is valid as an equality of elements of this space; and
(iii) conditions (2) are understood as the equalities of the elements of spaces W_{r+1} and W_r, respectively.

Theorem III.2.1. *Let* $f \in L^2(0, T; W_r)$, $\{y_0, y_1\} \in \mathcal{W}_{r+1}$, $\mathcal{W}_{r+1} := W_{r+1} \oplus W_r$. *Then problem (1), (2) has a unique solution, and the mapping*

$$\{f, y_0, y_1\} \mapsto \{y, \dot{y}\}$$

of space $L^2(0, T; W_r) \times \mathcal{W}_{r+1}$ *to space* $C(0, T; \mathcal{W}_{r+1})$ *is continuous.*

PROOF. As in the proof of Theorem 1.1, we use the Fourier method here. Let

$$
\begin{cases}
f(t) = \displaystyle\sum_{n=1}^{\infty} f_n(t)\varphi_n, & \displaystyle\sum_{n=1}^{\infty} \|f_n\|_{L^2(0,T)}^2 (\lambda_n + \alpha)^r < \infty \\[3mm]
y_0 = \displaystyle\sum_{n=1}^{\infty} y_n^0 \varphi_n, & \displaystyle\sum_{n=1}^{\infty} |y_n^0|^2 (\lambda_n + \alpha)^{r+1} < \infty \\[3mm]
y_1 = \displaystyle\sum_{n=1}^{\infty} y_n^1 \varphi_n, & \displaystyle\sum_{n=1}^{\infty} |y_n^1|^2 (\lambda_n + \alpha)^r < \infty.
\end{cases}
\tag{3}
$$

We search for function y in the form

$$
y(t) = \sum_{n=1}^{\infty} y_n(t)\varphi_n.
\tag{4}
$$

Substituting these expansions to (1), (2), we obtain

$$
\ddot{y}_n(t) + \lambda_n y_n(t) = f_n(t),
\tag{5}
$$

$$
y_n(0) = y_n^0, \qquad \dot{y}_n(0) = y_n^1, \qquad n \in \mathbb{N}.
\tag{6}
$$

The solutions of equations (5) with initial conditions (6) are

$$
y_n(t) = y_n^0 \cos\sqrt{\lambda_n}\,t + y_n^1 \frac{\sin\sqrt{\lambda_n}\,t}{\sqrt{\lambda_n}}
$$

$$
+ \int_0^t f_n(\tau)\frac{\sin\sqrt{\lambda_n}\,(t-\tau)}{\sqrt{\lambda_n}}\,d\tau \quad \text{for } \lambda_n > 0,
\tag{7}
$$

$$
y_n(t) = y_n^0 \cosh\sqrt{-\lambda_n}\,t + y_n^1 \frac{\sinh\sqrt{-\lambda_n}\,t}{\sqrt{-\lambda_n}}
$$

$$
+ \int_0^t f_n(\tau)\frac{\sinh\sqrt{-\lambda_n}\,(t-\tau)}{\sqrt{-\lambda_n}}\,d\tau \quad \text{for } \lambda_n < 0,
\tag{7'}
$$

$$
y_n(t) = y_n^0 + y_n^1 t + \int_0^t f_n(\tau)(t-\tau)\,d\tau \quad \text{for } \lambda_n = 0.
\tag{7''}
$$

From here

$$
\dot{y}_n(t) = -\sqrt{\lambda_n}\,y_n^0 \sin\sqrt{\lambda_n}\,t + y_n^1 \cos\sqrt{\lambda_n}\,t
$$

$$
+ \int_0^t f_n(\tau)\cos\sqrt{\lambda_n}\,(t-\tau)\,d\tau \quad \text{for } \lambda_n > 0,
\tag{8}
$$

$$\dot{y}_n(t) = -\sqrt{-\lambda_n}\, y_n^0 \sinh \sqrt{-\lambda_n}\, t + y_n^1 \cosh \sqrt{-\lambda_n}\, t$$

$$+ \int_0^t f_n(\tau) \cosh \sqrt{-\lambda_n}\, (t - \tau)\, d\tau \quad \text{for } \lambda_n < 0, \tag{8'}$$

$$\dot{y}_n(t) = y_n^1 + \int_0^t f_n(\tau)\, d\tau \quad \text{for } \lambda_n = 0. \tag{8''}$$

Treating functions $\cos \sqrt{\lambda}\, t$ and $\sin \sqrt{\lambda}\, t / \sqrt{\lambda}$ functions of complex argument and setting

$$\frac{\sin \sqrt{\lambda}\, t}{\sqrt{\lambda}} = t \quad \text{for } \lambda = 0,$$

one is able to combine formulas (7), (7'), (7'') into one formula (7), and (8), (8'), (8'') into one formula (8). These understandings are also used below.

Let us show that the function constructed by formulas (4), (7), and (8) is the desired solution to the problem (1), (2). At first, on the basis of equality

$$\|y(t)\|_{W_{r+1}}^2 = \sum_{n=1}^{\infty} |y_n(t)|^2 (\lambda_n + \alpha)^{r+1},$$

we check that $y(t) \in W_{r+1}$ for all $t \in [0, T]$. Using the estimate

$$\left| \frac{\sin \sqrt{\lambda}\, t}{\sqrt{\lambda}} \right|^2 (\lambda + \alpha) \prec 1$$

for $\lambda \in (-\alpha, \infty)$, $t \in [0, T]$, we find from (7) that

$$|y_n(t)|^2 (\lambda_n + \alpha)^{r+1} \prec |y_n^0|^2 (\lambda_n + \alpha)^{r+1} + |y_n^1|^2 (\lambda_n + \alpha)^r$$

$$+ \|f_n\|_{L^2(0, T)}^2 (\lambda_n + \alpha)^r, \quad n \in \mathbb{N}, t \in [0, T]. \tag{9}$$

Summing up these inequalities over n, we get

$$\|y(t)\|_{W_{r+1}}^2 \prec \|y_0\|_{W_{r+1}}^2 + \|y_1\|_{W_r}^2 + \|f\|_{L^2(0, T; W_r)}^2, \quad t \in [0, T]. \tag{10}$$

Moreover, equalities (8) provide

$$|\dot{y}_n(t)|^2 (\lambda_n + \alpha)^r \prec |y_n^0|^2 (\lambda_n + \alpha)^r |\lambda_n| + |y_n^1|^2 (\lambda_n + \alpha)^r$$

$$+ \|f_n\|_{L^2(0, T)}^2 (\lambda_n + \alpha)^r, \quad n \in \mathbb{N}, t \in [0, T]. \tag{11}$$

Since $\lambda_n + \alpha \geq \nu > 0$, $|\lambda_n| \prec \lambda_n + \alpha$. Therefore, function

$$\dot{y}(t) := \sum_{n=1}^{\infty} \dot{y}_n(t) \varphi_n$$

($\dot{y}_n(t)$ is determined by (8)) satisfies the estimate

$$\| \dot{y}(t) \|_{W_r}^2 \prec \| y_0 \|_{W_{r+1}}^2 + \| y_1 \|_{W_r}^2 + \| f \|_{L^2(0,T;W_r)}^2 \tag{12}$$

for all $t \in [0, T]$.

The other steps of the proof are similar to those carried out in Theorem 1.1. In particular, estimates (11), (12) imply

$$\| \{y, \dot{y}\} \|_{C(0,T;\mathscr{W}_{r+1})}^2 \prec \| y_0 \|_{W_{r+1}}^2 + \| y_1 \|_{W_r}^2 + \| f \|_{L^2(0,T;W_r)}^2. \tag{13}$$

Theorem 1 is proved.

Note that the theorem is precise in the same sense as Theorem 1.1: if $\lambda_n \to \infty$ then for any $p > r + 1$ it is possible to find $f \in L^2(0, T; W_r)$ such that inclusion $y(T) \in W_p$ is invalid.

It will be necessary to use Theorem 1 – as well as Theorems 1.1 and 1.2 concerning the solutions with values in the entire scale of spaces W_r – in our studies of control problems for systems of parabolic and hyperbolic types. Depending on the kind of control and the dimension of the domain, we apply these theorems with varying values of $r \in \mathbb{R}$.

2.2. Let U be a Hilbert space, $\mathscr{U} = L^2(0, T; U)$, B be a linear bounded operator from U to W_r. Consider a control system

$$\frac{d^2 y(t)}{dt^2} + Ay(t) = Bu(t), \qquad 0 < t < T, u \in \mathscr{U}, \tag{14}$$

with initial conditions

$$y(0) = y_0, \qquad \dot{y}(0) = y_1, \qquad \{y_0, y_1\} \in \mathscr{W}_{r+1}. \tag{15}$$

Theorem 1 allows us to define the reachability set $\mathscr{R}(T, y_0, y_1)$ of system (14) properly:

$$\mathscr{R}(T, y_0, y_1) = \{\{y(T), \dot{y}(T)\} \in \mathscr{W}_{r+1} \mid u \in \mathscr{U}\}.$$

Each element of set $\mathscr{R}(T, y_0, y_1)$ may be uniquely represented in the form

$$y(T) = \sum_{n=1}^{\infty} y_n(T) \varphi_n, \qquad \| y(T) \|_{W_{r+1}} = \| \{y_n(T)\} \|_{r+1},$$

$$\dot{y}(T) = \sum_{n=1}^{\infty} \dot{y}_n(T) \varphi_n, \qquad \| \dot{y}(T) \|_{W_r} = \| \{\dot{y}_n(T)\} \|_r.$$

Therefore, set $\mathcal{R}(T, y_0, y_1)$ is isometric to set $\hat{\mathcal{R}}(T, \{y_n^0\}, \{y_n^1\}) \subset \ell_{r+1}^2 \oplus \ell_r^2$ consisting of pairs of sequences $(\{y_n(T)\}, \{\dot{y}_n(T)\})$ corresponding to $\{y(T), \dot{y}(T)\} \in \mathcal{R}(T, y_0, y_1)$.

By analogy with (15), Section III.1, we introduce operators $\mathscr{S}(T)$ and $\mathscr{K}(T)$ such that

$$\begin{pmatrix} y(T) \\ \dot{y}(T) \end{pmatrix} = \mathscr{S}(T) \begin{pmatrix} y_0 \\ y_1 \end{pmatrix} + \mathscr{K}(T)u.$$

Here,

$$\mathscr{S}(T) \in \mathscr{L}(\mathscr{W}_{r+1}, \mathscr{W}_{r+1}), \qquad \mathscr{K}(T) \in \mathscr{L}(\mathscr{U}, \mathscr{W}_{r+1}).$$

The explicit form of these operators is easily presented by analogy with (16), (17) of Section III.1, starting from formulas (7), (8).

As in Section 1, we are interested first in the set $\mathcal{R}(T) := \mathcal{R}(T, 0, 0)$, which is the image of operator $\mathscr{K}(T)$, and in the isometric to it, set

$$\hat{\mathcal{R}}(T) := \hat{\mathcal{R}}(T, 0, 0).$$

We use $B \in \mathscr{L}(U, W_r)$ to define operator $B^* \in \mathscr{L}(W_{-r}, U)$ with the help of the relation

$$\langle Bv, \psi \rangle_* = (v, B^*\psi)_U, \qquad v \in U, \psi \in W_{-r}.$$

Formulas (7), (8) yield the elements of set $\hat{\mathcal{R}}(T)$ to be determined by control $u \in \mathscr{U}$ according to (compare with (21) of Section III.1)

$$y_n(T) = \int_0^T \left(u(t), \frac{\sin\sqrt{\lambda_n}(T - t)}{\sqrt{\lambda_n}} B^*\varphi_n \right)_U dt \qquad (16)$$

$$\dot{y}_n(T) = \int_0^T (u(t), \cos\sqrt{\lambda_n}(T - t) B^*\varphi_n)_U \, dt. \qquad (17)$$

Let us introduce in space $\mathscr{U}$ families of functions $\{\zeta_n^{(1)}\}$ and $\{\zeta_n^{(2)}\}$, $n \in \mathbb{N}$,

$$\zeta_n^{(1)}(t) = \frac{\sin\sqrt{\lambda_n}\, t}{\sqrt{\lambda_n}} B^*\varphi_n, \qquad (18)$$

$$\zeta_n^{(2)}(t) = \cos\sqrt{\lambda_n}\, t B^*\varphi_n. \qquad (19)$$

By changing variable $t' = T - t$ in the integrals in (16), (17), we reduce the problem of description of the system (14) reachability set $\mathcal{R}(T)$ to the problem of moments relative to the family of functions $\{\zeta_n^{(1)}\} \cup \{\zeta_n^{(2)}\}$. The obtained results may be formulated as the following theorem.

Theorem III.2.2. Reachability set $\mathcal{R}(T)$, $\mathcal{R}(T) \subset \mathcal{W}_{r+1}$, of system (14) is isometric to set $\hat{\mathcal{R}}(T) \subset \ell_{r+1}^2 \oplus \ell_r^2$, where $\hat{\mathcal{R}}(T)$ coincides with the set of pairs of sequences $(\{a_n\}, \{b_n\})$ for which the problem of moments

$$a_n = (u, \zeta_n^{(1)})_{\mathcal{U}}, \qquad b_n = (u, \zeta_n^{(2)})_{\mathcal{U}}, \quad n \in \mathbb{N}, \tag{20}$$

has the solution $u \in \mathcal{U}$.

For further investigations, it is convenient to represent moment relations (20) in a somewhat modified form. For this, we set $\mathbb{K} := \mathbb{Z} \setminus \{0\}$ and introduce space $\tilde{\ell}_r^2$ of sequences $\{c_k\}$, $k \in \mathbb{K}$, with the norm

$$\|\{c_k\}\|_r := \left(\sum_{k \in \mathbb{K}} |c_k|^2 (\lambda_{|k|} + \alpha)^r \right)^{1/2}.$$

We write

$$\kappa = \inf_{n \in \mathbb{N}} \{|\lambda_n| \mid \lambda_n \neq 0\},$$

$$\mathbb{K}_0 = \begin{cases} \{k \in \mathbb{K} \mid \lambda_{|k|} = 0\} & \text{if } \kappa > 0, \\ \{k \in \mathbb{K} \mid |\lambda_{|k|}| < 1\} & \text{if } \kappa = 0. \end{cases}$$

In the applications to control problems for differential equations of mathematical physics considered in this book, $\lambda_n \to +\infty$ and hence $\kappa > 0$. In the case $\kappa = 0$ (0 is the spectrum condensation point) in the definition of set $\mathbb{K}_0$, one may consider inequality $|\lambda_{|k|}| < \rho$, instead of $|\lambda_{|k|}| < 1$, with any positive number ρ.

Set

$$\omega_n = \begin{cases} \sqrt{\lambda_n}, & \lambda_n \geq 0 \\ -i\sqrt{-\lambda_n}, & \lambda_n < 0, \end{cases} \qquad \omega_{-n} = -\omega_n, n \in \mathbb{N}.$$

Associate with the pair of sequences $(\{a_n\}, \{b_n\})$ sequence $\{c_k\}$ according to formulas

$$\begin{cases} c_k = -i\omega_k a_{|k|} + b_{|k|} & \text{for } k \in \mathbb{K} \setminus \mathbb{K}_0 \\ c_{|k|} = a_{|k|}, c_{-|k|} = b_{|k|} & \text{for } k \in \mathbb{K}_0 \end{cases}. \tag{21}$$

One verifies directly that mapping $(\{a_n\}, \{b_n\}) \mapsto \{c_k\}$ given by formulas (21) is an isomorphism of spaces $\ell_{r+1}^2 \oplus \ell_r^2$ and $\tilde{\ell}_r^2$.

Instead of the families $\{\zeta_n^{(1)}\}$ and $\{\zeta_n^{(2)}\}$, let us consider in space $\mathcal{U}$ family $\mathscr{E} = \{e_k\}$, $k \in \mathbb{K}$:

$$\begin{cases} e_k(t) = e^{i\omega_k t} B^* \varphi_{|k|}, & k \in \mathbb{K} \setminus \mathbb{K}_0, \\ e_{|k|} = \zeta_{|k|}^{(1)}, e_{-|k|} = \zeta_{|k|}^{(2)}, & k \in \mathbb{K}_0. \end{cases} \tag{22}$$

Multiplying the first of the equalities (20) for $n \in \mathbb{K} \backslash \mathbb{K}_0$ by $\pm i\omega_n$ and adding the result to the second ones, we obtain moment equalities

$$c_k = (u, e_k)_{\mathcal{U}}, \qquad k \in \mathbb{K}. \qquad (23)$$

Thus from Theorem 2 and the subsequent arguments we can draw the following conclusion.

Theorem III.2.3. Reachability set $\mathcal{R}(T) \subset \mathcal{W}_{r+1}$ *of system* (14) *is isomorphic to the set of sequences* $\{c_k\}$ *in space* $\tilde{\ell}_r^2$ *for which the problem of moments* (23) *has the solution* $u \in \mathcal{U}$.

By multiplying both parts of equalities (23) by $(\lambda_{|k|} + \alpha)^{r/2}$, as in the case (24), (25) of Section III.1 we can move to the problem of moments for sequences from the space $\tilde{\ell}^2 := \tilde{\ell}_0^2$.

In comparison with Theorem 1, the peculiar properties of family $\mathscr{E}$ may sometimes provide an additional regularity for the solution to problem (14), (15). (This fact is often exploited in Chapters V and VII.)

Lemma III.2.4. Again, let $B \in \mathscr{L}(U, W_r)$ *and for some* $p > r$ *let an estimate be valid*

$$\sum_{k \in \mathbb{K}} |(u, e_k)_{L^2(0,t;\,U)}|^2 (\lambda_{|k|} + \alpha)^p \prec \|u\|^2_{L^2(0,t;\,U)}, \qquad (24)$$

$$t \in [0, T], \qquad u \in \mathcal{U} = L^2(0, T; U).$$

If $\{y_0, y_1\} \in \mathcal{W}_{p+1}$, *then the solution of problem* (14), (15) *satisfies the inclusion* $\{y, \dot{y}\} \in C(0, T; \mathcal{W}_{p+1})$ *and*

$$\|\{y, \dot{y}\}\|^2_{C(0,T;\,\mathcal{W}_{p+1})} \prec \|\{y_0, y_1\}\|^2_{\mathcal{W}_{p+1}} + \|u\|^2_{\mathcal{U}}. \qquad (25)$$

PROOF. By analogy with formula (21), set

$$\begin{cases} z_k(t) = -i\omega_k y_{|k|}(t) + \dot{y}_{|k|}(t), & k \in \mathbb{K} \backslash \mathbb{K}_0 \\ z_{|k|}(t) = y_{|k|}(t), \ z_{-|k|}(t) = \dot{y}_{|k|}(t), & k \in \mathbb{K}_0 \end{cases}$$

$$\begin{cases} z_k^0 = -i\omega_k y_{|k|}^0 + y_{|k|}^1, & k \in \mathbb{K} \backslash \mathbb{K}_0 \\ z_{|k|}^0 = y_{|k|}^0, \ z_{-|k|}^0 = y_{|k|}^1, & k \in \mathbb{K}_0. \end{cases}$$

Formulas (7), (8), (16), and (17) produce

$$|z_k(t)|^2 \prec |z_k^0|^2 + |(u^t, e_k)_{L^2(0,t;\,U)}|^2, \qquad k \in \mathbb{K}, \ u^t(\tau) := u(t - \tau).$$

Therefore,

$$\|\{y(t), \dot{y}(t)\}\|^2_{\mathcal{W}_{p+1}} \asymp \sum_{k\in\mathbb{K}} |z_k(t)|^2 (\lambda_{|k|} + \alpha)^{p+1} \prec \sum_{k\in\mathbb{K}} |z_k^0|^2 (\lambda_{|k|} + \alpha)^{p+1}$$

$$+ \sum_{k\in\mathbb{K}} |(u^t, e_k)_{L^2(0,t;\,U)}|^2 (\lambda_{|k|} + \alpha)^{p+1}$$

$$\prec \|\{y_0, y_1\}\|^2_{\mathcal{W}_{p+1}} + \|u\|^2_{L^2(0,t;\,U)}.$$

The continuity of $\{y(t), \dot{y}(t)\}$ in t in the norm of space $\mathcal{W}_{p+1}$ is established on the grounds of estimate (24), as in the case of the continuity of $x(t)$ in Theorem 1.1.

The analogous statement holds for problem (13), (14) of Section III.1 as well.

3. Controllability types and their relationship with exponential families

3.1. We now return to the question of the controllability of the system

$$\frac{dx(t)}{dt} + Ax(t) = Bu(t), \qquad 0 < t < T, \tag{1}$$

$$u \in \mathscr{U} = L^2(0, T; U), \qquad B \in \mathscr{L}(U, W_{r-1}),$$

with initial condition $x(0) = 0$.

Our aim is to analyze reachability set $R(T)$, which we determine to be the image of operator $K(T)$ (see (15), Section III.1). As shown in Section III.1, operator $K(T) \in \mathscr{L}(\mathscr{U}, W_r)$ is of the form

$$K(T)u = x(T) = \sum_{n=1}^{\infty} c_n(T)\varphi_n, \tag{2}$$

where

$$c_n(T) = (u^T, e_n)_{\mathscr{U}}$$

$$e_n(t) = e^{-\lambda_n t} B^* \varphi_n, \qquad u^T(t) = u(T - t).$$

Let H_0 be a Hilbert space densely and continuously embedded into W_r, which contains all the eigenfunctions φ_n; in particular, H_0 may coincide with W_r. Space H_0 figures in the definition of controllability types for system (1) presented below.

Definition III.3.1. System (1) is said to be

(a) *B*-controllable relative to H_0 in time T, if $R(T) = H_0$;
(b) *E*-controllable relative to H_0 in time T, if $R(T) \supset H_0$;
(c) *UM*-controllable relative to H_0 in time T, if for any $n \in \mathbb{N}$ there may be found a control $u_n \in \mathscr{U}$ such that $K(T)u_n = \varphi_n$ and $\|u_n\|_{\mathscr{U}} < \|\varphi_n\|_{H_0}$, $n \in \mathbb{N}$;
(d) *M*-controllable in time T, if for any $n \in \mathbb{N}$ there may be found a control $u_n \in \mathscr{U}$ such that $K(T)u_n = \varphi_n$; and
(e) *W*-controllable in time T, if $\mathrm{Cl}_{W_r} R(T) = W_r$.

Until now, two types of controllability have been most often considered in the literature. These types are usually called exact and approximate controllability. In our terms, they correspond to E and W controllability, respectively. The suggested definitions sharpen the existing classification. With respect to exact controllability, we recognize "the best" case when we have a complete description of reachability set $R(T)$ in the form of its coincidence with the space H_0. This is B controllability. For approximate controllability, it is physically meaningful to separate three cases:

(1) when for any $n \in \mathbb{N}$ it is possible to transfer the system in time T from the zero state to the state φ_n with the help of a control, whose norm is estimated uniformly in n by the "energy" norm of φ_n, and this is the *UM* controllability;
(2) when for any $n \in \mathbb{N}$ it is possible to transfer the system in time T from the zero state to the state φ_n, but without the guaranteed upper bound of the control norms (*M* controllability);
(3) when in time T it is still possible, up to the error of an arbitrarily small norm, to reach any state in W_r, but we cannot guarantee that all the states φ_n are reachable from the zero state (*W* controllability).

In the last case, the set $R(T)$, although it is dense in W_r, does not contain the linear span of the eigenfunctions of operator A. In our view, this case has to be considered a nonphysical one. For instance, under finite dimensional control of the process of heat propagation in a rectangle domain, W controllability takes place depending on whether the aspect ratio of a rectangle is a rational number.

Note also that the standard separation of controllability in the exact and the approximate case is somewhat conventional. For example, if a system has the *UM*-controllability property relative to some space H_0, one easily constructs space H_1 relative to which the system is

E-controllable (see Theorem 5 below). Moreover, if a system is W-controllable, it is B-controllable relative to some space H_0 (Theorem 6). However, one is far from being able to describe this space in standard terms at all times.

Further, we demonstrate (Theorem 3) that our classification of controllability types is closely related to the properties of vector families considered in Chapter I, which influence the solvability of the moment problem. This circumstance clarifies the notations used in the definitions. In the remaining chapters, we present many examples of parabolic and hyperbolic systems with various types of controllability.

Remark III.3.2. It is possible to consider controllability in relation to a space H_0 nondensely embedded into W_r. This makes sense, for instance, when $R(T)$ is a proper subspace of W_r. An example of this is the reachability set for string with the boundary control at small T (Avdonin and Ivanov 1983) (for details see Chapters V and VII; see also Remark VII.3.6).

3.2. Let us establish a relation between the introduced controllability types and the properties of vector exponential families $\mathscr{E} = \{e^{-\lambda_n t} B^* \varphi_n\}$ in space $\mathscr{U} = L^2(0, T; U)$. In doing this, we restrict ourselves to the space H_0 of a more specific kind.

Let $\rho = \{\rho_n\}_{n=1}^{\infty}$, $\rho_n > 0$. Let $W(\rho_n)$ denote a closure of finite sums $\sum c_n \varphi_n$ in the norm $(\sum |c_n|^2 \rho_n^2)^{1/2}$. Take H_0 in the form $W(\rho_n)$. To make inclusion $H_0 \subset W_r$ hold, let us require relation

$$\rho_n^2 \succ (\lambda_n + \alpha)^r, \ n \in \mathbb{N},$$

to be valid. Note that with the help of new notations space W_r may be written as $W(r_n)$, $r_n := (\lambda_n + \alpha)^{r/2}$.

Along with the family $\mathscr{E} = \{e_n\}$, let us introduce families $\tilde{\mathscr{E}} = \{r_n e_n\}$ and $\mathscr{E}_0 = \{\rho_n e_n\}$ differing from the first one by normalization, and consider isometric operators

$$\tilde{\mathfrak{U}}: \ell^2 \mapsto W_r, \qquad \tilde{\mathfrak{U}}(\{a_n\}_{n=1}^{\infty}) = \sum_{n=1}^{\infty} \frac{a_n}{r_n} \varphi_n,$$

$$\mathfrak{U}_0: \ell^2 \mapsto H_0, \qquad \mathfrak{U}_0(\{a_n\}_{n=1}^{\infty}) = \sum_{n=1}^{\infty} \frac{a_n}{\rho_n} \varphi_n.$$

Formula (2) implies immediately that operator $K(T)$ may be represented via the problem of moments operators $\mathscr{J}_{\tilde{\mathscr{E}}}$ and $\mathscr{J}_{\mathscr{E}_0}$ (for the definition of the problem of moments operator, see Section I.2):

$$K(T) = \tilde{\mathfrak{U}} \mathscr{I}_{\tilde{\mathscr{E}}} V \tag{3}$$

$$K(T)|_{V^{-1}\mathscr{D}_{\mathscr{E}_0}} = \mathfrak{U}_0 \mathscr{I}_{\mathscr{E}_0} V|_{V^{-1}\mathscr{D}_{\mathscr{E}_0}}, \tag{4}$$

where V is an isomorphism of space $\mathscr{U}: (Vf)(t) = f(T - t)$.

Theorem III.3.3. Let $H_0 = W(\rho_n)$. The following assertions are then true:

(a) *System* (1) *is B-controllable relative to H_0 in time T if and only if $\mathscr{E}_0 \in (LB)$ in space $\mathscr{U} = L^2(0, T; U)$.*

(b) *If system* (1) *is E-controllable relative to H_0 in time T, then $\mathscr{E}_0 \in (UM)$ in $\mathscr{U}$.*

(c) *System* (1) *is UM-controllable relative to H_0 in time T, if and only if $\mathscr{E}_0 \in (UM)$ in $\mathscr{U}$.*

(d) *System* (1) *is M-controllable in time T if and only if $\mathscr{E}_0 \in (M)$ in $\mathscr{U}$.*

(e) *System* (1) *is W-controllable in time T if and only if $\tilde{\mathscr{E}} \in (W)$ in $\mathscr{U}$.*

PROOF. The assertions of the theorem follow from Theorem I.2.1 and representations (3), (4).

(a) *B* controllability is equivalent, since $\mathfrak{U}$ and V are isometric, to the fact that operator $\mathscr{I}_{\mathscr{E}_0}$ is defined on the whole $\mathscr{U}$ and its image coincides with ℓ^2. By the closed graph theorem, operator $\mathscr{I}_{\mathscr{E}_0}|_{\bigvee \mathscr{E}_0}$ is then an isomorphism between $\bigvee \mathscr{E}_0$ and ℓ^2, which is equivalent (by Theorem I.2.1(a)) to the basis property of $\mathscr{E}_0$. The arguments are invertible.

(b) *E* controllability in relation to H_0 is equivalent to the equality $R_{\mathscr{E}_0} = \ell^2$, which, by Theorem I.2.1(b), yields $\mathscr{E}_0 \in (UM)$.

(c) *UM* controllability relative to H_0 means that for any $n \in \mathbb{N}$ there exists $u_n \in \mathscr{U}$ satisfying relations $Ku_n = \varphi_n$, $\|u_n\|_{\mathscr{U}} < \|\varphi_n\|_{H_0}$. Let $\{\zeta_n\}$ denote the standard basis in ℓ^2. Using representation (4) and equality $\mathfrak{U}_0^{-1}\varphi_n = \rho_n\zeta_n$, we obtain $\mathscr{I}_{\mathscr{E}_0}(v_n\rho_n^{-1}\} = \zeta_n$, $v_n := Vu_n$. Therefore, family $\{v_n\rho_n^{-1}\}_{n=1}^{\infty}$ is biorthogonal to $\mathscr{E}_0$. Since

$$\|v_n\|_{\mathscr{U}} = \|u_n\|_{\mathscr{U}} \quad \text{and} \quad \|\varphi_n\|_{H_0} = \rho_n, \quad \text{then} \quad \|v_n\rho_n^{-1}\| < 1$$

and hence $\mathscr{E}_0 \in (UM)$. These arguments are invertible.

The proof of assertion (d) is contained in the proof of assertion (c).

(e) Making use of representation (3), we find W controllability to be equivalent to $\mathrm{Cl}\, R_{\tilde{\mathscr{E}}} = \ell^2$, which, by Theorem I.2.1(d), is, in turn, equivalent to the inclusion $\tilde{\mathscr{E}} \in (W)$.

Remark III.3.4. From the proof, it is clear that B controllability of the system relative to H_0 implies the equivalence of the norms of the control and the state $x(T)$:

$$[K(T)u = x(T), u^T \in \bigvee \mathscr{E}] \Rightarrow \|u\|_{\mathscr{U}} \asymp \|x(T)\|_{H_0}.$$

(Note that the control orthogonal to all the elements of family $\mathscr{E}$ transfers the system from the zero state to the zero one.)

On the other hand, if estimate $\|u\|_{\mathscr{U}} \asymp \|x(T)\|_{H_0}$ takes place, operator $\mathscr{I}_{\mathscr{E}_0}$ is an isomorphism on its image. That is, $R(T)$ is a subspace of H_0. Together with W controllability, this provides B controllability relative to H_0.

Let us present a result concerning controllability in relation to different spaces H_0 of the form $W(\rho_n)$.

Theorem III.3.5.

(a) *Let* $\{\rho_n\}$ *and* $\{\hat{\rho}_n\}$ *be positive sequences such that* $\sum_{n=1}^{\infty} \rho_n^2 \hat{\rho}_n^{-2} < \infty$. *If system* (1) *is UM-controllable in time* T *relative to* $W(\rho_n)$, *then it is E-controllable in time* T *relative to* $W(\hat{\rho}_n)$.

(b) *Let* $\mathscr{E} \in (M)$ *in* $\mathscr{U}$; *the norms of the elements of the biorthogonal to* $\mathscr{E}$ *family* $\Theta = \{\theta_n\}$ *and sequence* $\{\rho_n\}$ *are such that* $\sup_{n \in \mathbb{N}} \|\theta_n\|_{\mathscr{U}} \rho_n^{-1} = \infty$. *Then system* (1) *is not E-controllable in time* T *relative to* $W(\rho_n)$.

PROOF.

(a) Let us show that $R(T) \supset W(\rho_n)$. From the proof of Theorem 3(c), it follows that the norms of the elements of the biorthogonal to $\mathscr{E}$ family Θ satisfy the estimate $\|\theta_n\|_{\mathscr{U}} \prec \rho_n$. Elements of family $\tilde{\Theta}$, biorthogonal to $\tilde{\mathscr{E}} = \{r_n e_n\}$, are of the form $\tilde{\theta}_n = r_n^{-1}\theta_n$, and hence $\|\tilde{\theta}_n\|_{\mathscr{U}} \prec \rho_n r_n^{-1}$. Let $x = \sum_{n=1}^{\infty} x_n \varphi_n \in W(\hat{\rho}_n)$. That is, $\sum_{n=1}^{\infty} |x_n|^2 \hat{\rho}_n^2 < \infty$. Let us prove that $x \in R(T)$. In view of (3), it is equivalent to the fact that the sequence $\{x_n r_n\}$ belongs to the image of operator $\mathscr{I}_{\mathscr{E}_0}$. By Corollary I.2.5(a), for the latter inclusion to be correct it is sufficient for series $\sum_{n=1}^{\infty} |x_n r_n| \|\tilde{\theta}_n\|_{\mathscr{U}}$ to be convergent. This is true, since

$$\sum_{n=1}^{\infty} |x_n r_n| \|\tilde{\theta}_n\|_{\mathscr{U}} \prec \sum_{n=1}^{\infty} |x_n| r_n \rho_n r_n^{-1} = \sum_{n=1}^{\infty} |x_n| \rho_n = \sum_{n=1}^{\infty} |x_n| \hat{\rho}_n \rho_n \hat{\rho}_n^{-1}$$

$$\leq \left(\sum_{n=1}^{\infty} |x_n|^2 \hat{\rho}_n^2 \right)^{1/2} \left(\sum_{n=1}^{\infty} \rho_n^2 \hat{\rho}_n^{-2} \right)^{1/2} < \infty.$$

(b) Since $\sup_{n \in \mathbb{N}} \|\theta_n\|_{\mathscr{U}} \rho_n^{-1} = \infty$, vector $a \in \ell^2$ may be found such that $\sum_{n=1}^{\infty} |a_n|^2 \|\theta_n\|_{\mathscr{U}}^2 \rho_n^{-1} = \infty$. In other words, $\sum_{n=1}^{\infty} |a_n|^2 \|\theta_n^0\|_{\mathscr{U}}^2 = \infty$, where $\{\theta_n^0\}_{n=1}^{\infty}$ is the family biorthogonal to $\mathscr{E}_0$. Then, by Corollary I.2.5(b), $R_{\mathscr{E}_0} \neq \ell^2$, and consequently $\tilde{R}(T) \neq W(\rho_n)$. The theorem is proved.

The following theorem implements for system (1) the Hilbert Uniqueness Method (HUM) suggested by J.-L. Lions (1986).

Theorem III.3.6. If system (1) is W-controllable in time T, then it is B-controllable in time T relative to a dense space H_0 in W_r, which is constructed in the proof.

PROOF. Consider, along with control system (1), the observation system dual to it:

$$\begin{cases} -\dfrac{d\varphi(t)}{dt} + A^*\varphi(t) = 0, & 0 < t < T, \\[2mm] \varphi(T) = \xi \in W_{-r}; & \upsilon(t) = B^*\varphi(t). \end{cases} \qquad (5)$$

(We are treating A and A^* as bounded operators in the scale of spaces W_r. Since $A \in \mathscr{L}(W_{r+1}, W_{r-1})$, then $A^* \in \mathscr{L}(W_{-r+1}, W_{-r-1})$.)

By Theorem 1.2, $\varphi \in L^2(0, T; W_{-r+1})$, $(d\varphi/dt) \in L^2(0, T; W_{-r-1})$; therefore, $A^*\varphi \in L^2(0, T; W_{-r-1})$.

Since $B^* \in \mathscr{L}(W_{-r+1}, U)$, then $\upsilon \in \mathscr{U} = L^2(0, T; U)$ and $\|\upsilon\|_{\mathscr{U}} \prec \|\xi\|_{W_{-r}}$.

Introduce space $\hat{H}$ as the closure of W_{-r} in the norm $\|\xi\|_{\hat{H}} := \|\upsilon\|_{\mathscr{U}}$. Let us check this equality to actually define the norm, namely, that $\upsilon = 0$ implies $\xi = 0$.

From (1), (5) we have

$$0 = \int_0^T \left\langle \frac{dx}{dt} + Ax - Bu, \varphi \right\rangle_* dt - \int_0^T \left\langle x, -\frac{d\varphi}{dt} + A^*\varphi \right\rangle_* dt$$

$$= -\int_0^T \langle Bu, \varphi \rangle_* \, dt + \langle x(T), \varphi(T) \rangle_*$$

and so

$$\langle x(T), \xi \rangle_* = (u, \upsilon)_{\mathscr{U}}. \qquad (6)$$

If $\upsilon = 0$, then the left side of (6) equals zero for all $u \in \mathscr{U}$. Since system (1) is W-controllable, the set of all $x(T)$ is dense in W_r, and therefore $\xi = 0$. (Dual system (5) is observable.) Hence, space $\hat{H}$ is defined properly and $\hat{H} \supset W_{-r}$.

Let us write H_0 for the space dual to $\hat{H}$. In view of this, H_0 is densely embedded into W_r. Let us demonstrate system (1) to be B-controllable in time T relative to H_0.

From (6), we have

$$|\langle x(T), \xi \rangle_*| \le \|u\|_{\mathcal{U}} \|v\|_{\mathcal{U}} = \|u\|_{\mathcal{U}} \|\xi\|_{\hat{H}}.$$

Then, $x(T) \in H_0$ for any $u \in \mathcal{U}$, i.e. $R(T) \subset H_0$. Let us show that inverse inclusion is also valid.

We connect system (1) with system (5) in the following way: put in (1), $u(t) = v(t)$. Then $x(T)$ depends on ξ. Define operator $\Lambda \colon \hat{H} \mapsto W_r$ according to the rule $\Lambda \xi = x(T)$. It is bounded by Theorem 1.1 and 1.2. From (6), we have

$$\langle x(T), \xi \rangle_* = (v, v)_{\mathcal{U}} = \|v\|_{\mathcal{U}}^2.$$

Hence,

$$\langle \Lambda \xi, \xi \rangle_* = \|\xi\|_{\hat{H}}^2.$$

Therefore, the Lax–Milgram theorem (Lax and Milgram 1954) makes operator Λ an isomorphism of space $\hat{H}$ on $\hat{H}' = H_0$.

Equation $\Lambda \xi = x_1$ has the solution $\xi \in \hat{H}$ for any $x_1 \in H_0$. The function $v = B^* \varphi$ constructed by this ξ (φ being the solution to problem (5)) belongs to space $\mathcal{U}$. One easily sees that function $u(t) = v(t)$ transfers system (1) from zero at $t = 0$ to x_1 at $t = T$. Therefore, $R(T) \supset H_0$. The theorem is proved.

In Theorem 6, we neither assume nor state that space H_0 has the form $W(\rho_n)$; it is usually difficult to describe. Some positive examples are presented in the following chapters.

Let us proceed now to describe set $\bigcup_{T > 0} R(T)$.

Consider, along with the family $\mathcal{E} = \{e^{-\lambda_n t} B^* \varphi_n\}$, family $\mathcal{E}_\alpha = \{e_n^\alpha\}$, $e_n^\alpha(t) = e^{-(\lambda_n + \alpha)t} B^* \varphi_n$. Since $\lambda_n + \alpha > 0$, $\mathcal{E}_\alpha \subset L^2(0, \infty; U) =: \mathcal{U}_\infty$. Now turn to set $\hat{R}_\infty$ of sequences $\{c_n\}$, $n \in \mathbb{N}$, determined by the equalities

$$c_n = (u, e_n^\alpha)_{\mathcal{E}_\infty}, \qquad u \in \mathcal{U}_\infty.$$

In the proof of Theorem 1.1, we checked that $\{x_n(T)\} \in \ell_r^2$. In exactly the same way, one is able to show that $\hat{R}_\infty \subset \ell_r^2$. Set

$$R_\infty = \left\{ \sum_{n=1}^{\infty} c_n \varphi_n \in W_r \,\middle|\, \{c_n\} \in \hat{R}_\infty \right\}.$$

Theorem III.3.7. Reachability sets of system (1) *satisfy the relations*

$$\bigcup_{T>0} R(T) \subset R_\infty, \tag{7}$$

$$\mathrm{Cl}_{W_r} \bigcup_{T>0} R(T) \supset R_\infty. \tag{8}$$

PROOF. Inclusions (7), (8) are equivalent to the following relations in the spaces of sequences

$$\bigcup_{T>0} \hat{R}(T) \subset \hat{R}_\infty, \tag{9}$$

$$\mathrm{Cl}_{\ell_r^2} \bigcup_{T>0} \hat{R}(T) \supset \hat{R}_\infty. \tag{10}$$

Since

$$(u, e_n^\alpha)_{L^2(0, T; U)} = (u\, e^{-\alpha t}, e_n)_{L^2(0, T; U)},$$

then for any finite T an equality takes place

$$\hat{R}(T) = \hat{R}_\alpha(T) := \{\{c_n\} \in \ell_r^2 \mid c_n = (u, e_n^\alpha)_{\mathscr{U}}\}. \tag{11}$$

Since $(u, e_n^\alpha)_{L^2(0, T; U)} = (u_T, e_n^\alpha)_{\mathscr{U}_\infty}$, where

$$u_T(t) = \begin{cases} u(t), & t \le T, \\ 0, & t > T, \end{cases}$$

then $R(T) \subset R_\infty$ for any T, and hence inclusion (9) is valid.

To demonstrate inclusion (10), let us take an arbitrary element $\{c_n\} \in \hat{R}_\infty$. It corresponds with the element $u \in \mathscr{U}_\infty$, which is the solution to the problem of moments

$$c_n = (u, e_n^\alpha)_{\mathscr{U}_\infty} =: \mathscr{I}_{\mathscr{E}_\alpha}^{(r)} u, \qquad \mathscr{I}_{\mathscr{E}_\alpha}^{(r)} : \mathscr{U}_\infty \mapsto \ell_r^2.$$

Operator $\mathscr{I}_{\mathscr{E}_\alpha}^{(r)}$ is defined on the whole space $\mathscr{U}_\infty$ and, as shown in Section I.2, is closed. So it is bounded, and therefore

$$\sum_{n=1}^{\infty} (\lambda_n + \alpha)^r |(u - u_T, e_n^\alpha)_{\mathscr{U}_\infty}|^2 < \|u\|_{L^2(T, \infty; U)}.$$

This implies that sequence $c_n^T := (u_T, e_n^\alpha)_{\mathscr{U}}$ is converging to $\{c_n\}$ in ℓ_r^2 as $T \to \infty$. By equality (11), $\{c_n^T\} \in \hat{R}(T)$, which just proves inclusion (10). $\blacksquare$

Corollary III.3.8. Theorem 7 *provides* $\mathrm{Cl}_{W_r} \bigcup_{T>0} R(T) = \mathrm{Cl}_{W_r} R_\infty$. *Now, on the grounds of Theorem* I.2.1(d), *we conclude that* $\mathrm{Cl}_{W_r} \bigcup_{T>0} R(T) = W_r$ *is equivalent to inclusion* $\mathscr{E}_\alpha \in (W)$ *in space* $\mathscr{U}_\infty$.

3.3. Up to now, the arguments of Section 3 were related to the control system (1). We now show that all the definitions and constructions, with only slight modifications, may be transferred to the system described by the second order in a time equation.

Let $u \in \mathcal{U} = L^2(0, T; U)$, $B \in \mathcal{L}(U, W_r)$. Consider system

$$\frac{d^2 y}{dt^2} + Ay(t) = Bu(t), \qquad 0 < t < T, \tag{12}$$

with initial conditions

$$y(0) = 0, \qquad \dot{y}(0) = 0.$$

Let us study the reachability set $\mathcal{R}(T)$ of system (12), which is the image of operator $\mathcal{K}(T)$. Operator $\mathcal{K}(T)$, $\mathcal{K}(T) \in \mathcal{L}(\mathcal{U}, \mathcal{W}_{r+1})$, $\mathcal{W}_{r+1} = W_{r+1} \oplus W_r$, is defined in Section 2.

Let $\mathcal{H}_0$ be a space densely embedded into $\mathcal{W}_{r+1}$ and containing all the functions of the form $\{\varphi_n, 0\}$ and $\{0, \varphi_n\}$, $n \in \mathbb{N}$. The analog of Definition 1 establishing the hierarchy of controllability types for system (12) is as follows.

Definition III.3.9. System (12) is said to be

(a) *B-controllable relative to $\mathcal{H}_0$ in time T*, if $\mathcal{R}(T) = \mathcal{H}_0$;
(b) *E-controllable relative to $\mathcal{H}_0$ in time T*, if $\mathcal{R}(T) \supset \mathcal{H}_0$;
(c) *UM-controllable relative to $\mathcal{H}_0$ in time T*, if for any $n \in \mathbb{N}$ there may be found controls u_n^0 and $u_n^1 \in \mathcal{U}$ such that

$$\mathcal{K}(T)u_n^0 = \{\varphi_n, 0\}, \qquad \mathcal{K}(T)u_n^1 = \{0, \varphi_n\}, \tag{13}$$

$$\|u_n^0\|_{\mathcal{U}} < \|\{\varphi_n, 0\}\|_{\mathcal{H}_0}, \qquad \|u_n^1\|_{\mathcal{U}} < \|\{0, \varphi_n\}\|_{\mathcal{H}_0};$$

(d) *M-controllable in time T*, if for any $n \in \mathbb{N}$ one is able to find controls $u_n^{0,1} \in \mathcal{U}$ for which equalities (13) hold;
(e) *W-controllable in time T*, if $\mathrm{Cl}_{\mathcal{W}_{r+1}} \mathcal{R}(T) = \mathcal{W}_{r+1}$.

To reveal the association between the introduced controllability types and the properties of exponentials defined by formulas (22), Section III.2, we confine ourselves to the space $\mathcal{H}_0$ of the form $\mathcal{W}(\rho_n)$,

$$\mathcal{W}(\rho_n) := W((\lambda_n + \alpha)^{1/2}\rho_n) \oplus W(\rho_n), \qquad \rho_n \succ r_n := (\lambda_n + \alpha)^{r/2}.$$

In these notations, $\mathcal{W}_{r+1} = \mathcal{W}(r_n)$.

Let us define isomorphism $L: \ell^2_{r+1} \oplus \ell^2_r \mapsto \tilde{\ell}^2_r$ by formulas (21), Section III.2. The inverse mapping is specified by the formulas

$$
\begin{cases}
a_n = \dfrac{c_n - c_{-n}}{2i\omega_n}, & b_n = \dfrac{c_n + c_{-n}}{2}, & n \in \mathbb{N} \cap (\mathbb{K} \setminus \mathbb{K}_0), \\
a_n = c_n, & b_n = c_{-n}, n \in \mathbb{N} \cap \mathbb{K}_0.
\end{cases}
\tag{14}
$$

We define isomorphisms $\tilde{\mathfrak{U}}: \tilde{\ell}^2 \mapsto \mathscr{W}_{r+1}$ and $\mathfrak{U}_0: \tilde{\ell}^2 \mapsto \mathscr{W}(\rho_n)$ (analogs of the operators introduced for system (1)) as

$$
\tilde{\mathfrak{U}}: \{c_k\}_{k \in \mathbb{K}} \mapsto \left\{ \sum_{n=1}^{\infty} \frac{a_n}{r_n} \varphi_n, \sum_{n=1}^{\infty} \frac{b_n}{r_n} \varphi_n \right\},
$$

$$
\mathfrak{U}_0: \{c_k\}_{k \in \mathbb{K}} \mapsto \left\{ \sum_{n=1}^{\infty} \frac{a_n}{\rho_n} \varphi_n, \sum_{n=1}^{\infty} \frac{b_n}{\rho_n} \varphi_n \right\},
$$

where coefficients a_n and b_n, $n \in \mathbb{N}$, are determined via c_k, $k \in \mathbb{K}$, by formulas (14).

Together with family $\mathscr{E} = \{e_k\}$, $k \in \mathbb{K}$, introduced in (22), Section III.2, we now consider families $\tilde{\mathscr{E}} = \{r_{|k|} e_k\}$ and $\mathscr{E}_0 = \{\rho_{|k|} e_k\}$, which differ from it by the normalization.

From the previous constructions and Section III.2, it follows that operator $\mathscr{K}(T)$ may be represented in a form identical with (3), (4):

$$
\mathscr{K}(T) = \mathfrak{U}_0 \mathscr{J}_{\tilde{\mathscr{E}}} V, \qquad \mathscr{K}(T)|_{V^{-1} \mathscr{D}_{\mathscr{E}^0}} = \mathfrak{U}_0 \mathscr{J}_{\mathscr{E}_0} V|_{V^{-1} \mathscr{D}_{\mathscr{E}^0}}.
$$

On the basis of this representation, one is able to obtain the analogs of assertions 3.3–3.6. We would like to emphasize two points in particlar.

Theorem III.3.10. *Let $\mathscr{H}_0 = \mathscr{W}(\rho_n)$. The following assertions are then true.*

(a) *System (12) is B-controllable relative to $\mathscr{H}_0$ in time T if and only if $\mathscr{E}_0 \in (LB)$ in space $\mathscr{U} = L^2(0, T; U)$.*

(b) *If system (12) is E-controllable relative to $\mathscr{H}_0$ in time T, then $\mathscr{E}_0 \in (UM)$ in $\mathscr{U}$.*

(c) *System (12) is UM-controllable relative to $\mathscr{H}_0$ in time T if and only if $\mathscr{E}_0 \in (UM)$ in $\mathscr{U}$.*

(d) *System (12) is M-controllable in time T if and only if $\mathscr{E}_0 \in (M)$ in $\mathscr{U}$.*

(e) *System (12) is W-controllable in time T if and only if $\tilde{\mathscr{E}} \in (W)$ in $\mathscr{U}$.*

Theorem III.3.11. *If system (12) is W-controllable in time T, then it is B-controllable in the same time relative to the dense space $\mathscr{H}_0$ in $\mathscr{W}_{r+1}$, whose construction is shown below.*

The proof of Theorem 3.11 follows the lines of the proof of Theorem 3.6. The system dual to (12) is

$$
\begin{cases}
\dfrac{d^2\varphi(t)}{dt^2} + A^*\psi(t) = 0, & \psi(T) = \psi_0 \in W_{-r}, \qquad \dot\psi(T) = \psi_1 \in W_{-r-1}, \\
v = B^*\psi, & B^* \in \mathscr{L}(W_{-r}, U).
\end{cases}
$$

Space $\mathscr{H}$ is defined as the closure of $\mathscr{W}_{-r} = W_{-r} \oplus W_{-r-1}$ in the norm $\|\{\psi_0, \psi_1\}\|_{\mathscr{H}} := \|v\|_{\mathscr{U}}$, while space H_0 is the dual one to $\mathscr{H}$. It is verified that this definition is correct, since system (12) is W-controllable in time T. Space $\mathscr{H}_0$ is then densely embedded into $\mathscr{W}_{r+1}$, since

$$
(\mathscr{W}_{-r})' = (W_{-r} \oplus W_{-r-1})' = (W_{-r-1})' \oplus (W_{-r})' = W_{r+1} \oplus W_r = \mathscr{W}_{r+1}.
$$

The remaining details are similar to those in Theorem 6.

To describe set $\bigcup_{T>0} \mathscr{R}(T)$, we take arbitrary $\delta > \sqrt{\alpha}$ and consider, along with the family $\mathscr{E} = \{e_k\}$, $k \in \mathbb{K}$, introduced by formulas (22), Section III.2, family $\mathscr{E}_\delta = \{e_k^\delta\}$, $e_k^\delta(t) = e^{-\delta t} e_k(t)$. Since $\lambda_n + \alpha > 0$, $\mathscr{E}_\delta \subset \mathscr{U}_\infty$, $\mathscr{U}_\infty = L^2(0, \infty; U)$. Let $\hat{\mathscr{R}}_\infty$ be a set of sequences $\{c_k\}$ determined by equalities

$$
c_k = (u, e_k^\delta)_{\mathscr{U}_\infty}, \qquad k \in \mathbb{K}, \, u \in \mathscr{U}_\infty.
$$

We write $\mathscr{R}_\infty$ for the set of pairs of functions of the form

$$
\left\{ \sum_{n=1}^{\infty} a_n \varphi_n, \; \sum_{n=1}^{\infty} b_n \varphi_n \right\},
$$

with coefficients a_n and b_n expressed via $\{c_k\} \in \hat{\mathscr{R}}_\infty$ by formulas (14).

The following theorem is proved in the same way as Theorem 3.7.

Theorem III.3.12. Inclusions are true:

$$
\bigcup_{T>0} \hat{\mathscr{R}}(T) \subset \hat{\mathscr{R}}_\infty, \qquad \mathrm{Cl}_{\tilde{\ell}_r^2} \bigcup_{T>0} \hat{\mathscr{R}}(T) \supset \hat{\mathscr{R}}_\infty;
$$

$$
\bigcup_{T>0} \mathscr{R}(T) \subset \mathscr{R}_\infty, \qquad \mathrm{Cl}_{\mathscr{W}_{r+1}} \bigcup_{T>0} \mathscr{R}(T) \supset \mathscr{R}_\infty.
$$

An analog of Corollary 8 takes place.

Corollary III.3.13. Theorems 12 *and* I.2.1(d), *imply the equivalence of* $\mathrm{Cl}_{\mathscr{W}_{r+1}} \bigcup_{T>0} \mathscr{R}(T) = \mathscr{W}_{r+1}$ *and inclusion* $\mathscr{E}_\delta \in (W)$ *in space* $\mathscr{U}_\infty$.

To conclude this chapter, we would like to present one statement about the relationship between the controllability of systems of the first and the second orders in time.

Theorem III.3.14. If system (12) *is M-controllable in time T, then system* (1) *with the same operators A and B is M-controllable in any time* $\tau > 0$.

This assertion immediately proves to be correct in view of Theorem II.5.14, Theorem 3(d), and Theorem 10(d).

IV

Controllability of parabolic-type systems

Let Ω be a bounded domain in $\mathbb{R}^N$ with the boundary Γ. We assume Γ to be regular enough to provide the existence of all the subjects considered below. In some cases, the requirements for the smoothness of Γ will be made more precise. Set $Q = \Omega \times (0, T)$, $\Sigma = \Gamma \times (0, T)$, $T > 0$. Let a_0, $a_{ij} \in L^\infty(\Omega)$, $(i, j = 1, \ldots, N)$, $a_{ij} = \overline{a_{ji}}$, and for some $\kappa > 0$ let an estimate

$$\sum_{i, j = 1}^N a_{ij}(x) \xi_i \xi_j \geq \kappa \sum_{i=1}^N \xi_i^2 \tag{0.1}$$

be valid almost everywhere in Ω for each $\{\xi_i\}_{i=1}^N \in \mathbb{R}^N$.

Recall the notations from Section III.1 and take space $L^2(\Omega)$ as H and $H_0^1(\Omega)$ as V. Then V' is $H^{-1}(\Omega)$ (Lions 1968). For φ, $\psi \in H_0^1(\Omega)$, we set

$$a[\varphi, \psi] = \sum_{i, j = 1}^N \int_\Omega a_{ij}(x) \frac{\partial \varphi}{\partial x_i} \frac{\overline{\partial \psi}}{\partial x_j} \, dx + \int_\Omega a_0(x) \varphi \overline{\psi} \, dx. \tag{0.2}$$

It is not difficult to see that the introduced bilinear form $a[\varphi, \psi]$ satisfies the conditions of Section III.1; one may choose any number larger than $\|a_0\|_{L^\infty(\Omega)}$ for α. We write A for the operator generated by form $a[\varphi, \psi]$. Its eigenfunctions φ_n, $n \in \mathbb{N}$, are the solutions to the homogeneous Dirichlet boundary-value problem for an elliptic equation

$$A\varphi_n = \lambda_n \varphi_n \quad \text{in } \Omega, \quad \varphi_n|_\Gamma = 0.$$

In what follows we may suppose that φ_n are real. From this point on, we use the same symbol A both for the operator and for the differential

173

expression

$$A\varphi = -\sum_{i,j=1}^{N} \frac{\partial}{\partial x_i}\left(a_{ij}(x)\frac{\partial \varphi}{\partial x_j}\right) + a_0(x)\varphi$$

corresponding to it.

We need the additional restrictions on Γ and coefficients a_{ij} and a_0 to make the eigenfunctions of operator A sufficiently smooth and its eigenvalue estimates

$$\lambda_n \asymp n^{2/N} \qquad (0.3)$$

large enough n to ensure the validity of some of the results in this and the next chapter. Conditions guaranteeing these properties may be found in Agmon (1965), Birman and Solomyak (1977), and Mikhailov (1979). Without special indications, we consider them true wherever necessary.

In Theorems 1.2 and 2.1, one also needs to demand coefficients a_{ij}, a_0 and boundary Γ to be of class C^1. This makes firm the result on the uniqueness of solutions of elliptic equations used there (Landis 1956; Hörmander 1976). The additional smoothness of coefficients is also required to demonstrate those assertions in Theorems 1.3, 1.7, 2.6, and 2.7 associated with the case $N = 1$, in which the asymptotics of λ_n are exploited.

In this chapter, we present the application of the approach developed in Chapter III concerning the controllability of systems described by parabolic equations. Section 1 treats the cases of control with spatial support in the domain Ω, both infinite dimensional and finite dimensional, including pointwise. In Section 2 we investigate boundary control with spatial support on the boundary Γ or on a part of it. For this we use the results of Chapter II about the properties of exponential families and a number of specific properties of equations of the parabolic type.

To explain determination, we cite the example of the Dirichlet problem and what is reflected in taking V as space $H_0^1(\Omega)$. The Neumann problem is treated similarly; one should only take V for $H^1(\Omega)$. The controllability for some other types of boundary conditions may be accomplished in the same way. The results of the studies for all the types of boundary-value problems are expressed similarly in terms of the W_r spaces introduced in Section III.1.

For various kinds of boundary conditions, spaces W_r correspond differently to the Sobolev spaces $H^r(\Omega)$ and their subspaces $H_0^r(\Omega)$. However, several common relations that also take place are described briefly.

Let $H_0^1(\Omega) \subset V \subset H^1(\Omega)$, A be a self-adjoint operator in $L^2(\Omega)$ generated by form (0.2) specified on V and $A_\alpha = A + \alpha I$, $\mathscr{D}(A_\alpha^{1/2}) = V$. As in Section III.1, let W_r, $r \in \mathbb{R}$, denote a scale of Hilbert spaces associated with operator A_α, $W_r = \mathscr{D}(A_\alpha^{r/2})$ for $r > 0$.

Depending on the smoothness of boundary Γ and coefficients a_{ij}, a_0 a number $\hat{r}$ is defined such that for all $r \in [0, \hat{r}]$ the embeddings

$$H_0^r(\Omega) \subset W_r \subset H^r(\Omega) \tag{0.4}$$

hold. For instance, for $V = W_1 = H_0^1(\Omega)$ we have

$$W_2 = \mathscr{D}(A) = H^2(\Omega) \cap H_0^1(\Omega),$$

and

$$W_4 = \mathscr{D}(A^2) = \{f \in H^4(\Omega) \mid f \in H_0^1(\Omega),\ Af \in H_0^1(\Omega)\}.$$

For all the functions $f \in W_r$, $f = \sum_{n=1}^\infty c_n \varphi_n$, relations

$$\|f\|_{H^r(\Omega)} \asymp \|f\|_{W_r}, \qquad \|f\|_{W_r} := \left[\sum_{n=1}^\infty |c_n|^2 (\lambda_n + \alpha)^r\right]^{1/2} \tag{0.5}$$

are correct. From this formula one should exclude the case of half-integer r for which the norms of $H^r(\Omega)$ and W_r are related in a more complicated manner (see, e.g., Lions and Magenes 1968: chap. 1).

For positive even r (0.4), (0.5) follow from formula (0.1) and the sufficient smoothness of the boundary and coefficients (see, for instance, Berezanskiĭ 1965). For an arbitrary positive r they are proved by means of the interpolation (Lions and Magenes 1968: chap. 1). Proceeding with the dual spaces, it is possible to show that (0.5) holds also for $r \in [-\hat{r}, 0]$, $r \neq$ a half-integer.

1. Control with spatial support in the domain

1.1. Infinite dimensional control

Let $H = L^2(\Omega)$, $V = H_0^1(\Omega)$, $V' = H^{-1}(\Omega)$, and $U = H = L^2(\Omega)$, and let B be an identity operator in H. State $y(\cdot, t)$ of the system is determined as a solution of the problem

$$\begin{cases} \dfrac{\partial y}{\partial t} + Ay = u & \text{in } Q, \\[2mm] y|_\Sigma = 0, \qquad u \in \mathscr{U} = L^2(0, T; U) = L^2(Q), \end{cases} \tag{1}$$

with the initial condition

$$y(x, 0) = 0 \qquad \text{in } \Omega. \tag{2}$$

The solution to problem (1), (2), as well as to other initial boundary-value problems in this and the next chapter is understood within the framework of the scheme described in Chapter III.

From Theorem III.1.1 (for the case $r = 1$) it follows that for every $u \in \mathcal{U}$ there exists a unique solution $y \subset C(0, T; H_0^1(\Omega))$ of problem (1), (2) (we used equality $W_1 = V$; see Section III.1).

For system (1), we have

$$\langle Bv, \varphi_n \rangle_* = (v, \varphi_n)_{L^2(\Omega)}, \qquad v \in U = L^2(\Omega).$$

Therefore, family $\mathscr{E} = \{e_n\}_{n=1}^{\infty} \subset \mathcal{U}$ introduced in Section III.1 by formula (23) in our case is of the form $e_n(x, t) = e^{-\lambda_n t} \varphi_n(x)$.

Theorem IV.1.1. System (1) *is B-controllable relative to space* $H_0^1(\Omega)$ *in any time* $T > 0$.

PROOF. By Theorem III.3.3(a), the assertion of the theorem being proved now is equivalent to condition $\mathscr{E}_0 := \{\rho_n e_n\} \in (LB)$, where $\rho_n = (\lambda_n + \alpha)^{1/2}$. In view of the orthogonality of $\{\varphi_n\}$ in $L^2(\Omega)$, families $\mathscr{E}$ and $\mathscr{E}_0$ are orthogonal in $\mathcal{U}$. It remains to check whether $\mathscr{E}_0$ is almost normed. Indeed,

$$\|\rho_n e_n\|_{\mathcal{U}}^2 = (\lambda_n + \alpha) \int_0^T e^{-2\lambda_n t}\, dt = \begin{cases} (\lambda_n + \alpha)(2\lambda_n)^{-1}(1 - e^{-2\lambda_n t}), & \lambda_n \neq 0 \\ \alpha T, & \lambda_n = 0. \end{cases}$$

Hence, $\|\rho_n e_n\|_{\mathcal{U}} \asymp 1$, $n \in \mathbb{N}$. The theorem is proved.

Consider now the case when the control acts not on the whole domain Ω but only on part of it. Let Ω' be an arbitrary nonempty subdomain of Ω, $U = L^2(\Omega')$, $\mathcal{U} = L^2(0, T; U)$.

Theorem IV.1.2. For $U = L^2(\Omega')$, *system* (1) *is W-controllable relative to space* $H_0^1(\Omega)$ *in any time* $T > 0$.

PROOF. Operator B in this case is determined by

$$\langle Bv, \varphi \rangle_* = (v, \varphi)_{L^2(\Omega')}, \qquad v \in L^2(\Omega'), \varphi \in L^2(\Omega).$$

Therefore, family $\mathscr{E}$ is written as in the previous example but with the other space $\mathcal{U}$, and family $\mathscr{E}$ is not orthogonal in it. Family $\widetilde{\mathscr{E}}$ (defined in Section III.3) coincides in form with $\mathscr{E}_0$: $\widetilde{\mathscr{E}} = \{\rho_n e^{-\lambda_n t} \varphi_n(x)\}$. By Theorem III.3.3(e), the required assertion is equivalent to the inclusion $\widetilde{\mathscr{E}} \in (W)$ in space $L^2(0, T; L^2(\Omega'))$ for any $T > 0$.

Let us recall that we write λ_n for the eigenvalues of operator A taking into account their multiplicity, so that different elements of sequence $\{\lambda_n\}$ may coincide in their numerical value. Along with this, we later use another notation for the spectrum of A as well. Let μ_n, $n \in \mathbb{N}$, be different eigenvalues of this operator. The multiplicity of eigenvalue μ_n is denoted by κ_n and corresponding eigenfunctions by φ_{nj}, $j = 1, \ldots, \kappa_n$.

Let us rewrite family $\tilde{\mathscr{E}}$ in the form $\{\tilde{e}_{nj}\}$, $\tilde{e}_{nj}(t, x) = r_n\, e^{-\mu_n t}\, \varphi_{nj}(x)$, $r_n = (\mu_n + \alpha)^{1/2}$. Suppose that for some sequence $\{c_{nj}\} \in \ell^2$

$$\sum_{n=1}^{R} \sum_{j=1}^{\kappa_n} c_{nj}\tilde{e}_{nj} \xrightarrow[R \to \infty]{} \quad \text{in } L^2(0, T; L^2(\Omega')).$$

Then for any function $f \in L^2(\Omega')$ and every $g \in L^2(0, T)$,

$$\sum_{n=1}^{\infty} \sum_{j=1}^{\kappa_n} c_{nj}r_n(\varphi_{nj}, f)_{L^2(\Omega')}(e^{-\mu_n t}, g)_{L^2(0, T)} = 0.$$

Thus

$$\sum_{n=1}^{R} e^{-\mu_n t}\, r_n \sum_{j=1}^{\kappa_n} c_{nj}(\varphi_{nj}, f)_{L^2(\Omega')} \xrightarrow[R \to \infty]{} 0 \quad \text{in } L^2(0, T).$$

By Theorem II.6.3 and estimates (0.3) it follows that

$$\sum_{j=1}^{\kappa_n} c_{nj}(\varphi_{nj}, f)_{L^2(\Omega')} = 0 \quad \text{for all } n \in \mathbb{N}.$$

Because f is arbitrary, we find

$$\psi_n = \sum_{j=1}^{\kappa_n} c_{nj}\varphi_{nj} = 0 \quad \text{in } L^2(\Omega'), n \in \mathbb{N}.$$

Together with conditions $A\psi_n = \mu_n\psi_n$ in Ω, $\psi_n|_\Gamma = 0$, this implies $\psi_n = 0$, in Ω, $n \in \mathbb{N}$ (see Landis 1956; Hörmander 1976). Since functions φ_{nj} are orthogonal in $L^2(\Omega)$, the latter is possible if only $\{c_{nj}\} = 0$.

We have demonstrated that $\tilde{\mathscr{E}} \in (W)$ in $L^2(0, T; L^2(\Omega'))$ for any $T > 0$ and thus proved Theorem 2.

As can be seen from the proof, the theorem is also valid for a class of so-called bilinear controls $u(x, t) = f(x)g(t)$; $f \in L^2(\Omega')$, $g \in L^2(0, T)$, f and g being controls, that is narrower than $L^2(0, T; L^2(\Omega'))$.

1.2. Finite dimensional control

In contrast with the preceding subsection, let

$$U = \mathbb{C}^m, \qquad Bu(t) = \sum_{p=1}^{m} b_p u_p(t), \qquad b_p \in L^2(\Omega), \, u_p \in L^2(0, T).$$

The system state is determined as the solution to the problem

$$\begin{cases} \dfrac{\partial y}{\partial t} + Ay = Bu & \text{in } Q, \\[2mm] y|_\Sigma = 0, & u \in \mathcal{U} = L^2(0, T; \mathbb{C}^m), \end{cases} \tag{3}$$

with initial condition (2).

In this case, $\langle Bu(t), \varphi_n \rangle_* = (u(t), \eta_n)_{\mathbb{C}^m}$, where η_n is a vector from $\mathbb{C}^m$ with the components $\overline{\beta_n^p}$, $\beta_n^p := (b_p, \varphi_n)_{L^2(\Omega)}$. Therefore, family $\mathscr{E}$ has the form

$$\{\eta_n \exp(-\lambda_n t)\}, \qquad n \in \mathbb{N}.$$

Accounting for the multiplicity of the eigenvalues (as in Theorem 2), we write family $\mathscr{E}$ as

$$\{\eta_{nj} \exp(-\mu_n t)\}, \qquad j = 1, \ldots, \kappa_n, \, n \in \mathbb{N}, \, \eta_{nj} \in \mathbb{C}^m,$$

Theorem IV.1.3. The following statements are correct.

(a) *If sequence $\{\kappa_n\}$ is unbounded, then system (3) is not W-controllable for every finite T.*

(b) *If* $\sup \kappa_n =: \kappa < \infty$, *then for any* $T > 0$, *system (3) is W-controllable if and only if* $m \geq \kappa$ *and for all* $n \in \mathbb{N}$

$$\operatorname{rank}[(b_p, \varphi_{nj})_{L^2(\Omega)}]_{j=1,\ldots,\kappa_n}^{p=1,\ldots,m} = \kappa_n. \tag{4}$$

(c) *For $N > 1$, system (3) is not M-controllable for every T.*

(d) *For $N = 1$, conditions (4) (which in this case, by the force of equalities $\kappa_n = 1$, take the form $\eta_n \neq 0$, $n \in \mathbb{N}$) constitute a necessary and sufficient condition of system (3) M controllability for any T.*

PROOF. (a), (b). By Theorem III.3.3(e), the W controllability of system (3) is equivalent to the property $\tilde{\mathscr{E}} \in (W)$. The linear independence of family $\mathscr{E}$ serves as a necessary condition for the latter and is equivalent to linear independence in $\mathbb{C}^m$ for any $n \in \mathbb{N}$, of vectors η_{nj}, $j = 1, \ldots, \kappa_n$, corresponding to a single exponential $\exp(-\mu_n t)$. If for some n $\kappa_n > m$, then these vectors are linearly dependent, and system (3) is not W-controllable for every T.

If $\kappa_n \leq m$, then linear independence of vectors η_{nj}, $j = 1, \ldots, \kappa_n$ is equivalent to relations (4). Thus, conditions (4) are necessary for system (3) to be W-controllable. Let us prove they are sufficient. For this we show conditions (4) to provide inclusion $\tilde{\mathscr{E}} \in (W)$. Suppose that for some sequence $\{a_{nj}\} \in \ell^2$ series $\sum_{n, j} a_{nj} \tilde{e}_{nj}$ converges weakly to zero in space

$$L^2(0, T; \mathbb{C}^m), \qquad \tilde{e}_{nj}(t) := (\mu_n + \alpha)^{1/2} \eta_{nj} \, e^{-\mu_n t}.$$

Introduce vectors

$$a_n := \sum_{j=1}^{\kappa_n} a_{nj}(\mu_n + \alpha)^{1/2} \eta_{nj}$$

and denote the components of vector a_n by a_n^p, $p = 1, \ldots, m$. Then series $\sum_n a_n^p \, e^{-\mu_n t}$ converges weakly to zero in $L^2(0, T)$ for all $p = 1, \ldots, m$. Therefore, by Theorem II.6.3, $a_n^p = 0$ for all $p = 1, \ldots, m$; $n \in \mathbb{N}$.

Under conditions (4) it follows that $a_{nj} = 0$ for all $i = 1, \ldots, \kappa_n$; $n \in \mathbb{N}$. Assertions (a) and (b) of the theorem are proved.

(c) Let $N > 1$ and system (3) be M-controllable in time T. Then by Theorem III.3.3(d), family $\mathscr{E}$ is minimal in $L^2(0, T; \mathbb{C}^m)$. Since mapping $f(t) \mapsto e^{-\alpha t} f(t)$ is an isomorphism of space $\mathscr{U}$, family $\mathscr{E}_\alpha := \{\eta_n e^{-(\lambda_n + \alpha)t}\}$ is also minimal in $\mathscr{U}$. As $\lambda_n + \alpha > 0$ for all n, $\mathscr{E}_\alpha \subset L^2(0, \infty, \mathbb{C}^m)$ and is minimal in this space. Theorem II.2.4 then implies set $\{i(\lambda_n + \alpha)\}$ to satisfy Blaschke condition (B) (see Subsection II.2.2.2). Therefore $\sum_n (\lambda_n + \alpha)^{-1} < \infty$, which is impossible for $N > 1$ owing to estimates (0.3).

(d) It is clear that conditions $\eta_n \neq 0$, $n \in \mathbb{N}$, are necessary for system (3) M controllability. Let us show that they are also sufficient. By the force of Theorem III.3.3(d) it is enough to demonstrate that for $\eta_n \neq 0$, $n \in \mathbb{N}$, family $\mathscr{E}$ is minimal in space $L^2(0, T; \mathbb{C}^m)$ for any $T > 0$. Since for $N = 1$, $\lambda_n \asymp n^2$, then $\sum_n (\lambda_n + \alpha)^{-1} < \infty$ and hence set $\{i(\lambda_n + \alpha)\}$ satisfies the Blaschke condition. Therefore by Theorem II.2.4, family $\mathscr{E}_\alpha$ is minimal in space $L^2(0, \infty, \mathbb{C}^m)$ and then, by Proposition II.4.28, family $\mathscr{E}_\alpha$, and with it $\mathscr{E}$, is minimal in $L^2(0, T; \mathbb{C}^m)$. The theorem is proved.

Remark IV.1.4. Results close to assertions (a) and (b) of the theorem are put forth by Sakawa (1974).

Remark IV.1.5. For the case $m = 1$, assertion (d) of the theorem has been obtained (Fattorini and Russell 1971).

Remark IV.1.6. If conditions (4) are not fulfilled, then assertions (a), (b) of Theorem 3 may be complemented by the statement

$$\text{Cl}_{L^2(\Omega)} \bigcup_{T>0} R(T) \neq L^2(\Omega).$$ (5)

To prove it, let us suppose that for some $n \in \mathbb{N}$ we have $\sum_{j=1}^{\varkappa_n} \alpha_j \eta_{nj} = 0$. The function $f = \sum_{j=1}^{\varkappa_n} \alpha_j \varphi_{nj}$ is orthogonal to $R(T)$ for all T. Indeed, let $g \in R(T)$, then it may be represented as (see formula (22) of Section III.1)

$$g = \sum_{n,j} c_{nj} \varphi_{nj}, \qquad c_{nj} = (u, \eta_{nj} e^{-\mu_n(T-t)})_{\mathcal{U}}.$$

Therefore,

$$(g, f)_{L^2(\Omega)} = \sum_{j=1}^{\varkappa_n} c_{nj} \bar{\alpha}_j = \left(u, \sum_{j=1}^{\varkappa_n} \alpha_j \eta_{nj} e^{-\mu_n(T-t)}\right)_{\mathcal{U}} = 0.$$

1.3. Pointwise control

Let the system state now be determined as the solution to the initial boundary-value problem

$$\begin{cases} \dfrac{\partial y(x, t)}{\partial t} + Ay(x, t) = \sum_{p=1}^{m} u_p(t)\delta(x - x_p) & \text{in } Q, \\ y|_{\Sigma} = 0; \quad x_p \in \Omega, \quad u_p \in L^2(0, T), \quad p = 1, \dots, m; \end{cases}$$ (6)

$$y(x, 0) = 0 \quad \text{in } Q.$$ (7)

First of all, we clarify which space can be used to define this solution properly. To make use of the results of Section III.1, we represent the right-hand side of the first of the equations (6) in the form $Bu(t)$, $u(t) \in U = \mathbb{C}^m$ and find out for which q operator $B: U \mapsto W_q$ is bounded. Operator B is specified by the equality

$$\langle Bu(t), \varphi \rangle_* = \sum_{p=1}^{m} u_p(t)\overline{\varphi(x_p)}.$$ (8)

Therefore,

$$|\langle Bu(t), \varphi \rangle_*| \prec \|u(t)\|_{\mathbb{C}^m} \|\varphi\|_{C(\bar{\Omega})}.$$ (9)

Let domain Ω satisfy conditions providing the validity of the embedding theorem (Nikol'skiĭ 1969): for $s > N/2$ we have $H^s(\Omega) \subset C(\bar{\Omega})$ and

$$\|\varphi\|_{C(\bar{\Omega})} \prec \|\varphi\|_{H_s(\Omega)}.$$

Then (9) and (0.5) imply that $B \in \mathscr{L}(U, W_q)$ for $q < -N/2$. From Theorem III.1.1 we now conclude that for any function $u \in \mathscr{U} = L^2(0, T; \mathbb{C}^m)$ there exists a unique solution of problem (6), (7) such that $y \in C(0, T; W_r)$ for $r < -N/2 + 1$.

Thus the control problem (6) may be treated within the framework of the general scheme formulated in Chapter III. To study its controllability, we exploit the results of Chapter II. In view of (8), family $\mathscr{E}$ has in this case the form $\{e_{nj}\}$, $e_{nj}(t) = \eta_{nj} e^{-\mu_n t}$, where μ_n, $n \in \mathbb{N}$, are different eigenvalues of operator A, similar to the preceding subsection, while η_{nj}, $j = 1, \ldots, \kappa_n$, are vectors from $\mathbb{C}^m$ with the components $\varphi_{nj}(x_p)$, $p = 1, \ldots, m$. Family $\tilde{\mathscr{E}}$ (see Section III.3) is of the form $\{r_n e_{nj}\}$, $r_n = (\mu_n + \alpha)^r$, $r < -N/2 + 1$.

Theorem IV.1.7. *For system (6) all the assertions of Theorem 3 are valid with the replacement of equalities (4) by*

$$\operatorname{rank}[\varphi_{nj}(x_p)]_{j=1,\ldots,\kappa_n}^{p=1,\ldots,m} = \kappa_n. \tag{10}$$

The proof of this theorem repeats exactly that of Theorem 3 and is based on Theorems III.3.3, II.2.4, and II.6.3

Remark IV.1.8. For system (6) the statement of Remark 6 remains correct, with conditions (4) replaced by (10). The proof runs smoothly without any modifications.

2. Boundary control

2.1. Infinite dimensional control

Let Γ_1 be a nonempty, relatively open subset of boundary Γ, $\Gamma_0 := \Gamma \backslash \Gamma_1$. Set $\Sigma_0 = \Gamma_0 \times (0, T)$, $\Sigma_1 = \Gamma_1 \times (0, T)$ and consider a system

$$\begin{cases} \dfrac{\partial y}{\partial t} + Ay = 0 & \text{in } Q, \\ y|_{\Sigma_0} = 0, \quad y|_{\Sigma_1} = u; \end{cases} \tag{1}$$

$$u \in \mathscr{U} = L^2(0, T; U) = L^2(0, T; L^2(\Gamma_1)) = L^2(\Sigma_1).$$

As earlier, we take the zero initial condition

$$y(x, 0) = 0 \quad \text{in } \Omega. \tag{2}$$

Initial boundary-value problem (1), (2) falls into the scheme of Chapter III if one specifies operator B as

$$\langle Bu(t), \varphi \rangle_* = - \int_{\Gamma_1} u(s, t) \frac{\overline{\partial \varphi(s)}}{\partial v} \, ds. \tag{3}$$

Here,

$$\frac{\partial \varphi(s)}{\partial v} = \sum_{i,j=1}^{N} a_{ij}(s) \frac{\partial \varphi(s)}{\partial x_i} \cos(n(s), x_j),$$

where $n(s)$ is the outward normal vector to boundary Γ at the point s.

Such a definition of the solution to problem (1), (2) has a simple justification. If the problem has a classical solution, then integrating by parts one can easily show that it satisfies an integral identity

$$\int_Q \left(-\frac{\partial f}{\partial t} + Af \right) y \, dx \, dt = - \int_{\Xi_1} u \frac{\partial f}{\partial v} \, ds \, dt \tag{4}$$

for any smooth function $f(x, t)$ such that $f(\cdot, T) = 0$, $f|_{\Xi_1} = 0$. For a generalized solution this identity gives operator B of the form (3).

Let us find out the value of q for which $B \in \mathcal{L}(U, W_q)$. Since, by the condition, $u(\cdot, t) \in L^2(\Gamma_1)$, and, by the trace theorem (Lions and Magenes 1968: chap. I) $\partial \varphi / \partial v \in L^2(\Gamma)$ for $\varphi \in H^p(\Omega)$, $p > 3/2$, then one concludes from (3) that $B \in \mathcal{L}(U, W_q)$ for $q < -3/2$. Theorem III.1.1 implies the existence, for any function $u \in \mathcal{U}$, of the unique solution to (1), (2) such that $y \in C(0, T; W_r)$ for $r < -1/2$.

According to the general scheme presented in Chapter III, the study of system (1) controllability reduces to the investigation of properties of a vector exponential family. Formulas (3) and (23) of Section III.1 yield family $\mathcal{E} = \{e_n\} \subset \mathcal{U}$ of the form

$$e_n(s, t) = \frac{\partial \varphi_n(s)}{\partial v} e^{-\lambda_n t}, \qquad n \in \mathbb{N} \tag{5}$$

(For convenience, we omit the sign $-$ as not affecting the properties of family $\mathcal{E}$ that we are interested in.) Family $\tilde{\mathcal{E}}$ (see Section III.3) has the form $\{r_n e_n\}$ with $r_n = (\lambda_n + \alpha)^{r/2}$, $r < -1/2$.

Theorem IV.2.1. *System* (1) *is W-controllable in any time* $T > 0$.

PROOF. By (e) of Theorem III.3.3, the assertion to verify is equivalent to the inclusion $\tilde{\mathcal{E}} \in (W)$ in space $L^2(0, T; L^2(\Gamma_1))$ for any $T > 0$. Accounting for the multiplicity of eigenvalues λ_n (as we did in Section 1), we represent family $\tilde{\mathscr{E}}$ as

$$\tilde{\mathscr{E}} = \{\tilde{e}_{nj}\}, \qquad \tilde{e}_{nj}(s, t) = (\mu_n + \alpha)^{r/2} \frac{\partial \varphi_{nj}(s)}{\partial v} e^{-\mu_n t}, \qquad j = 1, \dots, \kappa_n, n \in \mathbb{N}.$$

Suppose for some sequence $\{c_{nj}\} \in \ell^2$

$$\sum_{n=1}^{R} \sum_{j=1}^{\kappa_n} c_{nj} \tilde{e}_{nj} \xrightarrow[R \to \infty]{} 0 \qquad \text{in } L^2(0, T; L^2(\Gamma_1)).$$

Then for any function $f \in L^2(\Gamma_1)$ and any $g \in L^2(0, T)$

$$\sum_{n=1}^{\infty} \sum_{j=1}^{\kappa_n} c_{nj} (\mu_n + \alpha)^{r/2} \left(\frac{\partial \varphi_{nj}}{\partial v}, f \right)_{L^2(\Gamma_1)} (e^{-\mu_n t}, g)_{L^2(0, T)} = 0.$$

In other words,

$$\sum_{n=1}^{R} e^{-\mu_n t} (\mu_n + \alpha)^{r/2} \sum_{j=1}^{\kappa_n} c_{nj} \left(\frac{\partial v_{nj}}{\partial v}, f \right)_{L^2(\Gamma_1)} \xrightarrow[R \to \infty]{} 0 \qquad \text{in } L^2(0, T).$$

By Theorem II.6.3 we find from here that for all $n \in \mathbb{N}$

$$\sum_{j=1}^{\kappa_n} c_{nj} \left(\frac{\partial \varphi_{nj}}{\partial v}, f \right)_{L^2(\Gamma_1)} = 0.$$

Therefore, since f is arbitrary,

$$\sum_{j=1}^{\kappa_n} c_{nj} \frac{\partial \varphi_{nj}}{\partial v} = 0 \qquad \text{in } L^2(\Gamma_1), n \in \mathbb{N}. \tag{6}$$

It is known that conditions

$$A\varphi = \lambda\varphi \qquad \text{in } \Omega, \qquad \varphi|_{\Gamma_1} = \frac{\partial \varphi}{\partial v}\bigg|_{\Gamma_1} = 0,$$

imply $\varphi = 0$ in Ω. This result had been obtained by Landis (1956) for the case when coefficients a_{ij}, a_0 of operator A and boundary Γ belong to the class C^2 and was generalized later (Hörmander 1976) for class C^1. Hence, from (6) it follows that

$$\sum_{j=1}^{\kappa_n} c_{nj} \varphi_{nj} = 0 \qquad \text{in } \Omega, n \in \mathbb{N}.$$

This, in view of the orthogonality of φ_{nj} in $L^2(\Omega)$, provides $c_{nj} = 0$ for all j and n. Thus, we have demonstrated that $\tilde{\mathscr{E}} \in (W)$ in $L^2(0, T; L^2(\Gamma_1))$. The theorem is proved.

Remark IV.2.2. It is seen from the proof that the theorem holds even for a class of bilinear controls narrower than $\mathcal{U}$:

$$u(s, t) = f(s)g(t), \qquad f \in L^2(\Gamma_1),\ g \in L^2(0, T).$$

Remark IV.2.3. W controllability of system (1) is equivalent to the following uniqueness theorem: conditions

$$\begin{cases} -\dfrac{\partial z}{\partial t} + Az = 0 \qquad \text{in } Q \\[2mm] z|_{\Sigma} = 0, \qquad \dfrac{\partial z}{\partial v}\bigg|_{\Sigma_1} = 0, \qquad z|_{t=T} \in H^q(\Omega), \qquad q > 1/2, \end{cases} \tag{7}$$

imply $z = 0$ in Q.

This equivalence is demonstrated with the help of integration by parts (see Lions 1968: chap. 3). Indeed, together with (1), (2) consider an initial boundary-value problem

$$-\frac{\partial \xi}{\partial t} + A\xi = 0 \qquad \text{in } Q,$$

$$\xi|_{\Sigma} = 0, \qquad \xi|_{t=T} \in \psi, \qquad \psi \in H^q(\Omega), \qquad q > 1/2.$$

The following chain of equalities holds (all the further integrals are understood in the sense of action of functionals $\langle \cdot, \cdot \rangle_*$, and Theorems III.1.1 and III.1.2 are exploited):

$$0 = \int_Q \left(\frac{\partial y}{\partial t} + Ay \right) \xi \, dx \, dt = \int_Q \left(-\frac{\partial \xi}{\partial t} + A\xi \right) y \, dx \, dt$$

$$- \int_\Sigma y \frac{\partial \xi}{\partial v} \, ds \, dt + \int_\Omega y(x, T)\psi(x) \, dx.$$

So

$$\int_{\Sigma_1} u(s, t) \frac{\partial \xi(s, t)}{\partial v} \, ds \, dt = \int_\Omega y(x, T)\psi(x) \, dx. \tag{8}$$

Let $\int_\Omega y(x, T)\psi(x) \, dx = 0$ for all $u \in L^2(\Sigma_1)$. Then (8) provides

$$\frac{\partial \xi}{\partial v}\bigg|_{\Sigma_1} = 0.$$

If uniqueness theorem (7) is correct, then $\xi = 0$ in Q. Hence, $\psi = \xi|_{t=T} = 0$, and system (1) is W-controllable.

Now, let function z satisfy the conditions stated in (7). Then, instead of (8), we have

$$\int_\Omega y(x, T)z(x, T)\, dx = 0 \qquad \text{for all } u \in L^2(\Sigma_1).$$

If system (1) is W-controllable, this gives $z(\cdot, T) = 0$ and, consequently $z = 0$ in Q.

Thus, our approach to the study of controllability allows us to obtain uniqueness theorem (7) as well. Theorems of such a kind have been proved by Mizohata (1958).

Remark IV.2.4. For the case when boundary Γ and coefficients a_{ij}, a_0 belong to the class C^∞, the statement of Theorem 1 may be found in Schmidt and Weck (1978).

Remark IV.2.5. If set Γ_1 is large enough, one is able to prove a "stronger" controllability for system (1). In Chapter V, B controllability is independently proved for a hyperbolic system differing from (1) by the replacement of $\partial y/\partial t$ by $\partial^2 y/\partial t^2$. The result is valid if set Γ_1 and coefficients of operator A satisfy the conditions of Theorem V.2.6. The other form of conditions that provide this result was obtained by Bardos, Lebeau, and Rauch (1992) and formulated after Corollary V.2.7. In both cases, it follows from Theorem III.3.14 that system (1) is M-controllable in any time $T > 0$.

Remark IV.2.6. The result of Remark 5 can be formulated in another interesting form. Let us consider system (1) with nonzero initial condition

$$y(x, 0) = y_0 \qquad \text{in } \Omega, \qquad y \in L^2(\Omega).$$

System (1) is said to be null-controllable in time T if for any $y_0 \in L^2(\Omega)$ there exists control $u \in \mathcal{U} = L^2(0, T; L^2(\Gamma_1))$ such that $y(x, T) = 0$. A review of some results on null controllability can be found in Russell (1978), who also shows that the problem can be reduced to the problem of moments

$$c_n e^{-\lambda_n T} = (u, e_n)_\mathcal{U}, \qquad n \in \mathbb{N},$$

where

$$\{c_n\} \in \ell^2, \qquad e_n(s, t) = \frac{\partial \varphi_n(s)}{\partial \nu} e^{-\lambda_n t}, \qquad \mathscr{E} = \{e_n\} \subset \mathcal{U}.$$

The problem of moments has the formal solution

$$u = \sum_{n=1}^{\infty} c_n e^{-\lambda_n T} \theta_n,$$

where $\{\theta_n\}$ is the family biorthogonal to $\mathscr{E}$. If the conditions discussed in Remark 5 are satisfied and the corresponding hyperbolic system is B-controllable (or E-controllable relative to some $\mathscr{W_r}$), then it follows from Theorems III.3.10 and II.5.14 that the series converges in $\mathscr{U}$ and so system (1) is null controllable. Note that null controllability for the heat equation $y_t = \Delta y$ with control acting on a subdomain $\Omega' \subset \Omega$ or on a part of the boundary $\Gamma_1 \subset \Gamma$ has been proved by Lebeau and Robbiano (1994).

For some special cases results dual to null controllability (i.e., observability and prediction results) were obtained by Mizel and Seidman (1969, 1972); and by Seidman (1976, 1977).

2.2. Finite dimensional control

Consider a system described by the initial boundary-value problem

$$\begin{cases} \dfrac{\partial y}{\partial t} + Ay = 0 & \text{in } Q, \\[2mm] y(s, t)|_{\Sigma} = \displaystyle\sum_{p=1}^{m} u_p(t) g_p(s); & \end{cases} \tag{9}$$

$$y(x, 0) = 0 \quad \text{in } \Omega. \tag{10}$$

Here, $u_p \in L^2(0, T)$, $g_p \in L^2(\Gamma)$, $p = 1, \ldots, m$.

System (9), (10) is a particular case of system (1), (2) with $\Gamma_1 = \Gamma$, $U = \mathbb{C}^m$. Therefore, the statement of Subsection 2.1 concerning the existence of a unique solution $y \in C(0, T; W_r)$, $r < 1/2$, remains to be correct for problem (9), (10). Formula (3) for this case provides

$$\langle Bu(t), \varphi_n \rangle_* = -\sum_{p=1}^{m} u_p(t) \int_{\Gamma} g_p(s) \overline{\frac{\partial \varphi_n(s)}{\partial \nu}} \, ds = (u(t), \eta_n)_{\mathbb{C}^m}, \tag{11}$$

where η_n are vectors from $\mathbb{C}^m$ with the components $-\overline{\beta_n^p}$

$$\beta_n^p := \left(g_p, \frac{\partial \varphi_n}{\partial \nu} \right)_{L^2(\Gamma)}, \qquad p = 1, \ldots, m, \ m \in \mathbb{N}.$$

So vector families $\mathscr{E}$ and $\tilde{\mathscr{E}}$, on whose properties system (8) controllability depends, are

$$\mathscr{E} = \{\eta_n\, e^{-\lambda_n t}\}, \qquad \tilde{\mathscr{E}} = \{(\lambda_n + \alpha)^{r/2}\eta_n\, e^{-\lambda_n t}\},$$

or, with the account for the eigenvalue multiplicity (see Theorem 1.2),

$$\mathscr{E} = \{\eta_{nj}\, e^{-\mu_n t}\}, \qquad \tilde{\mathscr{E}} = \{(\mu_n + \alpha)^{r/2}\eta_{nj}\, e^{-\mu_n t}\}, \qquad j = 1, \ldots, \kappa_n, \; n \in \mathbb{N}.$$

Theorem IV.2.7. All the assertions of Theorem 1.3 and Remark 1.6 are valid for system ($\mathcal{S}$) *providing that conditions* (4) *of Section IV.1 are replaced by*

$$\operatorname{rank}\left[\left(g_p, \frac{\partial \varphi_{nj}}{\partial v}\right)_{L^2(\Gamma)}\right]_{j=1,\ldots,\kappa_n}^{p=1,\ldots,m} = \kappa_n. \tag{12}$$

The proof of this theorem is similar to the proof of Theorem 1.3 and Remark 1.6.

Remark IV.2.8. It is possible to repeat word by word Remarks 1.4 and 1.5 regarding Theorem 7.

2.3. Pointwise control

Let a system be described by the initial boundary-value problem

$$\begin{cases} \dfrac{\partial y}{\partial t} + Ay = 0 & \text{in } Q, \\[2mm] y(s, t)|_\Sigma = \displaystyle\sum_{p=1}^m u_p(t)\delta(s - s_p); \end{cases} \tag{13}$$

$$u_p \in L^2(0, T), \qquad s_p \in \Gamma, \qquad p = 1, \ldots, m, \qquad y(x, 0) \qquad \text{in } \Omega, \tag{14}$$

which is understood within the framework of the scheme of Chapter III. For this, we set (compare with (8) of Section IV.1 and (3))

$$\langle Bu(t), \varphi \rangle_* = -\sum_{p=1}^m u_p(t)\frac{\overline{\partial \varphi(s_p)}}{\partial v}, \qquad U = \mathbb{C}^m. \tag{15}$$

As in Subsection 1.3, one has to find out for which q the inclusion $B \in \mathscr{L}(U, W_q)$ holds. From (15) it follows that

$$|\langle Bu(t), \varphi \rangle_*| \le \|u(t)\|_{\mathbb{C}^m}\left\|\frac{\partial \varphi}{\partial v}\right\|_{C(\Gamma)} \le \|u(t)\|_{\mathbb{C}^m}\|\varphi\|_{C^1(\bar{\Omega})}.$$

Just as in Subsection 1.3, we make use of the embedding theorem (Nikol'skiĭ 1969) to estimate the right-hand side of this inequality by

$$\|u(t)\|_{C^m}\|\varphi\|_{H^l(\Omega)}, \qquad l > 1 + N/2.$$

Therefore, $B \in \mathscr{L}(U, W_q)$ for $q < -1 - N/2$. Hence, by Theorem III.1.1, for any function $u \in \mathscr{U} = L^2(0, T; \mathbb{C}^m)$ there exists a unique solution of problem (13), (14) such that $y \in C(0, T; W_r)$ for any $r < -N/2$.

Formula (15) implies also the reduction of the study of system (13) controllability to the investigation of the properties of families

$$\mathscr{E} = \{\eta_{nj} e^{-\mu_n t}\}, \qquad \tilde{\mathscr{E}} = \{(\mu_n + \alpha)^{r/2}\eta_{nj} e^{-\mu_n t}\},$$

$$r < -N/2, \qquad j = 1, \dots, \kappa_n, \qquad n \in \mathbb{N},$$

where η_{nj} are vectors from $\mathbb{C}^m$ with the components

$$\frac{\partial\varphi_{nj}(s_p)}{\partial v}, \qquad p = 1, \dots, m.$$

Theorem IV.2.9. All the statements of Theorem 1.3 and Remark 1.6, remain true for system (13), providing that conditions (4) of Section IV.1 are replaced by

$$\operatorname{rank}\left[\frac{\partial\varphi_{nj}(s_p)}{\partial v}\right]_{j=1,\dots,\kappa_n}^{p=1,\dots,m} = \kappa_n. \tag{16}$$

The proof is similar to that of Theorem 1.3 and Remark 1.6.

Remark IV.2.10. As mentioned at the beginning of this chapter, along with the Dirichlet problem, one is able to treat the Neumann problem or the third boundary-value problem (different types of boundary conditions on the different parts of the boundary are also admissible). The controllability study for these problems proceeds just as it does for the Dirichlet problem, and the analogs of the statements proved in this chapter are valid. In the case of Neumann boundary control, the form of family $\mathscr{E}$ changes: factor $\varphi_n(s)$ instead of $\partial\varphi_n(s)/\partial v$ becomes involved.

Remark IV.2.11. For the case of a one-dimensional domain Ω ($N = 1$), operator $(-A)$ is prescribed by the differential expression

$$\frac{\partial}{\partial x}\left(a(x)\frac{\partial y}{\partial x}\right) - a_0(x)y \tag{17}$$

(and homogeneous Dirichlet boundary conditions). It is also possible to consider an operator defined by the differential expression

$$a(x)\frac{\partial^2 y}{\partial x^2} - a_0(x)y,$$

which proves to be self-adjoint in space $L^2_{1/a}(\Omega)$. For such an operator, the assertions of this and the next chapter about the controllability of corresponding systems for $N = 1$ remain true; one should only substitute scalar products of $L^2(\Omega)$ for scalar products with the weight $1/a$.

In the same way, we can replace the "models" of systems

$$\frac{\partial y}{\partial t} + Ay \quad \text{and} \quad \frac{\partial^2 y}{\partial t^2} + Ay$$

by

$$\rho(x)\frac{\partial y}{\partial t} + Ay \quad \text{and} \quad \rho(x)\frac{\partial^2 y}{\partial t^2} + Ay$$

with $\rho \in C(\bar{\Omega})$ and $\rho(x) > 0$ in $\bar{\Omega}$ in all the problems considered in Chapters IV and V.

Remark IV.2.12. Controllability problems for systems described by parabolic-type equations with time delays were investigated in Avdonin and Gorshkova (1986, 1987, 1992), whereas optimal control problems for such system were investigated in Wang (1975).

V

Controllability of hyperbolic-type systems

In this chapter we study the controllability of systems described by partial differential equations of the hyperbolic type under various control actions. The basic notations, such as Ω, Γ, Q, A, φ_n, and λ_n bear the same meaning as in Chapter IV. We are mainly concerned with the homogeneous and nonhomogeneous Dirichlet problem: problems with other boundary conditions may be treated similarly.

1. Control with spatial support in the domain

1.1. Infinite dimensional control

Consider a system determined by the initial boundary-value problem

$$\begin{cases} \dfrac{\partial^2 y}{\partial t^2} + Ay = u & \text{in } Q, \\[2mm] y|_\Sigma = 0, & u \in \mathscr{U} = L^2(Q), \end{cases} \tag{1}$$

with zero initial conditions

$$y(x, 0) = y_t(x, 0) = 0 \qquad \text{in } \Omega. \tag{2}$$

Solutions to problem (1), (2), and to other problems of this chapter are understood as in Section III.2. Initial boundary-value problem (1), (2) becomes immersed in the scheme from Section III.2 if set $H = L^2(\Omega)$, $V = H_0^1(\Omega)$, $U = L^2(\Omega)$, B being an identity operator in H. As it follows from Theorem III.2.1, for any $u \in \mathscr{U}$ problem (1), (2) has a unique solution such that

$$\{y, y_t\} \in C(0, T; \mathscr{W}_1), \qquad \mathscr{W}_1 = H_0^1(\Omega) \oplus L^2(\Omega).$$

Corresponding to the system (1) family of exponentials

$$\mathscr{E} = \{e_k\} \subset \mathscr{U} = L^2(0, T; L^2(\Omega)), \qquad k \in \mathbb{K} = \mathbb{Z}\backslash\{0\},$$

has the form (see Section III.2 and Subsection IV.1.1)

$$e_k(x, t) = \varphi_{|k|}(x)\, e^{i\omega_k t}, \qquad (\omega_k^2 = \lambda_{|k|}, \omega_{-k} = -\omega_k).$$

The fact is taken into account here that the point zero may be, by the well-known property of elliptic operators, only an isolated eigenvalue of operator A. Throughout this chapter we assume that all the eigenvalues of operator A differ from zero. If $\lambda_j = 0$ for some $j \in \mathbb{N}$, then terms $\varphi_{|j|}(x)$ and $t\varphi_{|j|}(x)$ are present in family $\mathscr{E}$ (see formulas (18), (19), (22) of Section III.2). As mentioned in Section II.5 (Remark II.5.8), the investigation of the properties of such families may be accomplished within the framework of the described scheme, and the assumption made does not influence the degree of generality of conclusions about the controllability of the considered systems.

Theorem V.1.1. System (1) *is B-controllable relative to space $\mathscr{W}_1$ in any time $T > 0$.*

PROOF. By Theorem III.3.10, the assertion we are proving is equivalent to inclusion $\mathscr{E} \in (LB)$ in space $L^2(0, T; L^2(\Omega))$ for any $T > 0$. Since family $\{\varphi_n\}, n \in \mathbb{N}$, constitutes an orthonormal basis in space $L^2(\Omega)$, the elements of family $\mathscr{E}$ corresponding to different values of $|k|$ are orthogonal. For any function $f \in L^2(Q)$, representation

$$f(x, t) = \sum_{n=1}^{\infty} f_n(t)\varphi_n(x), \qquad \|f\|_{L^2(Q)}^2 = \sum_{n=1}^{\infty} \|f_n\|_{L^2(0, T)}^2,$$

takes place. Therefore inclusion $\mathscr{E} \in (LB)$ is equivalent to the statement: for any functions f_n of the form

$$f_n(t) = \alpha_n\, e^{i\omega_n t} + \alpha_{-n}\, e^{-i\omega_n t}$$

uniform in $n \in \mathbb{N}$ estimate $\|f_n\|_{L^2(0, T)}^2 \asymp |\alpha_n|^2 + |\alpha_{-n}|^2$ holds. This, in turn, is equivalent to the inequality

$$\sup_{n \in \mathbb{N}} \frac{|(\xi_n^+, \xi_n^-)_{L^2(0, T)}|}{\|\xi_n^+\|_{L^2(0, T)} \|\xi_n^-\|_{L^2(0, T)}} < 1, \qquad \xi_n^{\pm}(t) := e^{\pm i\omega_n t}. \tag{3}$$

We denote the expression under the sign sup by d_n. Since functions ξ_n^+ and ξ_n^- are linearly independent for any n then $d_n < 1$. It is easy to calculate that

$$d_n = \frac{\sin \omega_n T}{\omega_n T} \quad \text{for } \lambda_n > 0.$$

By the force of (0.3), Chapter IV, $\lambda_n \to \infty$ and we obtain inequality (3). Theorem I is proved.

Now let Ω' be an arbitrary nonempty subdomain of domain Ω, and control $u(x, t)$ differ from zero only for $x \in \Omega'$.

Theorem V.1.2. If $U = L^2(\Omega')$, then reachability sets $\mathscr{R}(T)$ of system (1) satisfy equality

$$\mathrm{Cl}_{\mathscr{W}_1} \bigcup_{T > 0} \mathscr{R}(T) = \mathscr{W}_1. \tag{4}$$

Moreover, this equality is valid for a class of bilinear controls $u(x, t) = f(x)g(t)$; $f \in L^2(\Omega')$, $g \in L^2(0, T)$ (f and g being controls), that is narrower than $L^2(0, T; L^2(\Omega'))$.

PROOF. For the proof of the theorem, we need notations μ_n, κ_n, and φ_{nj} for spectral characteristics of operator A (namely, the eigenvalues, their multiplicities, and eigenfunctions), introduced in Section IV.1 and regularly exploited in Chapter IV. Set $v_n^2 = \mu_n$, $v_{-n} = -v_n$, and represent family $\mathscr{E}$ corresponding to system (1) for $U = L^2(\Omega')$ in the form

$$\{e_{kj}\} \subset \mathscr{U} = L^2(0, T; L^2(\Omega')), \qquad e_{kj}(x, t) = \varphi_{|k|j}(x)\, e^{iv_k t},$$

$$j = 1, \ldots, \kappa_{|k|}, \qquad k \in \mathbb{K}.$$

Let us also introduce family $\mathscr{E}_\delta \subset \mathscr{U}_\infty = L^2(0, \infty; L^2(\Omega'))$ (see Theorem III.3.12) consisting of the elements $\varphi_{|k|j}(x)\, e^{i(v_k + i\delta)t}$, $\delta > \sqrt{\alpha}$ (α is defined in the beginning of Chapter IV). By Corollary III.3.13, (4) is equivalent to inclusion $\mathscr{E}_\delta \in (W)$ in space $\mathscr{U}_\infty$. Further, we follow the line of arguments from the proof of Theorem IV.1.2. Suppose that for some squarely summable sequence $\{c_{kj}\}$, any function $f \in L^2(\Omega')$ and any $g \in L^2(0, \infty)$ the equality

$$\sum_{k \in \mathbb{K}} \sum_{j=1}^{\kappa_{|k|}} c_{kj} (\varphi_{|k|j}, f)_{L^2(\Omega')} (e^{i(v_k + i\delta)t}, g)_{L^2(0, \infty)} = 0 \tag{5}$$

holds. Agmon (1965) has shown that κ_n increases no faster than some positive power of λ_n. Taking estimate $\lambda_n \asymp n^{2/N}$ into account, for some

$R > 0$ we obtain

$$\left| \sum_{j=1}^{\kappa_{|k|}} c_{kj}(\varphi_{|k|j}, f)_{L^2(\Omega')} \right| \prec |k|^R$$

(here f is fixed). Therefore, equality (5) and Theorem II.6.4 imply

$$\sum_{j=1}^{\kappa_{|k|}} c_{kj}(\varphi_{|k|j}, f)_{L^2(\Omega')} = 0$$

for all $k \in \mathbb{K}$. As in the proof of Theorem IV.1.2, from this we derive $\{c_{kj}\} = 0$. That is why $\mathscr{E}_\delta \in (W)$ in $\mathscr{U}_\infty$.

The proof of the assertion concerning bilinear controls can easily be obtained in the same way as the proof of Theorem III.3.7. Theorem 2 is proved.

Specific properties of the hyperbolic equations were almost not used in the proofs of Theorems 1 and 2. Let us briefly describe some results of other approaches that use these properties.

Sharp sufficient conditions of the B controllability of system (1) with control on Ω' have been obtained (Bardos, Lebeau, and Rauch 1988a, 1988b). Roughly speaking, the system is B-controllable in time T (in space $\mathscr{W}_1$) if every ray crosses $\Omega' \times (0, T)$. If there is a ray that does not cross $\overline{\Omega'} \times [0, T]$, then the system is not B-controllable in time T.

It is easy to see that the system may not be B-controllable in any time. For example, that is the case if $A = -\Delta$, Ω is the unit disk, and Ω' is a subdomain such that $\overline{\Omega'} \subset \Omega$.

As for approximate controllability, the system turns out to be W-controllable for any subdomain Ω' in time $T > 2T^*$. Here T^* is the time of filling of the domain Ω by the waves from Ω':

$$T^* = \inf\left\{ T \mid \bigcup_{u \in L^2(0, T; L^2(\Omega'))} \text{supp } y(\cdot, T) = \bar{\Omega} \right\}.$$

This result can be proved with the help of Russell's result (Russell 1978) about connections between approximate controllability and a uniqueness theorem for the hyperbolic equations. In the case of analytic boundary and coefficients, it is the well known Holmgreen–John theorem. For the nonanalytic case, it has been recently proved by D. Tataru (1993).

The problem of M controllability for control acting on a subdomain remains open.

Hypothesis. If system (1) with control on Ω' is W-controllable in time T, then it is M-controllable in the same time.

1.2. Finite dimensional control

Let

$$U = \mathbb{C}^m, \qquad Bu(t) = \sum_{p=1}^{m} b_p u_p(t), \qquad b_p \in L^2(\Omega), \qquad u_p \in L^2(0, T),$$

$$p = 1, \ldots, m.$$

The system under consideration is described by the boundary-value problem

$$\begin{cases} \dfrac{\partial^2 y}{\partial t^2} + Ay = Bu & \text{in } Q, \\[2mm] y|_\Sigma = 0, \qquad u \in \mathcal{U} = L^2(0, T; \mathbb{C}^m), \end{cases} \tag{6}$$

with zero initial conditions (2). For initial boundary-value problem (6), (2) the assertion of the previous Subsection about the existence of a unique solution

$$\{y, y_t\} \in C(0, T; \mathcal{W}_1)$$

is, of course, true.

We write the family $\mathscr{E}$ corresponding to system (6) either as $\{\eta_{|k|} e^{i\omega_k t}\}$, $k \in \mathbb{K}$, or in the form

$$\{\eta_{|k|j} e^{iv_k t}\}, \qquad j = 1, \ldots, \varphi_{|k|}, \qquad k \in \mathbb{K},$$

where vectors η_n, $\eta_{nj} \in \mathbb{C}^m$ are the same as in Subsection IV.1.2; the components of η_n are $(b_p, \varphi_n)_{L^2(\Omega)}$, $p = 1, \ldots, m$.

Theorem V.1.3. The following statements are correct.

(a) *For $N > 1$ and every $T > 0$ system (6) is not M-controllable.*
(b) *For $N = 1$ system (6) is not M-controllable if*

$$T < T_0/m, \qquad T_0 := 2 \int_\Omega \frac{dx}{\sqrt{a(x)}}$$

[*function $a(x)$ is defined in Section IV.2, formula (17)). If $T \geq T_0/m$, then for any $r > 3/2$ one is able to choose functions b_p, $p = 1, \ldots, m$, in such a way that system (6) becomes B-controllable relative to space $\mathcal{W}_r$ (for the definitions see Section III.3)*].

(c) *If conditions*

$$\text{rank}\,[(b_p, \varphi_{nj})_{L^2(\Omega)}]_{j=1,\ldots,\kappa_n}^{p=1,\ldots,m} = \kappa_n, \qquad n \in \mathbb{N}, \tag{7}$$

are not satisfied – in particular, if sequence $\{\kappa_n\}$ is not bounded – then system (6) is not W-controllable for any T. Moreover,

$$\text{Cl}_{\mathscr{W}_1} \bigcup_{T > 0} \mathscr{R}(T) \neq \mathscr{W}_1.$$

(d) *If equalities (7) are valid, then*

$$\text{Cl}_{\mathscr{W}_1} \bigcup_{T > 0} \mathscr{R}(T) = \mathscr{W}_1. \tag{8}$$

(e) *If equalities (7) are valid and vectors η_n obey the estimate,*

$$\|\eta_n\|_{\mathbb{C}^m} \prec \exp(-\varepsilon|\omega_n|), \qquad n \in \mathbb{N},$$

with some $\varepsilon > 0$, then system (6) is W-controllable in any time $T > 0$.

PROOF.

(a) Suppose system (6) to be M-controllable for some $T > 0$. Then, by Theorem III.3.10, family $\mathscr{E}$ is minimal in space $L^2(0, T; \mathbb{C}^m)$. Therefore, family $\mathscr{E}_\delta = \{\eta_{|k|}\, e^{i(\omega_k + i\delta)t}\}$, $\delta > \sqrt{\alpha}$, is minimal in the same space as well. Consequently, family $\mathscr{E}_\delta$ is also minimal in space $L^2(0, \infty; \mathbb{C}^m)$. Then, by Theorem II.2.4, set $\{\omega_k + i\delta\}$ satisfies the Blaschke condition (see the definition in Subsection II.1.2.2) $\sum \omega_n^{-2} < \infty$, which contradicts for $N > 1$ the estimate $\lambda_n \asymp n^{2/N}$.

(b) Let $N = 1$, $T < T_0/m$, and system (6) be M-controllable in time T. Then, in view of Theorem III.3.10, $\mathscr{E} \in (M)$ in space $L^2(0, T; \mathbb{C}^m)$. For $N = 1$ all the eigenvalues of operator A are simple, so by Theorem II.5.1, family $E := \{e^{i\omega_k t}\}$, $|k| > m/2$, is minimal in $L^2(0, mT)$.

On the other hand, eigenfrequencies ω_k are known (Naimark 1969: chap. 2, n. 4.9) to satisfy the relation

$$\omega_k = \frac{2\pi k}{T_0} + \mathcal{O}(1). \tag{9}$$

Therefore, by Corollary II.4.2 family E cannot be minimal in $L^2(0, mT)$ for $mT < T_0$.

Theorem II.4.16 provides us an even stronger statement: namely, family E can be represented as a unification $E_0 \cup E_1$, in which E_0 is a Riesz basis in $L^2(0, mT)$, card $E_1 = \infty$.

Now, let $T \geq T_0/m$. Take functions b_p in the form

$$b_p(x) = \sum_{l=0}^{\infty} \frac{1}{(lm + p)^\gamma} \varphi_{lm+p}(x); \qquad p = 1, \ldots, m,$$

where γ is an arbitrary number more than $1/2$. Vectors $n^\gamma \eta_n$ then run cycle after cycle over a standard basis $\{\xi_p\}$ in $\mathbb{C}^m$. Family $\mathscr{E}$ falls into m orthogonal series:

$$\mathscr{E} = \bigcup_{p=1}^{m} \mathscr{E}_p, \qquad \mathscr{E}_p = \left\{ \frac{1}{(lm + p)^\gamma} \xi_p e^{i\omega_{lm+p}t} \right\}_{l \in \mathbb{K}}. \tag{10}$$

The minimality of $\mathscr{E}$ in $L^2(0, T; \mathbb{C}^m)$ is equivalent to the minimality of each of the families $E_p = \{e^{i\omega_{lm+p}t}\}_{l \in \mathbb{K}}$ in $L^2(0, T)$.

By the force of equality (9), family E_p is asymptotically close to family

$$\left\{ \exp\left[i\frac{2\pi}{T_0}(lm + p)t \right] \right\}_{l \in \mathbb{K}}$$

The latter becomes an orthogonal basis in $L^2(0, T_0/m)$ after being complemented by function

$$\exp\left(i\frac{2\pi}{T_0} pt \right)$$

corresponding to $l = 0$. Then by Lemma II.4.11 and Proposition II.4.13, E_p becomes a Riesz basis in $L^2(0, T_0/m)$ after any function of the form $e^{i\lambda t}$, $\lambda \notin \{\omega_{lm+p}\}_{l \in \mathbb{K}}$ is added to it. Hence, by Theorem II.4.26, $E_p \in (LB)$ in $L^2(0, T)$. Therefore, with the account for representation (10) family $\mathscr{E}$ is minimal in space $L^2(0, T; \mathbb{C}^m)$. Moreover, from (10) it follows that family $\{|k|^\gamma e_k\}_{k \in \mathbb{K}}$, $e_k \in \mathscr{E}$, constitutes an $\mathscr{L}$-basis in $L^2(0, T; \mathbb{C}^m)$. So, according to Theorem III.3.10, system (6) is B-controllable in time T relative to space $\mathscr{W}(\rho_n)$ for $\rho_n = n^\gamma$. One easily checks that in the notations of Section III.2 this space is exactly $\mathscr{W}_r$, where $r = \gamma + 1$.

(c) If conditions (7) are not satisfied, family $\mathscr{E}$ is linearly dependent in $L^2(0, T; \mathbb{C}^m)$ for any $T > 0$ (see the proof of assertions (a) and (b) of Theorem IV.1.3). Just as when proving Remark IV.1.6, one constructs element $f \in \mathscr{W}_1$ orthogonal to $\mathscr{R}(T)$ for all $T > 0$.

(d) In view of Corollary III.3.13, (8) is equivalent to inclusion $\mathscr{E}_\delta \in (W)$ in space $L^2(0, \infty; \mathbb{C}^m)$, where $\mathscr{E}_\delta := \{\eta_{|k|j} \, e^{i(v_k + i\delta)t}\}$, $\delta > \sqrt{\alpha}$. The demonstration of this inclusion (just as in the proof of assertions (a) and (b) of Theorem IV.1.3) reduces to the verification of conditions (7) (which are necessary and sufficient for linear independence of the family) and the implication

$$\sum_{k=-n}^{n} \alpha_{kp} \, e^{i(v_k + i\delta)t} \xrightarrow[n \to \infty]{} 0 \qquad \text{in } L^2(0, \infty) \Rightarrow \{\alpha_{kp}\} = 0. \quad (11)$$

Here,

$$\alpha_{kp} = \sum_{j=1}^{\kappa_{|k|}} a_{kj}(b_p, \varphi_{|k|})_{L^2(\Omega)}; \qquad p = 1, \ldots, m; \; j = 1, \ldots, \kappa_{|k|}; \; k \in \mathbb{K};$$

$\{a_{kj}\}$ is an arbitrary squarely summable sequence. As in the proof of Theorem 2, one verifies that $|\alpha_{kp}| < |k|^R$ for some $R > 0$. Therefore, Theorem II.6.4 actually yields implication (11).

(e) Let us check the formulated conditions for vectors η_n to provide inclusion $\mathscr{E} \in (W)$ in $L^2(0, T; \mathbb{C}^m)$ for any $T > 0$. In the same manner as in proofs of assertion on (d) and Theorem IV.3.1(a), (b), the proof of this inclusion reduces to the direct application of Theorem II.6.5.

Theorem 3 is proved completely.

Note that assertion (e) was obtained earlier by means of some other method (Triggiani 1978; see also Tsujioka 1970, with a close result). Our proof stresses the role of the (W) property of a vector exponential family in the study of DPS controllability.

1.3. Pointwise control

Consider now a system described by the boundary-value problem

$$\begin{cases} \dfrac{\partial^2 y}{\partial t^2} + Ay = \displaystyle\sum_{p=1}^{m} u_p(t)\delta(x - x_p) & \text{in } Q, \\[2mm] y|_\Sigma = 0; \qquad x_p \in \Omega, \qquad u_p \in L^2(0, T), \end{cases} \quad (12)$$

with zero initial conditions (2). As in the previous Subsection, here $U = \mathbb{C}^m$, $\mathscr{U} = L^2(0, T; \mathbb{C}^m)$. Let us first elucidate the space in which initial boundary-value problem (12), (2) is correct. In Subsection IV.1.3 we demonstrated the right-hand side of the first equation (12) to be represented in the form $Bu(t)$, $B \in \mathscr{L}(U, W_q)$, with $q < -N/2$ and

$$\langle Bu(t), \varphi \rangle_* = \sum_{p=1}^{m} u_p(t) \overline{\varphi(x_p)}, \qquad \varphi \in W_{-q}. \tag{13}$$

Therefore, Theorem III.2.1 implies that there exists a unique solution to problem (12), (2) such that for $r < -N/2$

$$\{y, y_t\} \in C(0, T; \mathscr{W}_{r+1}), \qquad \mathscr{W}_{r+1} = W_{r+1} \oplus W_r.$$

We now demonstrate that, making use of the specific form of operator B, this result may be essentially sharpened. Note that the theorem stated below is proved by Y. Meyer (1985) for the case $N = 3$, and we follow the scheme invented by him.

Theorem V.1.4. There exists a unique solution to initial boundary-value (12), (2) such that $\{y, y_t\} \in C(0, T; \mathscr{W}_{(3-N)/2})$. This result cannot be improved in class $C(0, T; \mathscr{W}_r)$.

The proof exploits the transposition method and the Fourier method. Let $T' > 0$, set $Q' = \Omega \times (0, T')$, $\Sigma' = \Gamma \times (0, T')$. Let $w(x, t)$ denote a solution of initial boundary-value problem

$$\begin{cases} \dfrac{\partial^2 w}{\partial t^2} + Aw = 0 & \text{in } Q', \\[2mm] w|_{\Sigma'} = 0, \qquad w|_{t=T'} = w_0, \qquad w_t|_{t=T'} = w_1. \end{cases} \tag{14}$$

Lemma V.1.5. Let $w_0 \in W_{(N-1)/2}$, $w_1 \in W_{(N-3)/2}$, $x_0 \in \Omega$. Then

$$w(x_0, \cdot) \in L^2(0, T')$$

and estimate

$$\|w(x_0, \cdot)\|_{L^2(0,T')}^2 \leq C(T')[\|w_0\|_{(N-1)/2}^2 + \|w_1\|_{(N-3)/2}^2] \tag{15}$$

holds.

PROOF. Let us expand terminal data w_0, w_1 in series

$$w_0(x) = \sum_{n=1}^{\infty} a_n \varphi_n(x), \qquad w_1(x) = \sum_{n=1}^{\infty} b_n \varphi_n(x),$$

where

$$\sum_{n=1}^{\infty} |a_n|^2 (\lambda_n + \alpha)^{(N-1)/2} = \|w_0\|_{(N-1)/2}^2 < \infty,$$

$$\sum_{n=1}^{\infty} |b_n|^2 (\lambda_n + \alpha)^{(N-3)/2} = \|w_1\|_{(N-3)/2}^2 < \infty. \tag{16}$$

By means of the Fourier method we obtain

$$w(x_0, t) = \sum_{n=1}^{\infty} \left[a_n \cos \omega_n(T' - t) + b_n \frac{\sin \omega_n(T' - t)}{\omega_n} \right] \varphi_n(x_0). \quad (17)$$

Using Euler's formulas series (17) may be written as

$$w(x_0, t) = \sum_{n=1}^{\infty} [\alpha_n^+ (\lambda_n + \alpha)^{(1-N)/4} e^{i\omega_n t} + \alpha_n^- (\lambda_n + \alpha)^{(1-N)/4} e^{-i\omega_n t}] \varphi_n x_0)$$

and, by virtue of (16), inequality

$$\sum_{n=1}^{\infty} (|\alpha_n^+|^2 + |\alpha_n^-|^2) < \infty$$

is valid. Now from Meyer's theorem (Proposition II.1.22) we conclude that the assertion of the lemma is equivalent to estimate

$$\sup_{l \in \mathbb{N}} \sum_{l \le \omega_n < l+1} |\varphi_n(x_0)|^2 (\lambda_n + \alpha)^{(1-N)/2} < \infty.$$

Thus we have to show that for some positive constant C'

$$\sum_{l \le \omega_n < l+1} |\varphi_n(x_0)|^2 \le C' l^{N-1}. \quad (18)$$

Introduce a spectral function of operator A (see, e.g., Hörmander 1968):

$$e(x, x', \lambda) = \sum_{\lambda_n < \lambda} \varphi_n(x) \overline{\varphi_n(x')}.$$

Inequality (18) is rewritten as

$$e(x_0, x_0, (l+1)^2) - e(x_0, x_0, l^2) \le C' l^{N-1}. \quad (19)$$

It is known (Agmon 1965; Hörmander 1968; Birman and Solomyak 1977) that

$$e(x, x, \lambda) = K\lambda^{N/2} + \mathcal{O}(\lambda^{(N-1)/2}), \quad (20)$$

with constant K depending on the coefficients of operator A, domain Ω, and being independent of $x \in \Omega$. This immediately implies inequality (19) and thus proves the lemma.

Returning to the proof of Theorem 4, let us apply the Fourier method to problem (12), (2). As is easily seen from (16), (17), Section III.2, and

(13), for $T' \in [0, T]$ we have

$$y(x, T) = \sum_{n=1}^{\infty} y_n(T')\varphi_n(x), \qquad y_t(x, T') = \sum_{n=1}^{\infty} \dot{y}_n(T')\varphi_n(x), \quad (21)$$

$$y_n(T') = \sum_{n=1}^{m} \int_0^{T'} \varphi_n(x_p)u_p(t)\frac{\sin \omega_n(T' - t)}{\omega_n}\, dt, \quad (22)$$

$$\dot{y}_n(T') = \sum_{p=1}^{m} \int_0^{T'} \varphi_n(x_p)u_p(t) \cos \omega_n(T' - t)\, dt. \quad (23)$$

Relations (17), (21–23) together with the definition of spaces W_r lead to equality

$$\langle y(\cdot, T'), w_1 \rangle_* + \langle y_t(\cdot, T'), w_0 \rangle_* = \sum_{p=1}^{m} \int_0^{T'} u_p(t)\overline{w(x_p, t)}\, dt. \quad (24)$$

By Lemma 5, functions $w(x_p, \cdot)$ belong to $L^2(0, T')$ and thus the right-hand side of equality (24) is a linear continuous functional relation to $w_0 \in W_{(N-1)/2}$, $w_1 \in W_{(N-3)/2}$. Hence, inclusions $y(\cdot, T') \in W_{(3-N)/2}$, $y_t(\cdot, T') \in W_{(1-N)/2}$ are established. Estimate (15) also implies inequality

$$\|\{y(\cdot, T'), y_t(\cdot, T')\}\|_{(3-N)/2}^2 \le C(T')\|u\|_{L^2(0, T'; \mathbb{C}^m)}^2$$

with $u(t) = \{u_p(t)\}_{p=1}^{m}$.

Pair $\{y(\cdot, T'), y_t(\cdot, T')\}$ is easily checked in a standard way using (24), to be continuous with respect to T' in the $\mathscr{W}_{(3-N)/2}$ metrics. Similar arguments were given in detail in the proof of Theorem III.1.1.

Let us proceed with the study of system (12) controllability. Denote η_{nj}, $j = 1, \ldots, \kappa_n$, $n \in \mathbb{N}$, vectors from $\mathbb{C}^m$ with the components $\varphi_{nj}(x_p)$, $p = 1, \ldots, m$. Formula (13) provides corresponding family $\mathscr{E} = \{e_{kj}\} \subset \mathscr{U}$ for system (12) to have the form $e_{kj}(t) = \eta_{|k|j}\, e^{i\nu_k t}$, $j = 1, \ldots, \kappa_{|k|}$, $k \in \mathbb{K}$, while family $\tilde{\mathscr{E}}$ (see Section III.3 for the definition) is

$$\{r_{|k|}e_{kj}\}, \qquad r_n = (\mu_n + \alpha)^s, \qquad s = \frac{1-N}{4}$$

(compare with Subsection IV.1.3).

Theorem V.1.6. For $N > 1$ the following statements hold.

(a) *For any $T > 0$ system (12) is not M-controllable.*
(b) *If conditions*

$$\operatorname{rank}[\varphi_{nj}(x_p)]_{j=1,\ldots,\kappa_n}^{p=1,\ldots,m} = \kappa_n, \qquad n \in \mathbb{N}, \quad (25)$$

are not valid, then for system (12)

$$\text{Cl}_{\mathscr{W}_{(3-N)/2}} \bigcup_{T>0} \mathscr{R}(T) \neq \mathscr{W}_{(3-N)/2}.$$

(c) *If conditions* (25) *are fulfilled, then*

$$\text{Cl}_{\mathscr{W}_{(3-N)}} \bigcup_{T>0} \mathscr{R}(T) = \mathscr{W}_{(3-N)/2}.$$

PROOF. The proof of (a), (b), and (c) is similar to that of the corresponding assertions (a), (c), and (d) of Theorem 3.

Let us deal in more detail with the case $N = 1$, $\Omega = (0, 1)$.

Suppose first that $m = 1$, that is, that there is only one scalar control action u applied at the point $x_0 \in \Omega$. So, let the system be described by boundary-value problem

$$\begin{cases} \dfrac{\partial^2 y}{\partial t^2} = \dfrac{\partial}{\partial x}\left(a(x) \dfrac{\partial y}{\partial x} \right) - a_0(x)y + u(t)\delta(x - x_0), \\ y(0, t) = y(l, t) = 0, \qquad u \in L^2(0, T), \end{cases} \tag{26}$$

with zero initial conditions (2);

$$T_0 := 2 \int_0^1 \frac{dx}{\sqrt{a(x)}}.$$

Theorem V.1.7. The following statements are correct

(a) *For $T < T_0$ system* (26) *is not M-controllable.*

(b) *If $\varphi_n(x_0) = 0$ for at least one value of $n \in \mathbb{N}$, then system* (26) *is not W-controllable for any T and, more to the point, $\text{Cl}_{\mathscr{W}_1} \bigcup_{T>0} \mathscr{R}(T) \neq \mathscr{W}_1$.*

(c) *If $\varphi_n(x_0) \neq 0$ for all $n \in \mathbb{N}$, then for $T \geq T_0$ system* (26) *is M-controllable.*

(d) *System* (26) *is not UM-controllable relative to $\mathscr{W}_1$ for any $T > 0$.*

(e) *For system* (26) *$\mathscr{R}(T) = \mathscr{R}(T_0)$ for $T \geq T_0$.*

PROOF. If $\varphi_n(x_0) = 0$ for at least one value of n, then family

$$\mathscr{E} = \{\varphi_{|k|}(x_0) \, e^{i\omega_k t}\}_{k \in \mathbb{K}}$$

is, evidently, linearly dependent. In such a case it is not difficult to construct explicitly an element orthogonal to $\mathscr{R}(T)$ for all T (see Theorem 1.3(c) and Remark IV.1.6). This proves assertion (b).

If $\varphi_n(x_0) \neq 0$ for all $n \in \mathbb{N}$, then the minimality of family $\mathscr{E}$ is equivalent to the minimality of family $E = \{e^{i\omega_k t}\}$. By virtue of asymptotics (9) and Corollary II.4.2, family E is not minimal in $L^2(0, T)$ for $T < T_0$. By the force of Proposition II.4.13, family $E \cup \{e^{i\omega t}\}$ ($\omega \neq \omega_k$, $k \in \mathbb{K}$) is a basis in $L^2(0, T_0)$. Therefore, E is an $\mathscr{L}$-basis in $L^2(0, T)$ for $T \geq T_0$ and $\mathscr{E}$ is minimal in this space.

Family $\mathscr{E}' = \{e'_k\}$ biorthogonal to $\mathscr{E}$ has the form $e'_k(t) = [\varphi_n(x_0)]^{-1}\theta_k(t)$ where $\{\theta_k\}$ is the family biorthogonal to E. That is why

$$\|e'_k\|_{L^2(0, T)} \asymp |\varphi_{|k|}(x_0)|^{-1}.$$

By the force of the asymptotics (Naimark 1969: chap. 2, sec. 4.10),

$$\varphi_n(x) = \frac{2}{\sqrt{T_0}} \frac{1}{\sqrt[4]{a(x)}} \sin\left[\frac{2\pi n}{T_0} \int_0^x \frac{ds}{\sqrt{a(s)}}\right] + \mathcal{O}(1/n) \qquad (27)$$

sequence $\{|\varphi_n(x_0)|^{-1}\}_{n \in \mathbb{N}}$ is not bounded. To prove this, we apply for

$$\beta = \frac{2}{T_0} \int_0^{x_0} \frac{ds}{\sqrt{a(s)}}$$

the following assertion (Cassels 1957: sec. I.1): for any β, $\varepsilon > 0$ there may be found $m, n \in \mathbb{N}$, sucn that $|\beta n - m| < \varepsilon$. So family $\mathscr{E}$ is not *-uniformly minimal in $L^2(0, T)$ for any T. Assertions (a), (c), and (d) follow now from Theorem III.3.10.

Consider moment problem operator $\mathscr{J} : u \mapsto \{q_k\}_{k \in \mathbb{K}}$ acting from space $L^2(0, T)$ to $\tilde{\ell}^2$ according to the formula

$$q_k = \int_0^T u(t) e^{i\omega_k(T-t)} dt, \qquad k \in \mathbb{K}, T \geq T_0.$$

Since, as already mentioned, family $\{e^{i\omega_k t}\}$ is an $\mathscr{L}$-basis in $L^2(0, T)$ for $T \geq T_0$, by Theorem I.2.1(a), operator $\mathscr{J}$ maps $L^2(0, T)$ onto the whole $\tilde{\ell}^2$. Set $\hat{\mathscr{R}}(T)$ of coefficients $\{c_k(t)\}$ (defined in Section III.2) isomorphic to set $\mathscr{R}(T)$, has the form $c_k(T) = \varphi_{|k|}(x_0)q_k$. This just proves assertion (e).

Remark V.1.8. From the theory of Sturm–Liouville problems, each function φ_n is known to have a finite number of zeros on the interval $(0, 1)$. Therefore, conditions $\varphi_n(x_0) \neq 0$, $n \in \mathbb{N}$, may be violated for only a countable set of points $x_0 \in (0, 1)$.

We can complement assertions of Theorem 7 using the following results of the metric theory of Diophantine approximations.

Proposition V.1.9 (Cassels 1957: sec. VII.1).

(à)   The inequality $|\beta n - m| < (n \log n)^{-1}$ has infinitely many integer solutions (m, n) for almost all β.

(b)   The inequality $|\beta n - m| < (n \log^2 n)^{-1}$ has only a finite number of integer solutions (m, n) for almost all β.

Theorem V.1.10. For almost all $x_0 \in (0, 1)$, system (26) is not UM-controllable relative to $\mathcal{W}_r$ for any $r < 2$ and $T > 0$.

PROOF. The theorem can be proved in the same way as assertion (d) of Theorem 7 using asymptotics (27), Theorem III.3.10 and Proposition 9(a).

For the case of equations with constant coefficients, when we know the explicit form of $\varphi_n(x)$, some positive results concerning exact controllability of system (26) can be obtained. We give two examples in this direction.

Example V.1.11. For almost all $x_0 \in (0, 1)$ and any $r > 2$ system,

$$\begin{cases} \dfrac{\partial^2 y}{\partial t^2} = \dfrac{\partial^2 y}{\partial x^2} + u(t)\delta(x - x_0), \\[2mm] y(0, t) = y(l, t) = 0, \qquad u \in L^2(0, T), \end{cases}$$

is E-controllable relative to space $\mathcal{W}_r$ in time $T \geq T_0 := 21$.

PROOF. By the force of Theorem III.2.3, the assertion is equivalent to the fact that the problem of moments

$$c_k = (u, e_k)_{L^2(0, T)}, \qquad k \in \mathbb{K}, \tag{28}$$

has the solution $u \in L^2(0, T)$ for almost all x_0 if $\sum_{k \in \mathbb{K}} |c_k|^2 \omega_{|k|}^{2r-2} < \infty$. Here,

$$e_k \in \mathcal{E} = \{\varphi_{|k|}(x_0) e^{i\omega_k t}\}_{k \in \mathbb{K}}, \qquad \omega_k = \frac{2\pi k}{T_0},$$

$$\varphi_n(x) = \frac{2}{\sqrt{T_0}} \sin \frac{2\pi n}{T_0} x.$$

From Proposition 9(b) it follows that for almost all x_0 $|\varphi_{|k|}(x_0)|^{-1} < \omega_{|k|}^s$ for any $s > 1$. As mentioned in the proof of Theorem 7, family $\{e^{i\omega_k t}\}_{k \in \mathbb{K}}$ is an $\mathcal{L}$-basis in $L^2(0, T)$ for $T \geq T_0$. Therefore, the family $\{\theta_k\}$ biorthogonal to it is also an $\mathcal{L}$-basis in $L^2(0, T)$. Problem of moments (28) has the formal solution $u(t) = \sum_{k \in \mathbb{K}} c_k [\varphi_{|k|}(x_0)]^{-1} \theta_k(t)$. For $s < r - 1$ this series converges in $L^2(0, T)$, since

$$\sum_{k \in \mathbb{K}} |c_k|^2 |\varphi_k(x_0)|^{-2} \prec \sum_{k \in \mathbb{K}} |c_k|^2 |\omega_k|^{2s} \le \sum_{k \in \mathbb{K}} |c_k|^2 |\omega_k|^{2r-2} < \infty.$$

Notice that the example was studied in fact in Butkovskiĭ (1976).

Example V.1.12. The system

$$\begin{cases} \dfrac{\partial^2 y}{\partial t^2} = \dfrac{\partial^2 y}{\partial x^2} + u(t)\delta(x - 1/2), \\[2mm] y(0, t) = y_x(l, t) = 0, \qquad u \in L^2(0, T), \end{cases}$$

is *B*-controllable in space $\mathscr{W}_1$ in time $T \ge T_0 := 2l$.

PROOF. In this case

$$\omega_k = \frac{2\pi(k - 1/2)}{T_0}, \qquad \varphi_n(x) = \frac{2}{\sqrt{T_0}} \sin \frac{2\pi(n - 1/2)}{T_0} x$$

and so $|\varphi_{|k|}(l/2)| = \sqrt{2/T_0}$. Therefore, family $\{\varphi_{|k|}(x_0) e^{i\omega_k t}\}_{k \in \mathbb{K}}$ forms an $\mathscr{L}$-basis in $L^2(0, T)$ for $T \ge T_0$ that proves our assertion.

Let us now dwell on the case of several control actions. Let a system be described by boundary-value problem

$$\begin{cases} \dfrac{\partial^2 y}{\partial t^2} = \dfrac{\partial}{\partial x}\left(a(x)\dfrac{\partial y}{\partial x}\right) - a_0(x)y + \displaystyle\sum_{p=1}^{m} u_p(t)\delta(x - x_p), \\[2mm] y(0, t) = y(l, t) = 0; \qquad x_p \in (0, 1), \qquad u_p \in L^2(0, T). \end{cases} \tag{29}$$

Family $\mathscr{E} = \{e_k\} \subset L^2(0, T; \mathbb{C}^m)$ corresponding to system (29) is of the form $e_k(t) = \eta_k e^{i\omega_k t}$, $k \in \mathbb{K}$, where η_k is the vector from $\mathbb{C}^m$ with the components $\varphi_{|k|}(x_p)$, $p = 1, \ldots, m$.

Theorem V.1.13.

(a) *For $T < T_0/m$ system (29) is not M-controllable.*

(b) *System (29) is not UM-controllable relative to space $\mathscr{W}_1$ for any T.*

(c) *Assertions (b), (c), and (e) of Theorem 7 are true for system (29) as well with the replacement of scalars $\varphi_n(x_0)$ by the vectors η_n.*

The proof of assertion (a) is similar to the proof of (b) of Theorem 3. The proof of (b) is based on the following theorem (Cassels 1957: sec. I.5): for any positive ε, $\alpha_1, \ldots, \alpha_m$ there may be found natural numbers n, $q_1, \ldots, q_m$, such that $|\alpha_p n - q_p| < \varepsilon$. From asymptotics (27) it follows that $\inf_k \|e_k\| = 0$. Taking into account formula (10) of Section I.1 we conclude that family $\mathscr{E}$ is not *-uniformly minimal in $L^2(0, T; \mathbb{C}^m)$ for any T. The proof of assertion (c) is just like that of Theorem 7.

2. Boundary control

2.1. Infinite dimensional control

As in Section 2 of Chapter IV, let the boundary of set Ω be represented as a unification of two sets: $\Gamma = \Gamma_0 \cup \Gamma_1$, with Γ_1 being a nonempty, relatively open set, $\Gamma_0 = \Gamma \setminus \Gamma_1$. Set $\Sigma_0 = \Gamma_0 \times (0, T)$, $\Sigma_1 = \Gamma_1 \times (0, T)$ and consider system

$$
\begin{cases}
\dfrac{\partial^2 y}{\partial t^2} + Ay = 0 & \text{in } Q = \Omega \times (0, T), \\[2mm]
y|_{\Sigma_0} = 0, \qquad y|_{\Sigma_1} = u;
\end{cases}
\tag{1}
$$

$$
u \in \mathscr{U} = L^2(\Sigma_1) = L^2(0, T; U), \qquad U = L^2(\Gamma_1).
$$

As before, we choose zero initial conditions:

$$
y(x, 0) = y_t(x, 0) = 0 \qquad \text{in } \Omega.
\tag{2}
$$

An integral identity corresponding to problem (1), (2) reads

$$
\int_Q y \left(\frac{\partial^2 f}{\partial t^2} + Af \right) dx \, dt = - \int_{\Sigma_1} u \frac{\partial f}{\partial \nu} ds \, dt
$$

for any $f \in H^2(Q)$ such that $f(\cdot, T) = f_t(\cdot, T) = 0$, $f|_\Sigma = 0$. This provides us with the grounds to treat problem (1), (2) in accordance with the scheme from Section III.2, defining operator B by the relation

$$
\langle Bv, \varphi \rangle_* = - \int_{\Sigma_1} v(s) \overline{\frac{\partial \varphi(s)}{\partial \nu}} \, ds, \qquad v \in L^2(\Gamma_1).
\tag{3}
$$

It is known (Lasiecka, Lions, and Triggiani 1986) that the solution of (1), (2) satisfies $\{y, y_t\} \in C(0, T; \mathscr{W}_0)$, $\mathscr{W}_0 = L^2(\Omega) \oplus H^{-1}(\Omega)$. From formula (3) and Section III.2 we conclude that exponential family $\mathscr{E} = \{e_k\} \subset \mathscr{U}$

corresponding to system (1) has the form

$$e_k(s, t) = \frac{\partial \varphi_{|k|}(s)}{\partial v} e^{i\omega_k t}, \qquad k \in \mathbb{K}.$$

Family $\tilde{\mathscr{E}}$ (see Section III.3) prescribed by family $\mathscr{E}$ and space $\mathscr{W}_0$ (in the notations of Section III.2) has the form $\tilde{\mathscr{E}} = \{\tilde{e}_k\} = \{(\lambda_{|k|} + \alpha)^{-1/2} e_k\}$ (number α is determined in the beginning of Chapter IV). As in Section IV.2, for convenience we have changed the sign (see formula (3)) of all the members of the families $\mathscr{E}$ and $\tilde{\mathscr{E}}$, which certainly does not affect any of their properties that are pertinent to the controllability study.

For the analysis of these properties the following statement is required.

Lemma V.2.1. *For any function $f \in L^2(\Gamma)$ an estimate is valid*

$$\left| \int_\Gamma f(s) \frac{\partial \varphi_n(s)}{\partial v} ds \right| \prec (\lambda_n + \alpha) \| f \|_{L^2(\Gamma)};$$

the constant in this inequality is independent of both f and $n \in \mathbb{N}$.

PROOF. Consider the boundary-value problem for an elliptic equation

$$(A + \alpha I)z = 0 \qquad \text{in } \Omega, \, z|_\Gamma = f.$$

A unique solution to this problem $z \in L^2(\Omega)$ is known to exist (Lions 1968: chap. II) and $\|z\|_{L^2(\Omega)} \prec \|f\|_{L^2(\Gamma)}$.

Accounting for the equality $\varphi_n|_\Gamma = 0$, we have

$$0 = \int_\Omega (A + \alpha I)z\varphi_n \, dx = - \int_\Gamma z \frac{\partial \varphi_n}{\partial v} ds + \int_\Omega z(A + \alpha I)\varphi_n \, dx.$$

So that

$$\int_\Gamma f \frac{\partial \varphi_n}{\partial v} ds = (\lambda_n + \alpha) \int_\Omega z\varphi_n \, dx.$$

Therefore,

$$\left| \int_\Gamma f \frac{\partial \varphi_n}{\partial v} ds \right| \le (\lambda_n + \alpha)\|z\|_{L^2(\Omega)}\|\varphi_n\|_{L^2(\Omega)} \prec (\lambda_n + \alpha)\|f\|_{L^2(\Gamma)}.$$

Remark V.2.2. For the third boundary-value problem, a similar estimate was obtained in Plotnikov (1968).

Making use of the notations introduced in Theorem 1.2, we rewrite families $\mathscr{E}$ and $\tilde{\mathscr{E}}$ as

$$\mathscr{E} = \left\{ \frac{\partial \varphi_{|k|j}(s)}{\partial v} e^{iv_k t} \right\}, \qquad \tilde{\mathscr{E}} = \left\{ r_k \frac{\partial \varphi_{|k|j}(s)}{\partial v} e^{iv_k t} \right\},$$

$$j = 1, \dots, \kappa_{|k|}, \qquad k \in \mathbb{K}, \qquad r_k = (\mu_{|k|} + \alpha)^{-1/2}.$$

Theorem V.2.3. For system (1) an equality takes place

$$\mathrm{Cl}_{\mathscr{W}_6} \bigcup_{T > 0} \mathscr{R}(T) = \mathscr{W}_0.$$

Moreover, this equality is valid for a class of bilinear controls $u(x, t) = f(x)g(t)$; $f \in L^2(\Gamma_1)$, $g \in L^2(0, T)$ (f and g being controls), that is narrower than $L^2(\Sigma_1)$.

PROOF. In view of Remark III.3.13, the demonstrating equality is equivalent to inclusion $\tilde{\mathscr{E}}_\delta \in (W)$ in space $L^2(0, \infty; U)$ where

$$\tilde{\mathscr{E}}_\delta = \left\{ r_k \frac{\partial \varphi_{|k|j}}{\partial v} e^{i(v_k + i\delta)t} \right\}, \qquad \delta > \sqrt{\alpha}.$$

Suppose that for some sequence $\{c_{kj}\} \in \ell^2$

$$\sum_{k=1}^{R} \sum_{j=1}^{\kappa_{|k|}} c_{kj} r_k \frac{\partial \varphi_{|k|j}}{\partial v} e^{i(v_k + i\delta)t} \xrightarrow[R \to \infty]{} \quad \text{in } L^2(0, \infty; U).$$

Then for any function $f \in L^2(\Gamma_1)$ and any $g \in L^2(0, \infty)$

$$\sum_{k \in \mathbb{K}} \sum_{j=1}^{\kappa_{|k|}} c_{kj} r_k \left(\frac{\partial \varphi_{|k|j}}{\partial v}, f \right)_{L^2(\Gamma_1)} (e^{i(v_k + i\delta)t}, g)_{L^2(0, \infty)} = 0. \tag{4}$$

Moreover, we would like to use Theorem II.6.4 as applied to family $\{e^{i(v_k + i\delta)t}\}$. When proving Theorem 1.2, we have noted that the growth of $\kappa_{|k|}$ is not faster than some power of $|k|$. Since $|\lambda_n| \asymp n^{2/N}$ it can be shown that

$$|k|^{1/N} \prec |v_k| \prec |k|^Q \qquad \text{for some } Q > 0 \tag{5}$$

(recall that numbers v_k are different). By Lemma 1,

$$\left| \left(\frac{\partial \varphi_{|k|j}}{\partial v}, f \right)_{L^2(\Gamma_1)} \right| \prec (\mu_{|k|} + \alpha) \|f\|_{L^2(\Gamma_1)}.$$

Taking (5) into account, we arrive at

$$\left| \sum_{j=1}^{\kappa_{|k|}} c_{kj} r_k \left(\frac{\partial \varphi_{|k|j}}{\partial v}, f \right)_{L^2(\Gamma_1)} \right| \prec |k|^R$$

for some $R > 0$.

Thus we have checked that the conditions of Theorem II.6.4 hold. By the theorem, from equality (4) we obtain

$$\sum_{j=1}^{\kappa_{|k|}} c_{kj} \left(\frac{\partial \varphi_{|k|j}}{\partial v}, f \right)_{L^2(\Gamma_1)} = 0, \qquad k \in \mathbb{K}.$$

As function $f \in L^2(\Gamma_1)$ is arbitrary, this implies

$$\sum_{j=1}^{\kappa_{|k|}} c_{kj} \frac{\partial \varphi_{|k|j}}{\partial v} = 0 \qquad \text{in } L^2(\Gamma_1)$$

for all $k \in \mathbb{K}$. As already demonstrated in the proof of Theorem IV.2.1, the latter equality provides coefficients c_{kj} that turn to zero for all k and j. Hence, $\tilde{\mathscr{E}}_\delta \in (W)$ in $L^2(0, \infty; L^2(\Gamma_1))$.

The assertion about bilinear controls follows from the proof if we take into account the arguments used in Theorem III.3.7. Theorem 3 is proved.

As in the case of control on a subdomain the stronger result on approximate controllability takes place. The system (1) is W-controllable in time $T > 2T^*$, where T^* is the time of filling of the domain Ω by the waves from Γ_1:

$$T^* = \inf \left\{ T \mid \bigcup_{u \in L^2(0, T; L^2(\Gamma_1))} \text{supp } y(\cdot, T) = \bar{\Omega} \right\}.$$

This result can be obtained by Russell's (1978) approach using the uniqueness theorem proved in Tataru (1993) (cf. with the case of control on subdomain Ω' in Subsection 1.1).

M controllability in the case of control acting on a part of the boundary in general is still an open problem.

Hypothesis. If system (1) is W-controllable in time T, then it is M-controllable in the same time.

The hypothesis is supported by the following examples for the wave equation $y_{tt} = \Delta y$.

(a) Ω is a rectangle, $\Omega = (0, a) \times (0, b)$, $\Gamma_1 = (0, a) \times \{0\}$. Here, $T^* = b$.
   Fattorini (1979) proved that the system is M-controllable in time $2b$.

(b) Ω is annular,

$$\Omega = \{(x_1, x_2) \mid r^2 < x_1^2 + x_2^2 < 1\}, \qquad \Gamma = \{(x_1, x_2) \mid x_1^2 + x_2^2 = 1\}.$$

Here, $T^* = 1 - r$. Avdonin, Belishev, and Ivanov (1991b) have shown that the system is M-controllable in time $2T^*$. Note that the system is B-controllable in time $2\sqrt{1 - r^2}$ (Bardos, Lebeau, and Rauch 1992).

If set Γ_1 is large enough, a considerably stronger controllability of system (1) takes place. For the case when operator A is $(-\Delta)$, several interesting results have been obtained by D. L. Russell (see the survey in Russell 1978 and the references therein). Our exposition is based on the results obtained by the Hilbert Uniqueness Method suggested by J.-L. Lions (1986, 1988a, 1988b).

The system dual to (1) (see Theorem III.3.11) is

$$\begin{cases} \dfrac{\partial^2 \psi}{\partial t^2} + A\psi = 0 & \text{in } Q, \\[2mm] \psi|_\Sigma = 0, \\[2mm] \psi(x, 0) = \psi_0(x), \qquad \psi_t(x, 0) = \psi_1(x), \\[2mm] \psi_0 \in H_0^1(\Omega), \qquad \psi_1 \in L^2(\Omega). \end{cases} \tag{6}$$

For an observation v, one takes $\partial\psi/\partial v|_\Sigma$.

Proposition V.2.4 (Lions 1983: chap. 2). *There exists a positive constant C_1 depending on the coefficients of operator A and domain Ω such that the solution to problem (6) satisfies inequality*

$$\left\| \frac{\partial\psi}{\partial v} \right\|_{L^2(\Sigma)}^2 \le C_1(\|\psi_0\|_{H^1(\Omega)}^2 + \|\psi_1\|_{L^2(\Omega)}^2). \tag{7}$$

Ho (1986) has shown that the converse inequality is correct under some assumptions as well. Let the coefficients of operator A satisfy conditions (in addition to those formulated at the beginning of Chapter IV)

$$\frac{\partial a_{ij}}{\partial x_k} \in L^\infty(\Omega) \qquad \text{for all } i, j, k = 1, \ldots, N.$$

Set

$$\beta = \operatorname*{vrai\,sup}_{x \in \Omega} \left(\sum_{i,j,k=1}^{N} \left| \frac{\partial}{\partial x_k} a_{ij}(x) \right|^2 \right)^{1/2},$$

$$R = \sup_{x \in \Omega} |x - x_0|, \qquad \Gamma_+ = \{ x \in \Gamma \mid (x - x_0, n(x))_{\mathbb{R}^N} > 0 \},$$

where $n(x)$ is the outward normal vector to Γ at the point x, and x_0 is some arbitrary point of space $\mathbb{R}^N$.

Proposition V.2.5 (Ho 1986). Let $\beta < 2\kappa/R$ (number κ was introduced at the beginning of Chapter IV). Then positive constants $\hat{T}$ and C_2 depending on A and Ω may be found such that for all $T \geq \hat{T}$ the solution to problem (6) obeys inequality

$$\left\| \frac{\partial \psi}{\partial v} \right\|_{L^2(\Sigma_+)}^2 \geq C_2 (\|\psi_0\|_{H^1(\Omega)}^2 + \|\psi_1\|_{L^2(\Omega)}^2), \tag{8}$$

$$\Sigma_+ := \Gamma_+ \times (0, T),$$

So under condition $\beta < 2\kappa/R$ (note that it is obviously true for equations with constant coefficients) the relation

$$\left\| \frac{\partial \psi}{\partial v} \right\|_{L^2(\Sigma_+)}^2 \asymp \|\psi_0\|_{H^1(\Omega)}^2 + \|\psi_1\|_{L^2(\Omega)}^2, \tag{9}$$

$$\{\psi_0, \psi_1\} \in H_0^1(\Omega) \oplus L^2(\Omega),$$

is valid.

Theorem V.2.6. If coefficients of operator A and domain Ω are such that condition $\beta < 2\kappa/R$ holds; and $\Gamma_1 \supset \Gamma_+$, then system (1) is B-controllable in time T relative to space $\mathcal{W}_0 = L^2(\Omega) \oplus H^{-1}(\Omega)$ for any $T \geq \hat{T}$.

PROOF. Applying to the given case the assertion and the plan of the proof of Theorem III.3.11 and using relation (8), we conclude that system (1) is E-controllable relative to $\mathcal{W}_0$ for $T \geq \hat{T}$; that is, $\mathcal{R}(T) \supset \mathcal{W}_0$. Since, as already noted, $\mathcal{R}(T) \subset \mathcal{W}_0$ for any $T > 0$, the theorem is proved.

Corollary V.2.7. Family $\tilde{\mathcal{E}} = \{\tilde{e}_k\}$, $k \in \mathbb{K}$,

$$\tilde{e}_k(s, t) = (\lambda_{|k|} + \alpha)^{-1/2} \frac{\partial \varphi_{|k|}(s)}{\partial v} e^{i\omega_k t},$$

constitutes an $\mathcal{L}$-basis in $L^2(0, T; L^2(\Gamma_1))$ for $T \geq \hat{T}$, $\Gamma_1 \supset \Gamma_+$. This statement follows from the previous theorem and Theorem III.3.10.

The most complete results on exact boundary controllability have been obtained by Bardos, Lebeau, and Rauch (1992) with the help of microlocal analysis. Roughly speaking, system (1) is *B*-controllable in time T (in space $\mathcal{W}_0$) if every ray crosses $\Gamma_1 \times (0, T)$. If there is a ray that does not cross $\overline{\Gamma}_1 \times [0, T]$, then the system is not *B*-controllable in time T.

2.2. Finite dimensional control

Consider now a system described by boundary-value problem

$$
\begin{cases}
\dfrac{\partial^2 y}{\partial t^2} + Ay = 0 & \text{in } Q, \\[2mm]
y(s, t)|_\Sigma = \displaystyle\sum_{p=1}^{m} u_p(t) g_p(s),
\end{cases}
\tag{10}
$$

with zero initial conditions (2). Here, $u_p \in L^2(0, T)$, $p = 1, \ldots, m$, are the control functions while functions $g_p \in L^2(\Gamma)$ are specified.

Since we are dealing with a particular case of nonhomogeneity of the form $y|_\Sigma = u \in L^2(\Sigma)$, the solution to problem (10), (2) as well as the solution of problem (1), (2) satisfies inclusion $\{y, y_t\} \in C(0, T; \mathcal{W}_0)$.

For system (10), $U = \mathbb{C}^m$, $\mathcal{U} = L^2(0, T; \mathbb{C}^m)$. Formula (3) yields (compare with Subsections 1.2 and IV.2.2) the corresponding family $\mathscr{E} = \{e_k\}$, $k \in \mathbb{K}$, of the form $e_k(t) = \eta_k e^{i\omega_k t}$ where η_k is the vector from $\mathbb{C}^m$ with the components

$$
\left(\frac{\partial \varphi_{|k|}}{\partial \nu}, g_p \right)_{L^2(\Gamma)} \qquad p = 1, \ldots, m.
$$

Having accounted for the multiplicity of the eigenvalues (see Subsections IV.1.2, V.1.1, and V.1.2), we rewrite family $\mathscr{E}$ as $\{e_{kj}\}$, $j = 1, \ldots, \kappa_{|k|}$, $k \in \mathbb{K}$, $e_{kj}(t) = \eta_{kj} e^{i\nu_k t}$ with the vector η_{kj} from $\mathbb{C}^m$, whose components are

$$
\left(\frac{\partial \varphi_{|k|j}}{\partial \nu}, g_p \right)_{L^2(\Gamma)}
$$

Theorem V.2.8. Let $N > 1$. Then

(a) for any $T > 0$, system (10) is not M-controllable;
(b) if conditions

$$
\operatorname{rank}\left[\left(g_p, \frac{\partial \varphi_{nj}}{\partial \nu} \right) \right]_{j=1,\ldots,\kappa_n}^{p=1,\ldots,m} = \kappa_n, \qquad n \in \mathbb{N},
\tag{11}
$$

are not satisfied – in particular, if sequence $\{\kappa_n\}$ is unbounded – then system (10) is not W-controllable for any T. Moreover,

$$\mathrm{Cl}_{\mathscr{W}_6} \bigcup_{T>0} \mathscr{R}(T) \neq \mathscr{W}_0;$$

(c) *if equalities (11) are valid, then*

$$\mathrm{Cl}_{\mathscr{W}_6} \bigcup_{T>0} \mathscr{R}(T) = \mathscr{W}_0.$$

The proof of this theorem follows the corresponding statements of Theorem 1.3.

Now consider the case $N = 1$ in more detail. Domain Ω here is a segment $(0, l)$ while its boundary, Γ, consists of two points $x = 0$ and $x = l$. That is why we have two essentially different versions for system (10): either the control is applied at one of the end points or at both of them. Let us start with the first one. Thus, we have a system described by initial boundary-value problem

$$\begin{cases} \dfrac{\partial^2 y}{\partial t^2} = \dfrac{\partial}{\partial x}\left(a(x)\dfrac{\partial y}{\partial x} \right) - a_0(x)y & \text{in } Q, \\ y|_{x=0} = 0, \quad y|_{x=l} = u \in L^2(0, T), \quad y(x, 0) = y_t(x, 0) = 0. \end{cases} \tag{12}$$

As in Section 1, let

$$T_0 = 2 \int_0^l \frac{dx}{(\sqrt{a(x)})}.$$

Theorem V.2.9. The following assertions are true.

(a) *For $T \geq T_0$, system (12) is B-controllable relative to space $\mathscr{W}_0 = L^2(0, l) \oplus H^{-1}(0, l)$.*

(b) *For $T < T_0$, system (12) is not W-controllable. Reachability set $\mathscr{R}(T)$ is a subspace of infinite codimension in $\mathscr{W}_0$.*

PROOF. In this particular case, formula (3) looks like

$$\langle Bv, \varphi \rangle_* = -v(l)\overline{\frac{\partial \varphi(l)}{\partial x}},$$

and family $\mathscr{E}$ corresponding to system (12) is of the form

$$\left\{ \frac{\partial \varphi_{|k|}(l)}{\partial x} e^{i\omega_k t} \right\}.$$

Family $\mathscr{E}_0$ corresponding to family $\mathscr{E}$ and space $\mathscr{W}_0$ (for the definition see Section III.3) is

$$\left\{ (\lambda_{|k|} + \alpha)^{-1/2} \frac{\partial \varphi_{|k|}(l)}{\partial x} e^{i\omega_k t} \right\}.$$

Numbers $\partial \varphi_{|k|}(l)/\partial x$, $k \in \mathbb{K}$, differ from zero, and, by the force of the known asymptotics (Naimark 1969: chap. II, secs. 4.5, 4.10),

$$(\lambda_{|k|} + \alpha)^{-1/2} \frac{\partial \varphi_{|k|}(l)}{\partial x} \asymp 1, \qquad k \in \mathbb{K}. \tag{13}$$

As we noted while proving Theorem 1.5, family $\{e^{i\omega_k t}\}$ constitutes an $\mathscr{L}$-basis in space $L^2(0, T)$ for $T \geq T_0$. Formula (13) implies that family $\mathscr{E}_0$ possesses the same property. Assertion (a) now follows from Theorem III.3.10.

For $T < T_0$, Theorem II.4.16 allows us to choose a subfamily $\{e^{i\omega_k t}\}_{k \in \mathbb{K}'}$ from family $\{e^{i\omega_k t}\}_{k \in \mathbb{K}}$ in such a way that it forms a Riesz basis in $L^2(0, T)$, and set $\{\mathbb{K} \setminus \mathbb{K}'\}$ is infinite.

Assertion (b) now becomes a direct consequence of Theorem I.2.2(e) and Theorem III.3.10.

Remark V.2.10. Assertion (a) of Theorem 9 was proved, in fact, by Russell (1967). Systems of a more general type – described by a non-self-adjoint or nonregular boundary-value problem (in particular, systems with infinite optical length) – were studied by Avdonin (1975, 1980, 1982).

Suppose now that the control actions are applied to both boundary points; that is, the boundary conditions for system (12) are

$$y(0, t) = u_1(t), \qquad y(l, t) = u_2(t); \qquad u_{1,2} \in L^2(0, T).$$

For such a system, the analog of Theorem V.2.9 is valid with the replacement of T_0 by $T_0/2$,

$$T_0/2 = \int_0^l \frac{dx}{\sqrt{a(x)}}.$$

This statement is demonstrated in Section 4 of Chapter VII (there the equation of string vibrations is considered, but this alteration does not influence the essence of the problem).

2.3. Pointwise control

Let $N > 1$ and the system be described by the initial boundary problem

$$
\begin{cases}
\dfrac{\partial^2 y}{\partial t^2} + Ay = 0 & \text{in } Q, \\[2mm]
y(s, t)|_\Sigma = \displaystyle\sum_{p=1}^{m} u_p(t)\,\delta(s - s_p),
\end{cases}
\tag{14}
$$

with zero initial conditions (2).

As established in Subsection IV.2.3, such a nonhomogeneous boundary condition is immersed in the scheme of Section III.2 with operator $B \in \mathscr{L}(U, W_q)$ for $q < -1 - N/2$.

Therefore, by Theorem III.2.1 for any function $u \in \mathscr{U} = L^2(0, T; \mathbb{C}^m)$, there exists a unique solution of problem (14), (2) such that $\{y, y_t\} \in C(0, T; \mathscr{W}_r)$ for any $r < -N/2$.

Theorem V.2.11. *The following assertions are valid.*

(a) *For any $T > 0$, system (14) is not M-controllable.*
(b) *If conditions*

$$
\operatorname{rank}\left[\varphi_{nj}(s_p)\right]_{j=1,\ldots,\kappa_n}^{p=1,\ldots,m} = \kappa_n, \qquad n \in \mathbb{N},
\tag{15}
$$

do not hold, then for system (14)

$$
\operatorname{Cl}_{\mathscr{W}_r} \bigcup_{T>0} \mathscr{R}(T) \neq \mathscr{W}_r, \qquad r < -N/2.
$$

(c) *If conditions (15) are satisfied, then*

$$
\operatorname{Cl}_{\mathscr{W}_r} \bigcup_{T>0} \mathscr{R}(T) = \mathscr{W}_r, \qquad r < -N/2.
$$

The proof of these statements follows that of the corresponding statements of Theorem 1.3.

Remark V.2.12. As shown in the previous and the present chapters, parabolic- and hyperbolic-type equations with various kinds of finite dimensional control are not M-controllable if $\dim \Omega > 1$. The reason is that corresponding vector exponential families are not minimal in $L^2(0, T; \mathbb{C}^m)$ for any $m \in \mathbb{N}$ and $T > 0$, since the sets $\{i\lambda_n\}$ and $\{\omega_n + i\}$ do not satisfy the Blaschke condition. By the force of Theorems III.3.3

and III.3.10, the lack of M controllability implies a lack of E controllability relative to Sobolev space with any exponent. These results seem to have first been obtained by Avdonin and Ivanov (1989b).

Triggiani (1991) proved that the wave equation with finite dimensional Dirichlet boundary control is not exactly controllable in $L^2(\Omega) \oplus H^{-1}(\Omega)$.

Remark V.2.13. In Chapter VI we prove stronger results concerning the lack of approximate controllability for the wave equation in a rectangle with finite dimensional control. We expect these results to be valid for hyperbolic equations of the general type.

Remark V.2.14. Regarding the results of this chapter, an analog of Remark IV.2.8 proves to be correct: instead of the Dirichlet problem, one is able to study controllability in systems with other boundary conditions by the same approach.

VI

Control of rectangular membrane vibrations

In this chapter we examine the controllability of a system describing rectangular membrane oscillations under boundary and pointwise controls of some type. With the aid of the explicit form of the eigenfunctions and eigenvalues of the Laplace operator, we can now complement and sharpen several results of Chapter V for this particular system, namely, those pertaining to initial boundary-value problems. We also demonstrate that for some kinds of finite dimensional control the system proves to be not W-controllable in any finite time.

Section 1 is based on the work of Avdonin and Ivanov (1988, 1989b), Section 2 on a detailed exposition by Avdonin, Ivanov, and Joó (1990).

1. Boundary control

1.1. Regularity of the solution

Let

$$a, b \in \mathbb{R}_+; \qquad \Omega = (0, a) \times (0, b); \qquad \Gamma = \partial\Omega; \qquad \Gamma_1 = (0, a) \times \{0\};$$

$$\Gamma_0 = \Gamma \backslash \Gamma_1; \qquad \Sigma_j = \Gamma_j \times (0, T); \qquad j = 0, 1;$$

and v be the outward normal vector to Γ. Consider a system described by the boundary value problem

$$\begin{cases} \dfrac{\partial^2 z}{\partial t^2} = \dfrac{\partial^2 z}{\partial x^2} + \dfrac{\partial^2 z}{\partial y^2} & \text{in } Q, \\[2mm] z|_{\Sigma_0} = 0, \qquad \dfrac{\partial z}{\partial v}\bigg|_{\Sigma_1} = u \in L^2(\Sigma_1), \end{cases} \tag{1}$$

216

with zero initial conditions

$$z = z_t = 0 \qquad \text{for } t = 0. \tag{2}$$

A generalized solution to initial boundary-value problem (1), (2) of class $L^2(Q)$ is understood in the sense of an integral identity

$$\int_Q z(w_{tt} - \Delta w)\, dx\, dt = \int_{\Sigma_1} uw\, ds\, dt \tag{3}$$

valid for any function $w \in H^2(Q)$ such that

$$w|_{\Sigma_0} = 0, \qquad \frac{\partial w}{\partial v}\bigg|_{\Sigma_1} = 0, \qquad w = w_t = 0 \quad \text{for } t = T.$$

The existence and uniqueness of such a solution is proved easily by the transposition method (Lions and Magenes 1968). We now demonstrate that the solution to (1), (2) is essentially more regular.

According to the scheme of Chapter III, system (1) may be represented in the form

$$z_{tt} + Az = Bu. \tag{4}$$

Here, A is the $(-\Delta)$ operator in space $L^2(Q)$ with the domain

$$\mathscr{D}(A) = \left\{ \varphi \in H^2(\Omega) \,\bigg|\, \varphi|_{\Gamma_0} = 0, \frac{\partial \varphi}{\partial v}\bigg|_{\Gamma_1} = 0 \right\}.$$

Its eigenvalues and eigenfunctions are conveniently enumerated by means of two indices $m, n \in \mathbb{N}$. It is easy to check that they are

$$\lambda_{mn} = \left(\frac{\pi}{a} m\right)^2 + \left(\frac{\pi}{b}(n - 1/2)\right)^2,$$

$$\varphi_{mn}(x, y) = \frac{2}{\sqrt{ab}} \sin \frac{\pi}{a} mx \cos \frac{\pi}{b}(n - 1/2)y.$$

As in the previous chapters, we let W_r, $r \in \mathbb{R}$, denote the space of functions f on Ω such that

$$f(x, y) = \sum_{m,n} c_{mn} \varphi_{mn}(x, y), \qquad \|f\|_{W_r}^2 := \sum_{m,n} |c_{mn}|^2 \lambda_{mn}^r < \infty.$$

One is able to show that the relations

$$H_0^r(\Omega) \subset W_r \subset H^r(\Omega), \qquad \|f\|_{W_r} \asymp \|f\|_{H^r(\Omega)}, \qquad (f \in W_r) \tag{5}$$

hold for all $r \geq 0$ (see (0.4) and (0.5) of Chapter IV). In the latter relation expressing the equivalence of norms, one should exclude half-integer r.

Set $U = L^2(\Gamma_1)$, $\mathcal{U} = L^2(0, T; U)$. By the force of equality (3), operator B in formula (4) is given by

$$\langle Bv, \varphi \rangle_* = \int_{\Gamma_1} v(s) \overline{\varphi(s)} \, ds \tag{6}$$

(compare with (V.2.3)). From the embedding theorem, it follows that

$$\|\varphi|_{\Gamma_1}\|_{L^2(\Gamma_1)} \prec \|\varphi\|_{W_r} \qquad \text{for } \varphi \in W_r, r < 1/2.$$

Therefore, by Theorem III.2.1, the solution to initial boundary-value problem (1), (2) satisfies inclusion $\{z, z_t\} \in C(0, T; \mathcal{W}_q)$ for any $q < 1/2$ with $\mathcal{W}_q = W_q \oplus \mathcal{W}_{q-1}$. So, by (5), $z \in C(0, T; H^q(\Omega))$.

This regularity result is valid for arbitrary domains $\Omega \subset \mathbb{R}^N$ and elliptic operators A (under certain assumptions regarding the regularity of the operator coefficients and the domain boundary).

In the case $N = 1$, it is possible to show (Avdonin and Ivanov 1984) that

$$\{z, z_t\} \in C(0, T; H^1(\Omega) \oplus L^2(\Omega)).$$

If Ω is a ball in $\mathbb{R}^N$, $A = -\Delta$, then

$$\{z, z_t\} \in C(0, T; H^{2/3}(\Omega) \oplus H^{-1/3}(\Omega))$$

(Graham and Russell 1975). For our case, the following theorem is valid.

Theorem VI.1.1. For any $u \in L^2(\Sigma_1)$ there exists a unique solution to problem (1), (2) such that

$$\{z, z_t\} \in C(0, T; H^{3/4}(\Omega) \oplus H^{-1/4}(\Omega)).$$

PROOF. Set $\omega_{mk} = (\operatorname{sgn} k)\sqrt{\lambda_{m|k|}}$, $m \in \mathbb{N}$, $k \in \mathbb{K} = \mathbb{Z} \backslash \{0\}$. Formula (6) implies

$$\text{family } \mathscr{E} = \{e_{mk}\} = \{B^* \varphi_{m|k|} e^{i\omega_{mk}t}\}$$

to be of the form

$$e_{mk}(x, t) = \varphi_{m|k|}|_{\Gamma_1} e^{i\omega_{mk}t} = \frac{2}{\sqrt{ab}} \sin \frac{\pi}{a} m|k| \, e^{i\omega_{mk}t}.$$

We will establish, for all $u \in \mathcal{U}$, $t \in [0, T]$, the validity of the estimate

$$\sum_{m, k} |(u, e_{mk})_{L^2(0, t; U)}|^2 |\omega_{mk}|^{-1/2} \prec \|u\|^2_{L^2(0, t; U)}. \tag{7}$$

The assertion of the theorem will follow then from Lemma III.2.4 and relations (5). Note that since $H^r(\Omega) = H_0^r(\Omega)$ for $0 \le r < 1/2$ (Lions and Magenes 1968: chap. I), then $W_r = H^r(\Omega)$ for $-1/2 < r < 1/2$.

To continue the proof, we need the following assertion.

Lemma VI.1.2. For any $f \in L^2(0, T)$, $t \in [0, T]$, relation

$$\sum_{k \in \mathbb{K}} |(f, e^{i\omega_{mk}\tau})_{L^2(0,t)}|^2 \prec \sqrt{m} \|f\|^2_{L^2(0,t)}$$

holds.

PROOF OF THE LEMMA. Let us fix $m \in \mathbb{N}$ and decompose sequence $\{\omega_{mk}\}_{k \in \mathbb{K}} \cup \{\omega_{m0}\}$ to $M :=$ entier $(\sqrt{m})$ subsequences $\{\mu_n^j\}_{n \in \mathbb{Z}}$, $\mu_n^j := \omega_{m, j + nM}$, $j = 1, \ldots, M$. Elementary calculations give inequality

$$|\mu_{n+1}^j - \mu_n^j| \ge \frac{\pi}{3\sqrt{a^2 + a^4/b^2}}.$$

For a countable set $\sigma \subset \mathbb{C}_+$ and any function $g \in L^2(0, \infty)$ an estimate

$$\sum_{\mu \in \sigma} \left|(g, e^{i\mu\tau})_{L^2(0,\infty)} \sqrt{2 \operatorname{Im} \mu}\right|^2 \le \left(32 + 64 \log \frac{1}{\delta(\sigma)}\right) \|g\|^2_{L^2(0,\infty)}$$

is valid (in Subsection II.1.3.12, the formula was cited for the case of simple fractions instead of exponentials). Let us apply it to $\sigma = \{\mu_n^j + i/2\}_{n \in \mathbb{K}}$ and use Lemma II.1.19. Continuing function f by the zero value from $(0, t)$ to (t, ∞), we arrive at

$$\sum_{n \in \mathbb{Z}} |(f, e^{i(\mu_n^j + i/2)\tau})_{L^2(0,t)}|^2 \le C(a, b) \|f\|^2_{L^2(0,t)}$$

and so

$$\sum_{n \in \mathbb{Z}} |(f, e^{i\mu_n^j\tau})_{L^2(0,t)}|^2 \le C(a, b) e^T \|f\|^2_{L^2(0,t)}$$

for $f \in L^2(0, T)$, $t \in [0, T]$. Summing up these inequalities over $j = 1, \ldots, M$, we obtain the assertion of the lemma.

We are now able to complete the proof of Theorem 1. Let $\{u_m(\tau)\}_{m \in \mathbb{N}}$ be Fourier coefficients in the expansion of function $u(x, \tau)$ over family $\{\sqrt{2/a} \sin(\pi/a) mx\}$. Then

$$(u, e_{mk})_{L^2(0,t; U)} = \sqrt{2/b} (u_m, e^{i\omega_{mk}\tau})_{L^2(0,t)}. \tag{8}$$

Using an obvious inequality $|\omega_{mk}|^{-1/2} \prec m^{-1/2}$ and applying Lemma 3 to functions u_m, we find

$$\frac{2}{b} \sum_{m,k} |(u_m, e^{i\omega_{mk}\tau})_{L^2(0,t)}|^2 |\omega_{mk}|^{-1/2} \prec \sum_m m^{-1/2} \sum_{m,k} |(u_m, e^{i\omega_{mk}\tau})_{L^2(0,t)}|^2$$

$$\prec \sum_m m^{-1/2} m^{1/2} \|u_m\|_{L^2(0,t)}^2 = \|u\|_{L^2(0,t;\,U)}^2.$$

Thus, in accounting for equalities (6), we have demonstrated relation (7) and with it Theorem 1.

Let us now show that the assertion of Theorem 1 cannot be improved in the scale of Sobolev spaces. Namely, for any $p > -1/4$ there exists control $u \in \mathcal{U}$ such that relation

$$\{z(\cdot, T), z_t(\cdot, T)\} \notin H^{p+1}(\Omega) \oplus H^p(\Omega) \tag{9}$$

holds for the solution to problem (1), (2) corresponding to this control. We take the control in the form

$$u(x, t) = \sum_{m=1}^{\infty} \gamma_m \sin \frac{\pi}{a} mx \exp\left(i \frac{\pi}{a} m(T - t)\right), \qquad \sum_{m=1}^{\infty} |\gamma_m|^2 < \infty.$$

Then using formulas (16) and (17) of Section III.2 we have

$$\|(z(\cdot, T), z_t(\cdot, T))\|_{\mathcal{W}_{p+1}}^2 \asymp \sum_{m,k} |\omega_{mk}|^{2p} |(u(x, T - t), e_{mk}(x, t))_{L^2(0,T)}|^2$$

$$\asymp \sum_{m=1}^{\infty} |\gamma_m|^2 \sum_{k \in \mathbb{K}} |\omega_{mk}|^{2p} \left| \left(\exp\left(i \frac{\pi}{a} mt\right), \exp(i\omega_{mk}t) \right)_{L^2(0,T)} \right|^2. \tag{10}$$

Set

$$c_{mk} := \left(\exp\left(i \frac{\pi}{a} mt\right), \exp(i\omega_{mk}t) \right)_{L^2(0,T)}, \qquad \kappa_{mk} := \omega_{mk} - \frac{\pi}{a} m.$$

A simple calculation shows that

$$c_{mk} = T e^{-i\kappa_{mk}T/2} \frac{\sin(\kappa_{mk}T/2)}{\kappa_{mk}T/2}.$$

Since for $k > 0$

$$\kappa_{mk} = \frac{\pi}{a} m \left[\sqrt{1 + \left(\frac{a(k - 1/2)}{bm} \right)^2} - 1 \right] < \frac{\pi}{a} m \frac{1}{2} \left(\frac{a(k - 1/2)}{bm} \right)^2 = \frac{\pi a(k - 1/2)^2}{2b^2 m},$$

$\kappa_{mk}T < \pi$ if

$$k - 1/2 < \frac{b}{\sqrt{aT/2}}\sqrt{m}.$$

Now from an elementary inequality

$$\frac{\sin x}{x} > \frac{2}{\pi}, \qquad x \in \left(0, \frac{\pi}{2}\right),$$

we derive

$$|c_{mk}|^2 \geq \left(\frac{2T}{\pi}\right)^2, \qquad \text{for } 1 \leq k < \frac{b}{\sqrt{aT/2}}\sqrt{m} + 1/2. \tag{11}$$

To verify relation (9) it is sufficient to demonstrate that

$$(z(\cdot, T), z_t(\cdot, T)) \notin W_p, \qquad -1/4 < p < 0.$$

For such p and $k \prec m$ we have

$$|\omega_{mk}|^{2p} \succ m^{2p} \qquad (m, k \in \mathbb{N}). \tag{12}$$

Now (10)–(12) implies

$$\|(z(\cdot, T), z_t(\cdot, T))\|^2 \succ \sum_{m=1}^{\infty} |\gamma_m|^2 m^{2p} \sum_{k(m)} |c_{mk}|^2 \succ \sum_{m=1}^{\infty} |\gamma_m|^2 m^{2p}\sqrt{m}.$$

For $p > -1/4$, sequence $\{\gamma_m\} \in \ell^2$ may be chosen in such a way that

$$\sum_{m=1}^{\infty} |\gamma_m|^2 m^{2p+1/2} = \infty$$

and hence, $(z(\cdot, T), z_t(\cdot, T)) \notin W_p$.

Results close to Theorem 1 and other results concerning the regularity of solutions of nonhomogeneous boundary-value problems can be found in Lasiecka and Triggiani (1981, 1989a), and Lasiecka, Lions, and Triggiani (1986).

1.2. Lack of controllability

Let us now proceed with the question of system (1) controllability. From the results of Fattorini (1979), M controllability of system (1) follows for $T > 2b$. We confine ourselves to the case in which control u is $u(x, t) = b(x)v(t)$, $b \in L^2(0, a)$ being a specified function, $v \in L^2(0, T)$ being a control. Thus $U = \mathbb{C}$, $\mathcal{U} = L^2(0, T)$, and corresponding family $\mathcal{E}$ (cf.

Section II.2) is of the form $\{\beta_m \, e^{i\omega_{mk}t}\}$ with

$$\overline{\beta_m} = \frac{2}{\sqrt{ab}} \int_0^a b(x) \sin \frac{\pi}{a} mx \, dx.$$

Theorem VI.1.3. For $\mathcal{U} = L^2(0, T)$ and any function $b \in L^2(0, a)$, system (1) is not W-controllable in any finite time T; that is,

$$\mathrm{Cl}_{\mathscr{W}_{3/4}} \mathscr{R}(T) \neq \mathscr{W}_{-3/4}, \qquad \mathscr{W}_{3/4} := W_{3/4} \oplus W_{-1/4}.$$

Remark VI.1.4. In Section V.2, set $\bigcup_{T>0} \mathscr{R}(T)$ was shown to be dense in $\mathscr{W}_{1/2}$ if and only if the "rank criterion" (V.2.11) is fulfilled. If the eigenvalue multiplicity is unbounded, the criterion is broken and Theorem 3 follows directly from Theorem V.2.8. This is particularly the case for a square membrane. If the ratio a^2/b^2 for a rectangular membrane is irrational, the spectrum multiplicity equals unity and the rank criterion may be valid. In this case assertion (b) of Theorem 3 is not implied by the results of Chapter V.

This remark also applies to Theorem 7 and Theorems 2.1, 2.2, which are proved below.

PROOF OF THEOREM 3. Take arbitrary $T > 0$. By Theorem III.3.10, the assertion we are proving is equivalent to condition $\tilde{\mathscr{E}} \notin (W)$ in space $L^2(0, T)$. Family $\tilde{\mathscr{E}} = \{\tilde{e}_{mk}\}$ corresponding to family $\mathscr{E}$ and space $\mathscr{W}_{3/4}$ is $\tilde{e}_{mk}(t) = \beta_m |\omega_{mk}|^{-1/4} e_{mk}(t)$.

If at least one of the coefficients β_m equals zero, then family $\tilde{\mathscr{E}}$ is linearly dependent. Let $\beta_m \neq 0$ for all $m \in \mathbb{N}$. By Theorem II.6.9 there exist $m_0 \in \mathbb{N}$ and sequence $\{c_{mk}\}$ such that

(a) $\displaystyle\sum_{m,k} c_{mk} \, e^{i\omega_{mk}t} = 0 \qquad$ in $L^2(0, T)$;

(b) $\displaystyle\sum_{m,k} |c_{mk}|^2 \omega_{mk}^2 < \infty$;

(c) $c_{mk} = 0$ for $m \geq m_0$.

Set $a_{mk} = c_{mk} |\omega_{mk}|^{1/4} \beta_m^{-1}$. The above assertions (b), (c) imply

$$\sum_{m,k} |a_{mk}|^2 < \infty.$$

Assertion (a) provides us with

$$\sum_{m,k} a_{mk} \tilde{e}_{mk} = \sum_{m,k} c_{mk} |\omega_{mk}|^{1/4} \beta_m^{-1} \beta_m |\omega_{mk}|^{-1/4} \, e^{i\omega_{mk}t} = 0.$$

So, family $\tilde{\mathscr{E}}$ is ω-linearly dependent (see Subsection I.1.2) and the more so $\tilde{\mathscr{E}} \notin (W)$. Theorem 3 is proved.

1.3. Estimate of the Carleson constant

A closely related problem of some interest in itself is to find an estimate of the Carleson constant (see Definition II.1.17) δ_m of set $\{\omega_{mk} + i/2\}_{k \in \mathbb{K}}$ depending on m. Knowledge of δ_m allows us to estimate the norms of elements of the family biorthogonal to $\{e^{i(\omega_{mk} + i/2)t}\}$ in space $L^2(0, \infty)$ (see for formula (12), Section II.1).

Notice that the following lemma allows us to prove Lemma 2 without ·the use of Lemma II.1.19.

Lemma VI.1.5. The Carleson constant of set $\{\omega_{mk} + i/2\}_{k \in \mathbb{K}}$ allows us to estimate

$$\log \delta_m^{-1} \prec \sqrt{m}, \qquad m \in \mathbb{N}.$$

PROOF. To simplify the writing, let us assume that

$$\omega_{mk} = (\operatorname{sgn} k)\sqrt{m^2 + k^2}.$$

For the case

$$\omega_{mk} = (\operatorname{sgn} k)\sqrt{\left(\frac{\pi m}{a}\right)^2 + \left(\frac{\pi}{b}(|k| - 1/2)\right)^2}$$

the proof needs only a few obvious alterations. We write the Carleson constant δ_m of set $\{v_{mn}\} = \{\omega_{mn} + i/2\}$, $n \in \mathbb{Z}\backslash\{0\}$ as

$$\delta_m = \inf_n \prod_{k \neq n} \left|\frac{v_{mk} - v_{mn}}{v_{mk} - \bar{v}_{mn}}\right|.$$

Hence,

$$\log \delta_m^{-2} = \sup_n \sum_{k \neq n} \log \left|\frac{v_{mk} - \bar{v}_{mn}}{v_{mk} - v_{mn}}\right|^2$$

$$= \sup_n \sum_{k \neq n} \log[1 + ((\operatorname{sgn} k)\sqrt{m^2 + k^2} - (\operatorname{sgn} n)\sqrt{m^2 + n^2})^{-2}].$$

Specifically, let us consider $n > 0$ and estimate the sum

$$S_n = \sum_{k \neq n} \log[1 + ((\operatorname{sgn} k)\sqrt{m^2 + k^2} - (\operatorname{sgn} n)\sqrt{m^2 + n^2})^{-2}]$$

(it is evident that $S_{-n} = S_n$). We can then represent S_n in the form

$$S_n = \left(\sum_{k=-\infty}^{-1} + \sum_{k=1}^{n-1} + \sum_{k=n+1}^{\infty} \right)$$

$$\times \log[1 + ((\operatorname{sgn} k)\sqrt{m^2 + k^2} - (\operatorname{sgn} n)\sqrt{m^2 + n^2})^{-2}]$$

and estimate each sum in the right-hand side separately, denoting them by $S_n^{(1)}$, $S_n^{(2)}$, and $S_n^{(3)}$, respectively.

(1) The first sum can be estimated by a constant independing on m.

$$S_n^{(1)} = \sum_{k=-\infty}^{-1} \log[1 + (\sqrt{m^2 + n^2} + \sqrt{m^2 + k^2})^{-2}]$$

$$< \sum_{k=1}^{\infty} \log\left(1 + \frac{1}{k^2}\right) < \infty;$$

(2) To estimate the second sum

$$S_n^{(2)} = \sum_{k=1}^{n-1} \log[1 + (\sqrt{m^2 + n^2} - \sqrt{m^2 + k^2})^{-2}]$$

consider first the case $n < m$. Represent $S_n^{(2)}$ as

$$\sum_{p=1}^{n-1} \log[1 + (\sqrt{m^2 + n^2} - \sqrt{m^2 + (n-p^2)})^{-2}].$$

By means of elementary manipulations, one is able to check that inequalities

$$\sqrt{m^2 + n^2} - \sqrt{m^2 + (n-p)^2} \geq \frac{p^2}{2\sqrt{2m}}, \qquad 1 \leq p < n < m,$$

are correct. Indeed

$$\sqrt{m^2 + n^2} - \sqrt{m^2 + (n-p)^2} = \frac{2np - p^2}{\sqrt{m^2 + n^2} + \sqrt{m^2 + (n-p)^2}}$$

$$\geq \frac{p(2n - p)}{\sqrt{2m^2} + \sqrt{2m^2}} \geq \frac{p^2}{2\sqrt{2m}}.$$

Therefore

$$\sum_{p=1}^{n-1} \log[1 + (\sqrt{m^2 + n^2} - \sqrt{m^2 + (n-p)^2})^{-2}]$$

$$\le \sum_{p=1}^{n-1} \log\left(1 + \frac{8m^2}{p^4}\right) < \sum_{p=1}^{\infty} \log\left(1 + \frac{8m^2}{p^4}\right) < \int_1^{\infty} \log\left(1 + \frac{8m^2}{x^4}\right) dx$$

$$= \sqrt{m} \int_{1/\sqrt{m}}^{\infty} \log\left(1 + \frac{8}{t^4}\right) dt \le \sqrt{m} \int_0^{\infty} \log\left(1 + \frac{8}{t^4}\right) dt = c_1 \sqrt{m}$$

for some $c_1 \in (0, \infty)$.

Now let $n \ge m$. With the help of elementary calculations, one again checks that inequalities

$$\sqrt{m^2 + n^2} - \sqrt{m^2 + (n-p)^2} \ge \frac{p}{2\sqrt{2}}, \qquad 1 \le p \le n-1, n \ge m,$$

hold. So

$$\sum_{p=1}^{n-1} \log[1 + (\sqrt{m^2 + n^2} - \sqrt{m^2 + (n-p)^2})^{-2}] \le \sum_{p=1}^{\infty} \log\left(1 + \frac{8}{p^2}\right) < \infty.$$

Thus we have proved $\sup_{n \ne 0} S_n^{(2)} \prec \sqrt{m}$.

(3) For $n < m$ sum, represent $S_n^{(3)}$ as

$$\left(\sum_{k=n+1}^{m} + \sum_{k=m+1}^{\infty} \right) \log[1 + (\sqrt{m^2 + k^2} - \sqrt{m^2 + n^2})^{-2}].$$

Moreover,

$$\sum_{k=n+1}^{m} \log[1 + (\sqrt{m^2 + k^2} - \sqrt{m^2 + n^2})^{-2}]$$

$$= \sum_{p=1}^{m-n} \log[1 + (\sqrt{m^2 + (n+p)^2} - \sqrt{m^2 + n^2})^{-2}] \le c_2 \sqrt{m}$$

for some $c_2 > 0$. The latter inequality is justified in the same way as when estimating sum $S_n^{(2)}$ on the grounds of inequalities

$$\sqrt{m^2 + (n+p)^2} - \sqrt{m^2 + n^2} \ge \frac{p^2}{2\sqrt{2m}}, \qquad 1 \le p \le m-n, n < m,$$

valid. Now consider the sum

$$\sum_{k=m+1}^{\infty} \log[1 + (\sqrt{m^2 + k^2} - \sqrt{m^2 + n^2})^{-2}]$$

$$= \sum_{p=1}^{\infty} \log[1 + (\sqrt{m^2 + (m + p)^2} - \sqrt{m^2 + n^2})^{-2}].$$

It is estimated by a constant (independent of m and n) since, as can be easily checked,

$$\sqrt{m^2 + (m + p)^2} - \sqrt{m^2 + n^2} \ge \frac{p}{2\sqrt{2}}, \qquad p \ge 1, n < m.$$

It remains only to estimate sum $S_n^{(3)}$ for $n \ge m$:

$$\sum_{k=n+1}^{\infty} \log[1 + (\sqrt{m^2 + k^2} - \sqrt{m^2 + n^2})^{-2}]$$

$$= \sum_{p=1}^{\infty} \log[1 + (\sqrt{m^2 + (n + p)^2} - \sqrt{m^2 + n^2})^{-2}] < c_3$$

for some $c_3 > 0$. The latter inequality follows from

$$\sqrt{m^2 + (n + p)^2} - \sqrt{m^2 + n^2} \ge \frac{p}{2\sqrt{2}}, \qquad p \ge 1, n \ge m.$$

1.4. Pointwise boundary control

Consider now the case of the pointwise boundary control of rectangular membrane vibrations:

$$\begin{cases} z_{tt} = z_{xx} + z_{yy} & \text{in } Q, \\ z|_{\Sigma_0} = 0, \quad \dfrac{\partial z}{\partial v}\bigg|_{\Sigma_1} = v(t)\delta(x - x_0), \end{cases} \tag{13}$$

$$v \in L^2(0, T), \qquad x_0 \in (0, a), \qquad z|_{t=0} = z_t|_{t=0} = 0. \tag{14}$$

Theorem VI.1.6. Initial boundary-value problem (13), (14) has a unique solution $z \in C(0, T; W_{1/2})$ for which $z_t \in C(0, T; W_{-1/2})$.

Theorem VI.1.7. For $\mathcal{U} = L^2(0, T)$ and any $x_0 \in (0, a)$, system (11) is not W-controllable in any finite time T. That is,

$$\text{Cl}_{\mathscr{W}_{1/2}}\mathscr{R}(T) \ne \mathscr{W}_{1/2}, \qquad \mathscr{W}_{1/2} = W_{1/2} \oplus W_{-1/2}.$$

The proof of Theorem 6 is arranged according to the scheme of the proof of Theorem V.1.4. One introduces the initial boundary-value problem dual to (13), (14),

$$
\begin{cases}
w_{tt}(p, t) = \Delta w(p, t), \quad p = (x, y) \in \Omega, \quad t \in (0, T'), \\
w|_{\Sigma_0} = 0, \quad \dfrac{\partial w}{\partial v}\bigg|_{\Sigma_1} = 0,
\end{cases}
\tag{15}
$$

$$
w|_{t=T'} = w_0 \in W_{1/2}, \qquad w_t|_{t=T'} = w_1 \in W_{-1/2}.
$$

Moreover, with the help of the Fourier method, representations for z and w are obtained in the form of a series in φ_{mn}. We also have

$$
\langle z(\cdot, T'), w_1 \rangle_* + \langle z_t(\cdot, T'), w_0 \rangle_* = \int_0^{T'} v(t)\overline{w(p_0, t)}\, dt
$$

with $p_0 = (x_0, 0)$. Let us verify inequality

$$
\|w(p_0, \cdot)\|_{L^2(0, T)}^2 \le C(T')[\|w_0\|_{W_{1/2}}^2 + \|w_1\|_{W_{-1/2}}^2],
$$

which is valid if and only if (see the proof of Theorem V.1.4)

$$
\sup_{l \in \mathbb{N}} \sum_{l \le \omega_{mn} < l+1} \varphi_{mn}^2(p_0)\omega_{mn}^{-1} < \infty.
$$

Since numbers

$$
\varphi_{mn}(p_0) = \frac{2}{\sqrt{ab}} \sin \frac{\pi}{a} m x_0
$$

are bounded, it suffices to show that

$$
\sup_{l \in \mathbb{N}} \sum_{l \le \omega_{mn} < l+1} \omega_{mn}^{-1} < \infty.
$$

The latter inequality is easily established by the explicit expression for ω_{mn}:

$$
\omega_{mn} = \left[\left(\frac{\pi}{a} m \right)^2 + \frac{\pi^2}{b^2} (n - \tfrac{1}{2})^2 \right]^{1/2}
$$

The remaining stages are completed as in the proof of Theorem V.1.4.

The proof of Theorem 7 repeats that of Theorem 3 except that it replaces β_m by

$$
\gamma_m := \frac{2}{\sqrt{ab}} \sin \frac{\pi}{a} m x_0.
$$

Remark VI.1.8. By the same method, one can prove the lack of approximate controllability for more powerful kinds of control than illustrated in Theorems 3 and 7. Avdonin and Ivanov (1995) investigated the wave equation in N-dimensional parallelepiped with the boundary control equal to zero everywhere except for an edge of dimension $N - 2$. In another case, the boundary control was supposed to be acting on a face of dimension $N - 1$ and depending on $N - 1$ independent variables (including t). It was proved that in both cases the system is not W-controllable for any $T > 0$.

Model problems of this kind allow us to put forward the following hypothesis concerning controllability of hyperbolic equations of the second order.

Hypothesis. If a control acts on an m-dimensional part of the boundary and/or on an m-dimensional part of domain Ω and $m < N - 1$, or if (more general formulation) a control function depends on less than N independent variables including t, then the system is not approximately controllable in any finite time.

2. Initial and pointwise control

In this section we consider vibrations of a rectangular membrane with homogeneous Dirichlet boundary conditions. We demonstrate that, roughly speaking, for any finite number of membrane points one is able to obtain arbitrary trajectories by choosing the appropriate initial conditions.

In a sense the problem is dual to the pointwise control problem. The reachability set of the system under the action of any finite number of pointwise controls is proved to be not dense in the phase space.

2.1. Principal results

Let $\Omega = (0, a) \times (0, b)$ and A be operator $(-\Delta)$ with the domain $\mathscr{D}(A) = H^2(\Omega) \cap H_0^1(\Omega)$. Let ω_{mn}^2 denote the eigenvalues of operator A,

$$\omega_{mn} = \sqrt{\left(\frac{\pi}{a} m\right)^2 + \left(\frac{\pi}{b} n\right)^2}; \qquad m, n \in \mathbb{N},$$

and φ_{mn} the corresponding eigenfunctions normalized in $L^2(\Omega)$

$$\varphi_{mn}(p) = \frac{2}{\sqrt{ab}} \sin \frac{\pi m}{a} x \sin \frac{\pi n}{a} y, \qquad p = (x, y).$$

As before, we introduce spaces $W_r = D(A^{r/2})$, $r \geq 0$; $W_{-r} = W_r'$.

Let $p_1, \ldots, p_N$ be arbitrary (distinct) points of Ω, $p_k = (x_k, y_k)$, T be any positive number, $Q = \Omega \times (0, T)$, and $\Sigma = \partial\Omega \times (0, T)$. Consider the initial boundary-value problem

$$\begin{cases} v_{tt} = \Delta v & \text{in } Q, \\ v|_{\Sigma} = 0, \\ v|_{t=0} = v_0, \quad v_t|_{t=0} = v_1. \end{cases} \tag{1}$$

Use $\Phi(t, v_0, v_1)$ to denote vector function

$$(v(p_1, t), \ldots, v(p_N, t)), \qquad \mathcal{W}_r := W_r \oplus W_{r-1}.$$

Theorem VI.2.1. Let r be a nonnegative integer. Then

(a) *for any vector-function $F \in H^r(0, T; \mathbb{C}^N)$ there exists initial state $(v_0, v_1) \in \mathcal{W}_r$ such that $\Phi(t, v_0, v_1) = F(t)$, $t \in [0, T]$;*

(b) *the dimension of the set of initial states $(v_0, v_1) \in \mathcal{W}_r$ for which $\Phi(t, v_0, v_1) = 0$, $t \in [0, T]$, is infinite.*

Let us write $z(p, t)$ for the solution of initial boundary-value problem

$$\begin{cases} z_{tt}(p, t) = \Delta z(p, t) + \displaystyle\sum_{k=1}^{N} \delta(p - p_k)u_k(t) & \text{in } Q, \\ z|_{\Sigma} = 0, \\ z|_{t=0} = z_t|_{t=0} = 0. \end{cases} \tag{2}$$

If $u_k \in L^2(0, T)$, $k = 1, \ldots, N$, then inclusion $(z, z_t) \in C(0, T; \mathcal{W}_{1/2})$ holds. This assertion follows immediately from Theorem V.1.4 for $N = 2$. Note that in our case it is possible to avoid the use of asymptotic properties of the spectral function by exploiting instead the explicit form of φ_{mn} as was done while proving Theorem 1.6.

Theorem VI.2.2. The reachability set $\mathcal{R}(T)$,

$$\mathcal{R}(T) := \{(z(\cdot, T), z_t(\cdot, T)) \mid u_k \in L^2(0, T), k = 1, \ldots, N\},$$

is not dense in $\mathcal{W}_{1/2}$ and $\operatorname{codim} \mathcal{R}(T) = \infty$.

2.2. Initial controllability

We prove here Theorem 1 concerning initial controllability of system (1). Let initial data v_0, v_1 belong to W_r and W_{r-1}, respectively. Then expansions

$$v_0 = \sum_{m,n} a_{mn}\varphi_{mn}, \qquad v_1 = \sum_{m,n} b_{mn}\varphi_{mn} \tag{3}$$

take place, and

$$\sum_{m,n} |a_{mn}|^2 \omega_{mn}^{2r} < \infty, \qquad \sum_{m,n} |b_{mn}|^2 \omega_{mn}^{2(r-1)} < \infty. \tag{4}$$

The following formula is obtained by means of the Fourier method:

$$v(p, t) = \sum_{m,n} \left[a_{mn} \cos \omega_{mn} t + \frac{b_{mn}}{\omega_{mn}} \sin \omega_{mn} t \right] \varphi_{mn}(p).$$

Introduce vector functions

$$c_{mn}(t) = \eta_{mn} \cos \omega_{mn} t, \qquad s_{mn}(t) = \eta_{mn} \sin \omega_{mn} t,$$

where

$$\eta_{mn} = \omega_{mn}^{-r}(\varphi_{mn}(p_1), \dots, \varphi_{mn}(p_N)) \in \mathbb{R}^N$$

and let $\mathscr{F}$ denote family $\{c_{mn}, s_{mn}\}$, $m, n \in \mathbb{N}$. Then

$$\Phi(t, u_0, u_1) = \sum_{m,n} [a_{mn}\omega_{mn}^r c_{mn}(t) + b_{mn}\omega_{mn}^{r-1} s_{mn}(t)]. \tag{5}$$

Lemma VI.2.3. If there exists subset $\mathscr{M} \subset \mathbb{N} \times \mathbb{N}$ such that family $\mathscr{F}_* = \{c_{mn}, s_{mn}\}_{(m,n)\in\mathscr{M}} \subset \mathscr{F}$ constitutes a Riesz basis in $H^r(0, T; \mathbb{C}^N)$, then assertion (a) of Theorem 1 holds.

PROOF. If $\mathscr{F}_*$ is a Riesz basis, then function F in $H^r(0, T; \mathbb{C}^N)$ may be expanded in a series

$$F(t) = \sum_{(m,n)\in\mathscr{M}} [f_{mn}^0 c_{mn}(t) + f_{mn}^1 s_{mn}(t)], \tag{6}$$

$$\sum_{(m,n)\in\mathscr{M}} [(f_{mn}^0)^2 + (f_{mn}^1)^2] < \infty. \tag{7}$$

Choose coefficients in expansions (3) in the following way:

$$a_{mn} = b_{mn} = 0, \qquad (m, n) \notin \mathscr{M}, \tag{8}$$

$$a_{mn} = f_{mn}^0 \omega_{mn}^{-r}, \qquad b_{mn} = f_{mn}^1 \omega_{mn}^{1-r}, \qquad (m, n) \in \mathscr{M}. \tag{9}$$

Inequality (7) implies inequalities (4), and inclusions $v_0 \in W_r$, $v_1 \in W_{r-1}$ are true. By comparing (5) and (6), we complete the proof of Lemma 3.

To construct basis $\mathscr{F}_*$, it is convenient to consider, instead of $\mathscr{F}$, family $\mathscr{E}$ of vector exponentials:

$$\mathscr{E} = \{e_{mn}^\pm\}_{m,n\in\mathbb{N}}, \qquad e_{mn}^\pm(t) = \eta_{mn} e^{\pm i\omega_{mn} t}.$$

Lemma VI.2.4. Family $\mathscr{F}_* = \{c_{mn}, s_{mn}\}_{(m,n)\in\mathscr{M}}$, $\mathscr{M} \subset \mathbb{N} \times \mathbb{N}$, *constitutes a Riesz basis in* $H^r(0, T; \mathbb{C}^N)$ *if and only if family* $\mathscr{E}_* = \{e_{mn}^\pm\}_{(m,n)\in\mathscr{M}}$ *is a Riesz basis in* $H^r(0, T; \mathbb{C}^N)$.

PROOF. Let $\mathscr{E}_0$ be a Riesz basis in $H^r(0, T; \mathbb{C}^N)$. Define operator $\mathfrak{U}$ over the linear span of $\mathscr{E}_*$ by formulas

$$\mathfrak{U}e_{mn}^+ = c_{mn}, \qquad \mathfrak{U}e_{mn}^- = s_{mn}, \qquad (m, n) \in \mathscr{M}.$$

If operator $\mathfrak{U}$ can be extended to an isomorphism of the whole $H^r(0, T; \mathbb{C}^N)$, then family $\mathscr{F}_* = \mathfrak{U}\mathscr{E}_*$ is also a Riesz basis in $H^r(0, T; \mathbb{C}^N)$.

Let $\{a_{mn}^\pm\}$ be a finite sequence of complex numbers and

$$g = \sum_{(m,n)\in\mathscr{M}} (a_{mn}^+ e_{mn}^+ + a_{mn}^- e_{mn}^-).$$

Since $\mathscr{E}_*$ is a Riesz basis, then

$$\|g\|_{H^r(0, T; \mathbb{C}^N)}^2 \asymp \sum_{(m,n)\in\mathscr{M}} (|a_{mn}^+|^2 + |a_{mn}^-|^2). \tag{10}$$

With the help of the Euler formula, function $\mathfrak{U}g$ may be expanded over the basis $\mathscr{E}_*$

$$\mathfrak{U}g = \sum_{(m,n)\in\mathscr{M}} (a_{mn}^+ c_{mn} + a_{mn}^- s_{mn}) = \sum_{(m,n)\in\mathscr{M}} (\tilde{a}_{mn}^+ e_{mn}^+ + \tilde{a}_{mn}^- e_{mn}^-).$$

Moreover,

$$|\tilde{a}_{mn}^+|^2 + |\tilde{a}_{mn}^-|^2 = \tfrac{1}{2}(|a_{mn}^+|^2 + |a_{mn}^-|^2). \tag{11}$$

Again, using the Riesz basis property of $\mathscr{E}_*$, we have

$$\|\mathfrak{U}g\|_{H^r(0, T; \mathbb{C}^N)}^2 \asymp \sum_{(m,n)\in\mathscr{M}} (|\tilde{a}_{mn}^+|^2 + |\tilde{a}_{mn}^-|^2). \tag{12}$$

From relations (10)–(12) we conclude that

$$\|\mathfrak{U}g\|_{H^r(0, T; \mathbb{C}^N)}^2 \asymp \|g\|_{H^2(0, T; \mathbb{C}^N)}^2.$$

Therefore, operator $\mathfrak{U}$ may be extended to an isomorphism in $H^r(0, T; \mathbb{C}^N)$.

The converse statement – if $\mathscr{F}_*$ is a Riesz basis in $H^r(0, T; \mathbb{R}^N)$, then $\mathscr{E}_*$ is a Riesz basis in $H^r(0, T; \mathbb{C}^N)$ – is proved in a similar way.

Theorem VI.2.5. For any nonnegative integer r and any $T > 0$ there exists family $\mathscr{E}_r \subset \mathscr{E}$, which forms a Riesz basis in $H^r(0, T; \mathbb{C}^N)$, and set $\mathscr{E}\setminus\mathscr{E}_r$ is infinite.

The proof of this theorem is given below in Subsection 2.4.

Statement (a) of Theorem 1 follows from Lemmas 3, 4 and Theorem 5. To prove statement (b), we introduce notation $\mathscr{F}_r$ for family

$$\mathscr{F}_r = \{c_{mn} s_{mn}\}_{(m,n) \in \mathcal{M}}, \qquad \mathcal{M} := \{(m,n) \mid e_{mn}^{\pm} \in \mathscr{E}_r\}.$$

By virtue of Lemma 3, family $\mathscr{F}_r$ forms a Riesz basis in $H'(0, T; \mathbb{R}^N)$. It is evident that the set of elements of $\mathscr{F}$ that do not belong to $\mathscr{F}_r$ is also infinite.

Let c_{kl} be an arbitrary element of $\mathscr{F} \backslash \mathscr{F}_r$. It may be represented by series

$$c_{kl}(t) = \sum_{(m,n) \in \mathcal{M}} [\alpha_{mn} c_{mn}(t) + \beta_{mn} s_{mn}(t)]. \tag{13}$$

By analogy with (8), (9), we choose coefficients a_{mn}, b_{mn} in expansions (3) in the following way:

$$a_{mn} = b_{mn} = 0, \qquad (m,n) \notin \mathcal{M} \cup (k,l),$$

$$a_{mn} = \alpha_{mn} \omega_{mn}^{-r}, \qquad b_{mn} = \beta_{mn} \omega_{mn}^{1-r}, \qquad (m,n) \in \mathcal{M},$$

$$a_{kl} = -1, \qquad b_{kl} = 0.$$

Then formulas (13), (5) imply $\Phi(t, v_0, v_1) = 0$, $t \in [0, T]$. The initial data (v_0, v_1) constructed in this manner for distinct elements from $\mathscr{F} \backslash \mathscr{F}_r$ are linearly independent. Theorem 1 is proved.

2.3. Lack of pointwise controllability

In order to prove Theorem 2 consider initial boundary-value problem

$$\begin{cases} \omega_{tt}(p,t) = \Delta \omega(p,t), \qquad p \in \Omega, \qquad t \in (0,T), \\ \omega|_{\Sigma} = 0, \qquad \omega|_{t=T} = \omega_0, \qquad \omega_t|_{t=T} = \omega_1, \end{cases} \tag{14}$$

coinciding with (1) where variable t is changed to $T - t$. Hence, on the grounds of assertion (b) of Theorem 1, linear set

$$\{(\omega_0, \omega_1) \in \mathscr{W}_1 \mid \omega(p_j, t) = 0, j = 1, \ldots, N; t \in [0, T]\}$$

has an infinite dimension. At the same time, an equality takes place (see formula (24) of Section V.1 with $T' = T$)

$$\langle z(\cdot, T), \omega_1 \rangle_* + \langle z_t(\cdot, T), \omega_0 \rangle_* = \sum_{j=1}^{N} \int_0^T u_j(t) \overline{\omega(p_j, t)} \, dt.$$

This completes the proof of Theorem 2.

2.4. Construction of basis subfamily

Let us start the proof of Theorem 5 with the verification of completeness in $\mathbb{C}^N$ for the family of vectors $\{\eta_{mn}\}$, which is obviously a necessary condition for a basis subfamily $\mathscr{E}_r$ to exist.

Lemma VI.2.6. Vector family $\{\eta_{mn}\}_{m,n=1}^{2N}$ is complete in $\mathbb{C}^N$.

PROOF. Suppose vector $\gamma = \{\gamma_k\}_{k=1}^N$ may be found to be orthogonal to all the vectors η_{mn}; $m, n \leq 2N$:

$$\sum_{k=1}^N \gamma_k \sin \frac{\pi}{a} mx_k \sin \frac{\pi}{b} ny_k = 0; \qquad m, n = 1, \ldots, 2N. \tag{15}$$

Introduce notations

$$\tilde{x}_k = \frac{\pi}{a} x_k, \qquad \tilde{y}_k = \frac{\pi}{a} y_k, \qquad \tilde{\gamma}_k = \gamma_k \sin \tilde{x}_k \sin \tilde{y}_k.$$

Function $\sin qx / \sin x$ is a polynomial of the degree $q - 1$ in $\cos x$. Therefore, a relation may be derived by induction from (15)

$$\sum_{k=1}^N \tilde{\gamma}_k \cos^{n-1} \tilde{x}_k \cos^{m-1} \tilde{y}_k = 0; \qquad m, n = 1, \ldots, 2N \tag{16}$$

(one first checks this equality for $m = 1, n = 1, \ldots, N$; then for $m = 2$, $n = 1, \ldots, N$, and so on).

Let $P(\xi, \zeta)$ be an arbitrary polynomial of a degree not larger than $2N - 1$ in each variable. From (16) it follows that

$$\sum_{k=1}^N \tilde{\gamma}_k P(\cos \tilde{x}_k, \cos \tilde{y}_k) = 0.$$

Making use of this relation for

$$P(\xi, \zeta) = \prod_{k=2}^N [(\xi - \cos \tilde{x}_k)^2 + (\zeta - \cos \tilde{y}_k)^2],$$

we obtain $\tilde{\gamma}_1 = 0$ and hence $\gamma_1 = 0$. Equalities

$$\gamma_2 = \gamma_3 = \cdots = \gamma_N = 0$$

are obtained in a similar way. Lemma 6 is proved.

Let us choose in family $\{\eta_{mn}\}_{m,n=1}^{2N}$ subfamily $\{\eta^j\}_{j=1}^N$, which forms a basis in $\mathbb{C}^N$.

Lemma VI.2.7. For any $\varepsilon > 0$ and $T > 0$, family $\mathscr{E}$ contains subfamily $\tilde{\mathscr{E}}$, which can be represented as

$$\tilde{\mathscr{E}} = \mathscr{E}^1 \cup \mathscr{E}^2 \cup \cdots \cup \mathscr{E}^N, \qquad \mathscr{E}^j = \{\eta_m^j \, e^{\pm i\mu_{mj}t}\}_{m=R}^\infty,$$

and the equalities hold

$$\langle\langle \eta_m^j - \eta^j \rangle\rangle_N < \varepsilon, \qquad j = 1, \ldots, N; \; m = R, R+1, \ldots. \qquad (17)$$

$$\left| \mu_{mj} - \frac{2\pi}{T} m \right| < \varepsilon, \qquad j = 1, \ldots, N; \; m = R, R+1, \ldots. \qquad (18)$$

To prove the lemma we use the following form of the Kronecker theorem.

Proposition VI.2.8. Let $\xi_1, \ldots, \xi_N, \zeta_1, \ldots, \zeta_N$ be real numbers such that for any integer $n_1, \ldots, n_N$ equality

$$n_1 \xi_1 + \cdots + n_N \xi_N \equiv 0 \qquad (\mathrm{mod} \; 2\pi)$$

implies

$$n_1 \zeta_1 + \cdots + n_N \zeta_N \equiv 0 \qquad (\mathrm{mod} \; 2\pi).$$

Then for any $\varepsilon > 0$ there exists a number $Q_\varepsilon = Q_\varepsilon \, (\xi_1, \ldots, \xi_N) > 0$ (independent of $\{\zeta_1, \ldots, \zeta_N\}$) such that on any segment of the length Q_ε an integer n_0 may be found for which

$$\||n_0 \xi_1 - \zeta_1\|| < \varepsilon, \ldots, \||n_0 \xi_N - \zeta_N\|| < \varepsilon,$$

$$(\||x\|| := \mathrm{dist}(x, 2\pi\mathbb{Z})).$$

The principal difference between this statement and the one presented in Cassels (1957: sec. III.5) is that there the number n_0 satisfies estimate $|n_0| < Q_\varepsilon$. Passing from ζ_j to $\zeta_j - n\xi_j$, $n \in \mathbb{Z}$, we arrive at Proposition 8.

PROOF OF LEMMA 7. Specify $\varepsilon > 0, j \, (1 \le j \le N)$ and define m', n' according to the conditions

$$\eta_{m'n'} = \eta^j, \qquad 1 \le m', n' \le 2N.$$

Sets

$$\{\tilde{x}_1, \ldots, \tilde{x}_N\}, \qquad \{m' \tilde{x}_1, \ldots, m' \tilde{x}_N\} \qquad (19)$$

satisfy the conditions of Proposition 8. The same also goes for sets

$$\{\tilde{y}_1, \ldots, \tilde{y}_N\}, \qquad \{n' \tilde{y}_1, \ldots, n' \tilde{y}_N\} \qquad (20)$$

Let $Q_\varepsilon = \max\{Q_\varepsilon(\tilde{x}_1, \ldots, \tilde{x}_N), Q_\varepsilon(\tilde{y}_1, \ldots, \tilde{y}_N)\}$.

Take some integer m and define $m_0 \in \mathbb{N}$ from the conditions

$$\frac{2a}{T} m \le m_0 \le \frac{2a}{T} m + Q_\varepsilon, \tag{21}$$

$$\||m_0 \tilde{x}_k - m' \tilde{x}_k\|| < \varepsilon, \qquad k = 1, \dots, N. \tag{22}$$

Proposition 8 applied to sets (19) show that such m_0 really exists.

Introduce function

$$g(\tau) = \left| \frac{2\pi}{T} m - \sqrt{\left(\frac{\pi}{a} m_0\right)^2 + \left(\frac{\pi}{b} \tau\right)^2} \right|$$

and define integer $q = q(m, n)$ in such a way that $g(q) = \min_{n \in \mathbb{N}} g(n)$. Further, let us find $n_0 \in \mathbb{N}$ satisfying conditions

$$q \le n_0 \le q + Q_\varepsilon, \tag{23}$$

$$\||n_0 \tilde{y}_k - n' \tilde{y}_k\|| < \varepsilon, \qquad k = 1, \dots, N. \tag{24}$$

Its existence follows from Proposition 8 applied to sets (20).

Let us describe the procedure for all j. Thus, there is a pair of numbers $m_0, n_0 \in \mathbb{N}$ that can be assigned to each $j \in \{1, \dots, N\}$ and every $m \in \mathbb{N}$. Let us check vector functions

$$\eta_m^j e^{\pm i\mu_{mj}t} := \eta_{m_0 n_0} e^{\pm i\omega_{m_0 n_0} t}$$

to satisfy conditions (17), (18) for large enough m. Inequalities (22), (24) imply inequalities (17) (for all m), where ε should be replaced by $4\varepsilon\sqrt{(N/ab)}$.

To prove inequalities (18), we note that

$$
\left| \mu_{mj} - \frac{2\pi}{T} m \right| = \left| \sqrt{\left(\frac{\pi}{a} m_0\right)^2 + \left(\frac{\pi}{b} n_0\right)^2} - \frac{2\pi}{T} m \right|
$$

$$
= \left| \frac{\left(\frac{\pi}{a} m_0\right)^2 + \left(\frac{\pi}{b} n_0\right)^2 - \left(\frac{2\pi}{T} m\right)^2}{\sqrt{\left(\frac{\pi}{a} m_0\right)^2 + \left(\frac{\pi}{b} n_0\right)^2} + \frac{2\pi}{T} m} \right|
$$

$$
\le \frac{T}{2\pi m} \left| \left(\frac{\pi}{a} m_0\right)^2 + \left(\frac{\pi}{b} n_0\right)^2 - \left(\frac{2\pi}{T} m\right)^2 \right|. \tag{25}
$$

Let $q_1(m)$ be a positive root of function $g(\tau)$. Using (21), one can easily demonstrate that $q_1(m) = \mathcal{O}(\sqrt{m})$. It is evident that $|q_1(m) - q(m, m_0)| < 1$.

Therefore, (23) implies $n_0 = q_1(m) + \mathcal{O}(1)$. Substituting this into (25), we find

$$\left| \mu_{mj} - \frac{2\pi}{T} m \right| \le \frac{T}{2\pi m} \left| \left(\frac{\pi}{a} m_0 \right)^2 + \left(\frac{\pi}{b} n_0 \right)^2 - \left(\frac{2\pi}{T} m \right)^2 + \mathcal{O}(\sqrt{m}) \right|$$

$$= \frac{T}{2\pi m} |g(q_1) + \mathcal{O}(\sqrt{m})| = \mathcal{O}(1/\sqrt{m}).$$

Hence, inequalities (18) are valid for large enough m. Lemma 7 is proved.

Note that for small enough ε, all the elements of the constructed family $\tilde{\mathscr{E}}$ are different, as it follows from inequalities (17), (18).

Consider now family

$$\Xi = \{\eta^j \, e^{i(2\pi m/T)t}\}_{j=1, m \in \mathbb{Z}}^N$$

in space $L^2(0, T; \mathbb{C}^N)$. This family constitutes a Riesz basis, since it transfers to an orthogonal almost normed basis

$$\{h_j \, e^{i(2\pi m/T)t}\}_{j=1, m \in \mathbb{Z}}^N$$

under a mapping in $\mathbb{C}^N$, which transforms basis $\{\eta^j\}_{j=1}^N$ to some orthogonal basis $\{h_j\}_{j=1}^N$.

Let $\varepsilon > 0$ be so small that any family being an ε perturbation of family Ξ (in the sense of Theorem II.5.5) conserves the basis property. Using Lemma 7, we construct family $\tilde{\mathscr{E}}$ for such ε. Then, in accordance with the mentioned theorem, family

$$\tilde{\Xi} = \{\eta^j \, e^{i(2\pi m/T)t}\}_{j=1, |m| < R}^N \cup \tilde{\mathscr{E}} \tag{26}$$

forms a Riesz basis in $L^2(0, T; \mathbb{C}^N)$. We state this result in the form of a lemma.

Lemma VI.2.9. For any $T > 0$ there exists family $\tilde{\Xi}$ of the form (26) constituting a Riesz basis in $L^2(0, T; \mathbb{C}^N)$.

We are now able to prove Theorem 5 for $r = 0$ if it is known that family $\mathscr{E}$ is complete in this case in $L^2(0, T; \mathbb{C}^N)$. Indeed, we are able to complement family $\tilde{\mathscr{E}}$ up to a basis $\mathscr{E}_0$ in the following way. Let us take an element $e \in \mathscr{E}$, $e \notin \bigvee \tilde{\mathscr{E}}$. Since $\tilde{\mathscr{E}}$ is an $\mathscr{L}$-basis, family $\tilde{\mathscr{E}} \cup \{e\}$ is also an $\mathscr{L}$-basis. Continuing this process and taking into account Lemma 9, we obtain, in a finite number of steps, basis family $\mathscr{E}_0$.

Let us check that $\mathscr{E}$ is complete. Lemma 9 and theorem II.5.9(b) imply that $\tilde{\mathscr{E}}$, is complete in $L^2(0, T_1; \mathbb{C}^N)$ for any $T_1 < T$. Moreover, it is true for family $\mathscr{E}$. Since T is arbitrary, family $\mathscr{E}$ is complete in L^2 on any interval of $\mathbb{R}_+$.

Thus we have proved Theorem 5 for the case $r = 0$. The proof of the theorem in the general case runs along similar lines, but because the constructions are cumbersome, they are not given here.

Remark VI.2.10. Lebeau (1992) has shown that the lack of approximate controllability for the wave equation with pointwise control takes place for arbitrary domain Ω (dim $\Omega \geq 2$) with analytic boundary under certain geometrical conditions.

Remark VI.2.11. Questions similar to those discussed in this section have been studied by Haraux and Komornik (1985) and Komornik (1989a). In addition, similar problems Haraux and Jaffard (1991) and Haraux and Komornik (1991) investigated in the realm of plate vibrations.

VII

Boundary control of string systems

In this chapter we consider boundary control in systems described by hyperbolic equations for vector functions of one spatial variable. The study of controllability reduces to the investigations of families of vector exponentials $\eta_n\, e^{-i\omega_n t}$ with ω_n being eigenfrequencies of the system and η_n being the traces of eigenfunction derivatives at the boundary points where control is applied. Section 1 shows how the explicit form of ω_n and η_n can be used to study controllability in a system of homogeneous strings. In the following sections, the "strings" are nonhomogeneous, so that only the asymptotics of ω_n and η_n are known, which is not always accurate enough to separate the exponentials (Riekstyn'sh 1991). In the vector case, as pointed out in Remark II.5.8, the basis property (and then minimality) cannot be assessed on the basis of the asymptotics alone. In fact, there is practically only one way to examine vector exponential families, that is, by constructing and studying the generating (matrix) function GF. Remarkably, the problems under consideration have some physical foundation and naturally give rise to the GF, since it is expressed in the fundamental solution to ordinary differential equations of the Helmholtz type. Since such solutions are known to act as functions of the spectral parameter, and the eigenfunctions of the elliptic operator of the system forms an orthogonal basis, one is able to arrive at a conclusion about the $\mathscr{L}$-basis property of the corresponding exponential family.

1. System of connected homogeneous strings controlled at the ends

Let us consider a connected network of homogeneous strings to which control actions are applied in the nodal points. Such a network may be drawn as a graph (for convenience, we assume that it is an oriented one), whose edges correspond to strings and vertices – to the nodes of the network. The strings are enumerated by index s, $s = 1, 2, \ldots, M$, and the nodes by index p, $p = 1, 2, \ldots, N$. During the initial period, the system remains in some excited state. We are interested in the possibility of quieting all the string vibrations in some finite time T, independent of the initial data, by means of the controls $u_p(t)$, $u_p \in L^2(0, T)$, $p = 1, 2, \ldots, N$. It is convenient to consider control actions as components of vector function u of the space $\mathscr{U} := L^2(0, T; \mathbb{C}^N)$.

Oscillations of the sth string starting at the node $p(s)$ and ending at the node $p'(s)$ are described by the equation

$$\rho_s \frac{\partial y_s^2(x, t)}{\partial t^2} = \frac{\partial y_s^2(x, t)}{\partial x^2}, \qquad \rho_s > 0, 0 < x < l_s, 0 < t < T, \qquad (1)$$

where ρ_s is the constant density and l_s is the length of the string. The boundary conditions read

$$y_s(0, t) = u_{p(s)}(t), \qquad y_s(l_s, t) = u_{p'(s)}(t), \qquad (2)$$

while the initial ones are

$$y_s(x, 0) = \frac{\partial}{\partial t} y_s(x, 0) = 0. \qquad (3)$$

S. Rolewicz (1970) has studied the control problem associated with this system under the assumption that the optical lengths $L_s = l_s \sqrt{\rho_s}$ of all the strings are commensurable (ratios L_{s_i}/L_{s_j} are rational numbers). His solution depends on the presence of cycles in the graph representing the network. If there are no cycles – that is, if the graph is a tree – the system is controllable. When two or more cycles are involved, the system is uncontrollable. Rolewicz (1970) did not clarify the question of system controllability when exactly one cycle is present, however. His statement about controllability in this case can be proved incorrect by a simple counterexample of the cyclic network of two (or any number) of identical strings (see below). In this section we give the solution to the problems of B and M controllability of a string system without the restriction that their optical lengths are commensurable.

Let us treat system (1), (2) within the framework of the general scheme of Chapter III. To do this we introduce spaces $L_{\rho_s}^2(0, s)$, $s = 1, 2, \ldots, M$, of functions squarely integrable over interval $(0, l_s)$ with the norm

$$\left[\int_0^{l_s} \rho_s |\varphi_s(x)|^2 \, dx \right]^{1/2}$$

and set

$$H := \bigoplus_1^M L_{\rho_s}^2(0, l_s).$$

We also introduce the space $V \subset H$

$$V := \bigoplus_1^M H_0^1(0, l_s)$$

and bilinear form

$$a[\varphi, \psi] = \sum_{s=1}^M \int_0^{l_s} \varphi_s'(x)\bar{\psi}_s'(x)\, dx; \qquad \varphi, \psi \in V.$$

The form generates operator A (see Chapter III, Section 1):

$$(A\varphi, \psi)_H = a[\varphi, \psi]; \qquad \varphi \in \mathscr{D}(A), \varphi \in V.$$

$$\varphi = \{\varphi_s\}_{s=1}^M \overset{A}{\mapsto} \left\{ -\frac{1}{\rho_s}\frac{d^2}{dx^2}\varphi_s \right\}_{s=1}^M$$

with the domain $\mathscr{D}(A) = \{\varphi_s \mid \varphi_s \in H^2(0, l_s) \cap H_0^1(0, l_s), s = \{1, 2, \ldots, M\}$. The spectrum of operator A falls into M series $\{\lambda_{r,s}\}_{r\in\mathbb{N}, s=1,2,\ldots,M}$. Orthonormal eigenfunctions $\Phi_{r,s}$ corresponding to the eigenvalue λ have the form

$$\Phi_{r,s} = (0, 0, \ldots, \varphi_{r,s}, 0, \ldots, 0),$$

where in the sth place the eigenfunction of a single string

$$\varphi_{r,s}(x) = \sqrt{2/(\rho_s l_s)}\, \sin(\omega_{r,s}x), \qquad \omega_{r,s} := \sqrt{\lambda_{r,s}} = \pi r/L_s$$

is standing.

Let us write initial boundary-value problem (1)–(3) as an integral identity. Let $f_s \in C^2([0, l_s] \times [0, T])$, $s = 1, 2, \ldots, M$, and

$$f_s(\cdot, T) = \frac{\partial}{\partial t} f_s(\cdot, T) = 0.$$

Multiply each equation in (1) by $\bar{f}_s$. Then, integrating by parts and taking (2), (3) into account, we find

$$\sum_{s=1}^M \int_0^T \int_0^{l_s} y_s \left(\rho_s \frac{\partial^2}{\partial t^2} \bar{f}_s - \frac{\partial^2}{\partial t^2} \bar{f}_s \right) dx\, dt$$

$$= \sum_{s=1}^M \int_0^T \left[u_{p(s)}(t) \frac{\partial}{\partial x} \bar{f}_s(0, t) - u_{p'(s)}(t) \frac{\partial}{\partial x} \bar{f}_s(l_s, t) \right] dt. \quad (4)$$

In order to write problem (1)–(3) in the form (12) in Section III.3, we only need to define operator B. Equality (4) makes it clear that operator B may be defined by the formula

$$\langle Bv, \varphi \rangle_* = \sum_{s=1}^{M} (v_{p(s)} \bar{\varphi}'(0) - v_{p'(s)} \bar{\varphi}'_s(l_s)).$$

Here, $v \in \mathbb{C}^N$, $\varphi \in W_2 = D(A)$, and operator B acts from $\mathbb{C}^N$ to the space W_{-2}. (For the definition of spaces W_r see Section 1 of Chapter III.) Note that the space W_1 coincides with $\oplus_1^M H_0^1(0, l_s)$. The space H is topologically equivalent to $\oplus_1^M L^2(0, l_s)$. Therefore space W_{-1} dual to W_1 with respect to the scalar product in H is also topologically equivalent to the space $\oplus_1^M H^{-1}(0, l_s)$. Later in the discussion, we do not recognize the latter from the space W_{-1}.

To construct a family of exponentials, we must now find vector $B^*\Phi_{rs} \in \mathbb{C}^N$. For any element $v \in \mathbb{C}^N$, we have

$$\langle Bv, \Phi_{rs} \rangle_* = v_{p(s)} \bar{\varphi}'_{rs}(0) - v_{p'(s)} \bar{\varphi}'_{rs}(l_s).$$

This linear (in v) form may be written as

$$\langle Bv, \Phi_{rs} \rangle_* = \omega_{rs} \langle v, \eta_{rs} \rangle_{\mathbb{C}^N},$$

where vector η_{rs} has only two nonzero coordinates: the $p(s)$th coordinate equals d_s, the $p'(s)$th one equals $(-1)^{r+1}d_s$, $d_s := \sqrt{2/(\rho_s l_s)}$.

We now have everything we need to construct exponential family $\mathcal{E}_T$ (family $\mathcal{E}$ in the notations of Chapter III; see formula (22) in Section III.2):

$$\mathcal{E}_T = \{|\omega_{rs}|e_{rs}\}_{r \in \mathbb{K}, s = 1, 2, \ldots, M},$$

$$e_{rs} = \eta_{rs} \exp(i\omega_{rs}t), \qquad \omega_{rs} = \operatorname{sgn}(r)\sqrt{\lambda_{|r|s}} = \pi r/L_s,$$

$$\eta_{-rs} := \eta_{rs}, \qquad r > 0.$$

We let $\sigma^{(s)}$ denote the spectrum of the sth string, $\sigma^{(s)} := \{\pi r/L_s\}_{r \in \mathbb{K}}$.

The explicit form of the set of frequencies implies that two strings have an equal frequency if and only if their optical lengths are commensurable (i.e., their ratio is a rational number). Accordingly, we split the set of all the strings into R classes S_j, $j = 1, 2, \ldots, R$, of strings with pairwise commensurable optical lengths. We write σ_j for the family spectrum corresponding to the strings of the jth class,

$$\sigma_j = \{\pi r/L_s\}_{s \in S_j, r \in \mathbb{K}} = \bigcup_{s \in S_j} \sigma^{(s)}.$$

(By "spectrum," we mean here the set of points on the plane not accounting for the multiplicity.)

Lemma VII.1.1. Solution $y = (y_1, \ldots, y_N)$ *to problem* (1)–(3) *exists, is unique and*

$$\left\{ y, \frac{\partial}{\partial t} y \right\} \in C(0, T, \mathscr{W}_0), \tag{5}$$

$$\mathscr{W}_0 = W_0 \oplus W_{-1} = \left[\bigoplus_{s=1}^{M} L^2_{\rho_s}(0, l_s) \right] \oplus \left[\bigoplus_{s=1}^{M} H^{-1}(0, l_s) \right].$$

PROOF OF THE LEMMA. System (1), (2) is understood as equation (12) in Section III.3 and, since $B \in \mathscr{L}(\mathbb{C}^N, W_{-2})$, by Theorem II.2.1 solution y exists, is unique, and

$$\left\{ y, \frac{\partial}{\partial t} y \right\} \in C(0, T, \mathscr{W}_{-1}), \qquad \mathscr{W}_{-1} = W_{-1} \oplus W_{-2}.$$

Let us show that in fact the smoothness of the solution is one unit greater. Indeed, spectrum σ of the family is a unification of separable sets σ_j. Therefore, Lemma II.1.21 implies an estimate

$$\sum_{r,s} |(f, e_{rs})_{L^2(0, t; \mathbb{C}^N)}|^2 \leq C_T \|f\|^2_{L^2(0, t; \mathbb{C}^N)}, \qquad t \leq T.$$

What we want to prove now follows from Lemma III.2.4.

Relation (5) provides

$$\mathscr{R}(T) \subset \mathscr{W}_0. \tag{6}$$

Note that similar statements about the smoothness of solutions are valid for all string systems considered in this chapter.

Having established relation (6), we can study the controllability of system (1), (2). Family $\tilde{\mathscr{E}}_T$, corresponding to phase space $\mathscr{W}_0$, is obtained from family $\mathscr{E}_T$ by multiplying by $|\omega_{rs}|^{-1}$ and has the form

$$\tilde{\mathscr{E}}_T = \{e_{rs}\}_{r \in \mathbb{K}, s = 1, 2, \ldots, M}.$$

This corresponds to the transition from family $\mathscr{E}$ to $\tilde{\mathscr{E}}$, which in the abstract form is performed in Section III.3. For the space $\mathscr{H}_0$ of the abstract form we choose space $\mathscr{W}_0$, so that in this case families $\mathscr{E}_0$ and $\tilde{\mathscr{E}}$ coincide (this coincidence takes place for all the problems treated in this chapter). Properties of family $\tilde{\mathscr{E}}_T \subset L^2(0, T, \mathbb{C}^N)$ determine the controllability type of system (1), (2). Families of exponentials on the semiaxis are easier to study than those on a segment (there is no parameter T). That is why we reduce the study of family $\tilde{\mathscr{E}}_T$ to the investigation of an exponential family over the semiaxis. Since the spectrum of family $\tilde{\mathscr{E}}_T$ is real, $e_{rs} \notin L^2(0, \infty; \mathbb{C}^N)$. To consider exponentials on the semiaxis, we proceed – without changing the notations of family $\tilde{\mathscr{E}}_T$ and sets σ, σ_j, $\sigma^{(s)}$ – with the exponential family obtained from $\tilde{\mathscr{E}}_T$ by the shift of the spectrum in $\mathbb{C}^N$:

$$e_{rs} \mapsto \eta_{rs}\, e^{iv_{rs}t}, \qquad v_{rs} = \omega_{rs} + i, \qquad r \in K, s = 1, 2, \ldots, M.$$

This transition corresponds to the mapping $f(t) \mapsto f(t)\, e^{-t}$ being an isomorphism of space $L^2(0, T; \mathbb{C}^N)$ and hence does not change the properties of family $\tilde{\mathscr{E}}_T$ governing the type of controllability (Theorem I.1.25).

Lemma VII.1.2. For some T_0 orthoprojector P_{T_0} from space $L^2(0, \infty)$ to $L^2(0, T_0)$ restricted to $\bigvee \mathscr{E}_{sc}$, $\mathscr{E}_{sc} := \{\exp(i\lambda t)\}_{\lambda \in \sigma}$, $(\sigma := \bigcup_{r,s} \{v_{r,s}\})$ is an isomorphism on its image.

PROOF OF THE LEMMA. GF of an exponential family corresponding to the spectrum of one single string (shifted by i to the upper half-plane) is of the form $f_s(k) = \sin(L_s(k - i)) \exp(ikL_s)$ (this is the STF). Consider a function $f := \prod_1^M f_s$. Together with f^{-1}, it is bounded on $\mathbb{R}$ and allows a factorization

$$f = bf_e^+ = e^{2ikT_0} f_e^-,$$

where b is the Blaschke product turning to zero at the spectrum points, $T_0 = 2L_1 + \cdots + 2L_M$. Then the projector from K_b to K_{T_0} is an isomorphism (Theorem II.3.14). Let $\tilde{b}$ denote the Blaschke product having simple zeros at σ ($b(k)$ possess multiple zeros as soon as optical lengths of at least one pair of strings are commensurable). Then, obviously, operator $P_{T_0}|_{K_{\tilde{b}}}$ is an isomorphism on its image. Passing from the family of simple fractions to the family of exponentials, we obtain the assertion of the lemma.

Theorem VII.1.3.

(a) *If system (1), (2) is E-controllable relative to $\mathscr{W}_0$ in some time T, then it is B-controllable relative to $\mathscr{W}_0$ in T;*

(b) *system (1), (2) is M-controllable in time T_0 if and only if family $\tilde{\mathscr{E}}_T$ is linearly independent;*

(c) *family $\tilde{\mathscr{E}}_T$ is linearly independent if and only if any system of strings formed by any connected component of any class is B-controllable relative to $\mathscr{W}_0$ in time T_0.*

Note that the linear independence of an exponential family does not depend on interval $(0, T)$.

PROOF OF THE THEOREM.

(a) Inclusion (6) implies that under the condition $\mathscr{R}(T) \supset \mathscr{W}_0$ (i.e., under E controllability), equality $\mathscr{R}(T) = \mathscr{W}_0$ holds (B controllability).

(b) Recall that M controllability of system (1), (2) in time T_0 is equivalent to the fact that family $\tilde{\mathscr{E}}_{T_0}$ is minimal.

For $\lambda \in \sigma$, introduce family $\mathscr{M}_\lambda$ of vectors η_{rs} such that $\pi r/L_s + i = \lambda$. Then family $\tilde{\mathscr{E}}_{T_0}$ may be represented as

$$\tilde{\mathscr{E}}_{T_0} = \{\tilde{\mathscr{E}}_{T_0, \lambda}\}_{\lambda \in \sigma}, \qquad \tilde{\mathscr{E}}_{T_0, \lambda} = e^{i\lambda t}\{\eta\}_{n \in \mathscr{M}_\lambda}.$$

From Lemma 2 and the fact that $\sigma \in (\text{CN})$, it follows that scalar family $\{\exp(i\lambda t)\}_{\lambda \in \sigma}$ is minimal in space $L^2(0, T_0)$. By the force of Corollary II.2.2(a), the family of subspaces $\{e^{i\lambda t}\mathbb{C}^N\}_{\lambda \in \sigma}$ is then minimal in $L^2(0, T_0; \mathbb{C}^N)$. According to Corollary II.2.2(b), the family of subspaces

$$\{e^{i\lambda t}\mathfrak{N}_\lambda\}_{\lambda \in \sigma}, \qquad \mathfrak{N}_\lambda := \bigvee_{\eta \in \mathscr{M}_\lambda} \eta$$

is also minimal. Now, two situations are possible: either for any λ the family $\mathscr{M}_\lambda$ is linearly independent, and then family $\tilde{\mathscr{E}}_{T_0}$ is minimal, or for some λ_0, family $\mathscr{M}_{\lambda_0}$ proves to be linearly dependent, in which case family $\tilde{\mathscr{E}}_{T_0, \lambda_0}$ and hence $\tilde{\mathscr{E}}_{T_0}$ are also linearly dependent.

(c) From the considerations mentioned above it follows that family $\tilde{\mathscr{E}}_T$ is linearly independent if and only if each of the vector families $\mathscr{M}_\lambda$ is linearly independent. It is evident that any $\mathscr{M}_\lambda$ contains vectors corresponding to the same class. On the other hand, vectors corresponding to different connected components are orthogonal. If some family $\mathscr{M}_{\lambda_0}$ is linearly dependent, then such is the case for some subfamily $\tilde{\mathscr{M}}_{\lambda_0} \subset \mathscr{M}_{\lambda_0}$ corresponding to one connected component. In this case a subfamily of exponentials corresponding to the same component is linearly dependent, too. It is quite clear that the corresponding subsystem of strings is not B-controllable.

Suppose that $\mathscr{M}_\lambda$ is linearly independent for any λ. Let $\sigma_{j,q}$ denote the spectrum of the qth connected component of the jth class. Since

σ_j is a separable set lying on the line $Im\,k = 1$, σ_j is Carlesonian. It is even more so for $\sigma_{j,q}$. Then family $\mathscr{E}_{j,q}$ of subspaces $\{\exp(i\lambda t)\mathfrak{N}_\lambda\}_{\lambda \in \sigma_{j,q}}$ constitutes an $\mathscr{L}$-basis in $L^2(0, \infty; \mathbb{C}^N)$. Since $\mathscr{M}_\lambda$ is linearly independent for all $\lambda \in \sigma_{j,q}$, in $\mathfrak{N}_\lambda$ there exist skew projectors $\mathscr{P}_{rs}$ on every vector $\eta_{rs} \in \mathscr{M}_\lambda$ parallel to all the other ones. The whole set $\{\eta_{rs}\}_{r \in \mathbb{K}, s=1,2,...,M}$ is finite. Therefore the norms of $\mathscr{P}_{rs}$ are uniformly bounded. Together with the basis property of $\mathscr{E}_{j,q}$ this implies the basis property of not only the family of subspaces, but the family of elements: family $\{\exp(i\lambda t)\eta_{rs} \mid \lambda = \pi r/L_s + i \in \sigma_{j,q}\}$ forms an $\mathscr{L}$-basis in $L^2(0, \infty; \mathbb{C}^N)$. Lemma 2 and Corollary II.2.2 imply that the family forms an $\mathscr{L}$-basis in $L^2(0, T_0; \mathbb{C}^N)$. The theorem is proved.

Now we are able to deal directly with the controllability of the string network. We examine three cases: the number of strings is less than, equal to, and larger than the number of nodes.

(1) The number of strings is less than the number of nodes $M < N$.

In this case, the solution to the problem of system controllability can be seen more clearly in the primary formulation. The following arguments have already been presented (Rolewicz 1970). Note that they are valid for nonhomogeneous strings as well.

The system graph is a tree. Let us call the strings starting from its root the first-order strings; those branching from the first-order strings are second-order ones, and so on. The nodes between the strings of the q and $q + 1$ order are called nodes of the order q. We consider the root of the tree to be the node of the order zero.

The string of length l and optical length L is B controllable in $L^2(0, l) \oplus H^{-1}(0, l)$ in time $2L$ by control $u \in L^2(0, 2L)$ applied at some string end. The result may be established exactly in the same way as the controllability of system (12) of Section V.2 (see also Russell 1967).

In view of the invertibility of the hyperbolic-type equations, B controllability is equivalent to the fact that an arbitrary initial state may be transformed into zero by an admissible control. We prove that the system of a system of strings may be quieted down exactly in this sense.

Let the system stay at some initial state $\{y_s^0, y_s^1\}$. We perform the following procedure to quiet the strings. First, set the controls equal to zero in all the nodes except the first-order ones and quiet down all the first-order strings. Second, using only the controls at the second-order nodes, we quiet all the second-order strings without exciting the first-order ones. Since the graph is a tree, we will quiet the entire system during a finite number of steps. Thus, E controllability is proved, which implies B controllability (Theorem 3(a)). Let us write this result accurately.

Theorem VII.1.4. If the number of strings is less than the number of nodes, then system (1), (2) is B-controllable in $\mathcal{W}_0$ in some time period T.

(2) The numbers of strings and nodes are equal, $M = N$.

In this case, the system graph consists of one cycle (extracted uniquely) and perhaps of one or more trees with roots on the cycle.

Theorem VII.1.5. If the number of strings is equal to the number of nodes, then system (1), (2) is not E-controllable in space $\mathcal{W}_0$ in any time.

PROOF. Clearly, the system is not controllable if its subsystem forming the cycle is not controllable.

Consider the cyclic string system. We first assume that the optical lengths of all the strings are commensurable; that is, they have the form $L_s = n_s \mu$, where $n_s \in \mathbb{N}$ and the highest common divisor of $\{n_s\}$ equals unity. Let us prove linear dependence of family $\tilde{\mathcal{E}}_T$ (for any $T > 0$).

The answer to the question about linear independence of vectors $\eta_{r_s s}$ is given by the following lemma.

Lemma VII.1.6. Family $\{\eta_{r_s s}\}_{s=1}^{Q}$ for $Q < N$ is linearly independent while for

$$\det\{\eta_{r_s s}\}_{s=1}^{N} = C[1 - (-1)^{\sum_{s=1}^{N} r_s}], \qquad C := \sum_{s=1}^{N} d_s.$$

The proof is produced by elementary methods.

Consider the set of exponentials of family $\tilde{\mathcal{E}}_T$ of the form e_{r_s} for $r_s = 2n_s$, $s = 1, 2, \ldots, N$. These exponentials read as $\eta_{r_s s} \exp(i\alpha t)$ for

$$\alpha = \pi r_s / (\mu n_s) + i = 2\pi/\mu + i.$$

By Lemma 6, vectors $\{\eta_{rs}\}_{s=1}^{N}$ are linearly dependent; hence, so is family $\tilde{\mathcal{E}}_T$. Consequently, the cyclic system with commensurable optical lengths is not E-controllable nor even W-controllable.

Now let us deal with the general cycle and define for the jth class, $j = 1, 2, \ldots, R$, the number μ_j and integers n_s, $s \in S_j$, in such a way that $L_s = n_s \mu_j$ while the highest common divisor of $\{n_s\}_{s \in S_j}$ equals unity for any j.

Suppose that the cyclic system is E-controllable. Then, by Theorem 3(a), it is B-controllable (in space $\mathcal{W}_0$ in time T_0). Hence $\tilde{\mathcal{E}}_{T_0} \in (LB)$ and, more so, $\tilde{\mathcal{E}}_{T_0} \in (UM)$. Then by Lemma I.1.28, family $\tilde{\mathcal{E}}_\infty = \{e_{rs}\} \subset L^2(0, \infty; \mathbb{C}^N)$ is also *-uniformly minimal. The latter is almost normed and so is uniformly minimal (see Definition I.1.15). Therefore, by Proposition II.2.8, $\tilde{\mathcal{E}}_\infty \in (LB)$. Using the basis property criterion (Theorem II.2.12 and Corollary II.2.14), we now show that family $\tilde{\mathcal{E}}_\infty$ cannot be an $\mathscr{L}$-basis.

Let us split the system spectrum into sets $\Lambda_m(r)$ of its close points (see the definitions in Subsection II.2.2). Later, we consider r small enough to provide the absence in a group $\Lambda_m(r)$ of two different points belonging to the same class (this is possible because the spectrum of strings of a given class is separable). The group of exponentials $\eta_{r_s s} \exp(iv_{r_s s}t)$ such that $v_{r_s s} \in \Lambda_m(r)$ corresponds to the set $\Lambda_m(r)$.

Lemma VII.1.7. For any $r > 0$ there may be found set $\Lambda_m(r)$ such that the group of N exponentials with even r_s corresponds to it.

In view of Lemma 6, any N vectors $\eta_{r_s s}$ with even r_s are linearly dependent. The assertion of Theorem 5 follows immediately from Lemma 7 and Theorem II.2.12.

PROOF OF LEMMA 7. Let us cite one result (a consequence of the Dirichlet theorem) on the mutual approximation of homogeneous forms by integers.

Proposition VII.1.8 (Cassels 1957: sec. I.5). For any real $\beta_1, \beta_2, \ldots, \beta_R$ and any $\varepsilon > 0$, there exists integer n such that

$$\min_{m \in \mathbb{Z}} |\beta_j n - m| < \varepsilon, \qquad j = 1, 2, \ldots, R.$$

PROOF. Set $\beta_j := \mu_j/2\pi$ and take some $\varepsilon > 0$. Then, by Proposition 8, integers n and $q_j, j = 1, 2, \ldots, R$, may be found such that $|\mu_j n/2\pi - q_j| < \varepsilon$. Therefore the inequalities hold

$$|n - 2\pi q_j/\mu_j| < c\varepsilon, \qquad c := \max_{j=1, 2, \ldots, R} \{2\pi/\mu_j\}.$$

Now set $r_s := 2n_s q_j, s \in S_j, j = 1, 2, \ldots, R$. Group $\Lambda_m(r)$ formed by spectrum points $\pi r_s/L_s + i, s = 1, 2, \ldots, N$ satisfies Lemma 7 for $\varepsilon < r/c$. Lemma 7, and with it Theorem 5, is proved.

Let us demonstrate the lack of W controllability for a circle of two identical strings. Let the strings be at rest at $t = 0$. Then for any controls

$f_1, f_2 \in L^2(0, T)$ acting at the ends, the state of each string at $t = T$ is the same:

$$y^1(\cdot, T) = y^2(\cdot, T), \qquad y_t^1(\cdot, T) = y_t^2(\cdot, T).$$

Thus the reachability set is

$$\left\{ \begin{pmatrix} \{g, h\} \\ \{g, h\} \end{pmatrix} \mid g \in L^2(0, l),\, h \in H^{-1}(0, l) \right\}.$$

The orthogonal complement of $\mathcal{R}(T)$ is

$$\left\{ \begin{pmatrix} \{g, h\} \\ -\{g, h\} \end{pmatrix} \mid g \in L^2(0, l),\, h \in H^{-1}(0, l) \right\}.$$

(3) The number of strings is greater than the number of nodes, $M > N$.

Theorem VII.1.9. If $M > N$, then system (1), (2) is not E-controllable in space $\mathcal{W}_0$ in any time.

PROOF. For any $r > 0$ group, $\Lambda_m(r)$ may be found such that M elements $\eta_{r_{rs}} \exp(iv_{r_{rs}}t)$ of family $\tilde{\mathscr{E}}_\infty$ correspond to it (this statement, in fact, is contained in the proof of Lemma 7). The $\mathscr{L}$-basis property criterion then provides $\tilde{\mathscr{E}}_\infty \notin (LB)$ and, hence, $\tilde{\mathscr{E}}_T \notin (LB)$ for any T, since the set of vectors η_{rs} corresponding to $\Lambda_m(r)$ is linearly dependent (the number of vectors is larger than the space dimension). Therefore the system is not B-controllable. In view of Theorem 3(a) the system is also not E-controllable. The theorem is proved.

With the help of Theorem 3(b), (c), the question of M controllability of the systems can be reduced to B controllability of any connected component of any class. It follows from Theorems 4, 5, and 9 that system (1), (2) is M-controllable in some time T_0 if and only if the graph of any connected component of any class is a tree.

2. System of strings connected elastically at one point

A number of recent studies have dealt with controllability problems in multilinked flexible systems (see, e.g., Chen et al. 1987; Leugering and Schmidt 1989; Schmidt 1992; Lagnese, Leugering, and Schmidt 1993)). The main tool used in these studies is the Hilbert Uniqueness Method. In this section we use the example of strings linked at one point to show that the method of moments can be efficiently applied to such systems.

Consider a system described by the following initial boundary problem:

$$\rho_s(x)\frac{\partial^2 y_s(x, t)}{\partial t^2} = \frac{\partial^2 y_s(x, t)}{\partial x^2}, \qquad \rho_s > 0, \; 0 < x < l_s, \; 0 < t < T. \qquad (1)$$

$$y_s(0, t) = u_s(t), \qquad u_s \in L^2(0, T), \qquad (2)$$

$$y_1(l_1, \cdot) = y_2(l_2, \cdot) = \cdots = y_N(l_N, \cdot), \qquad (3)$$

$$\frac{\partial}{\partial x} y_1(l_1, \cdot) + \frac{\partial}{\partial x} y_2(l_2, \cdot) + \cdots + \frac{\partial}{\partial x} y_N(l_N, \cdot) = 0, \qquad (4)$$

$$y_s(x, 0) = \frac{\partial}{\partial t} y_s(x, 0) = 0, \qquad s = 1, 2, \ldots, N. \qquad (5)$$

It is assumed that $\rho_s \in C^2[0, l_s]$, $\rho_s > 0$. Functions u_s, $s = 1, 2, \ldots, N$, are the control actions; we consider them to be components of vector functions of $L^2(0, T; \mathbb{C}^N)$.

Conditions (3), (4) at the linking point are implied by the requirement that the system's energy be conserved. In Section 4, one particular string is treated as a system of two strings joined at one point, and we prove that (3) can be used as a continuity condition at the linking point for solution $y(x, t)$ to the string equation. Condition (4) may be considered the requirement for function $(\partial/\partial x)y(x, t)$ to be continuous at the linking point.

Let us embed the classically formulated problem (1)–(5) in the scheme of Chapter III and thus attach a strict meaning to it. Let $L^2_{\rho_s}(0, l_s)$ denote spaces of functions squarely summable on the segment $[0, l_s]$ with the weight $\rho_s(x)$ and, as in Section 1, set

$$H := \bigoplus_1^M L^2_{\rho_s}(0, l_s).$$

Introduce space V

$$V = \{v = (v_1, v_2, \ldots, v_N) \mid v_s \in H^1(0, l_s), \; v_s(0) = 0, \; s = 1, 2, \ldots, N,$$

$$v_1(l_1) = v_2(l_2) = \cdots = v_N(l_N)\},$$

and specify a bilinear form on V

$$a[\varphi, \psi] = \sum_{s=1}^N \int_0^{l_s} \varphi_s'(x)\bar{\psi}_s'(x)\, dx; \qquad \varphi, \psi \in V.$$

Integrating this by parts for $\varphi_s \in H^2(0, l_s)$ and taking boundary conditions for φ, ψ into account, we have

$$a[\varphi, \psi] = \sum_{s=1}^{N} \int_{0}^{l_s} \left(-\frac{1}{\rho_s} \varphi_s''(x) \right) \bar{\psi}_s(x) \, dx + \left(\sum_{s=1}^{N} \varphi_s'(l_s) \right) \bar{\psi}_0(l_s), \quad (6)$$

where $\psi_0(l_s)$ is the common value of functions ψ_s at points l_s.

According to Section III.1, bilinear form has to be represented as $a[\varphi, \psi] = (A\varphi, \psi)_H$; that is, it has to be a continuous functional in H for a given φ from the domain of operator A. From equality (6) we then conclude that

$$\varphi_1'(l_2) + \varphi_2'(l_2) + \cdots + \varphi_N'(l_N) = 0. \quad (7)$$

Therefore operator A acts according to the rule

$$\{\varphi_s(x)\}_{s=1}^{M} \mapsto \left\{ -\frac{1}{\rho_s(x)} \frac{d^2}{dx^2} \varphi_s(x) \right\}_{s=1}^{M}$$

and its domain consists of sets $\{\varphi_s\}_{s=1}^{N}$ of functions such that $\varphi_s \in H^2(0, l_s)$, $\varphi_s(0) = 0$, and condition (7) and $\varphi_1(l_2) = \varphi_2(l_2) = \varphi_N(l_N)$ hold at the linking point. For the eigenvalues and the eigenfunctions of A normed in H, we write λ_n and Φ_n, $n \in \mathbb{N}$, respectively. It is easy to see that $\lambda_n \neq 0$ for all n. From the scheme of Chapter III, operator B here acts from the space $\mathbb{C}^N$ to the space W_{-2} by the rule

$$\langle Bv, \varphi \rangle_* = \sum_{s=1}^{M} v_s(0) \bar{\varphi}_s'(0), \quad \varphi \in W_2 = D(A). \quad (8)$$

(The relation of operator B with problem (1)–(5) is clarified in the same manner as in Section 1; for the definition of spaces W_r see Section III.1.)

Lemma VII.2.1. Solution $y = \{y_s\}_{s=1}^{N}$ to the problem (1)–(5) exists, is unique, and

$$\left\{ y, \frac{\partial}{\partial t} y \right\} \in C(0, T, \mathcal{W}_0),$$

$$\mathcal{W}_0 = W_0 \oplus W_{-1} = H \oplus \left[\bigoplus_{s=1}^{N} H^{-1}(0, l_s) \right].$$

We prove this lemma after proving Theorem 6. For the time being, note its apparent consequence: $\mathcal{R}(T) \subset \mathcal{W}_0$.

Let us introduce the optical lengths L_s of the strings:

$$L_s := \int_0^{l_s} \sqrt{\rho_s(x)}\, dx$$

and the notation T_0 for $2\max\{L_s\}$.

Theorem VII.2.2. System (1)–(4) *is B-controllable in space* $\mathcal{W}_0$ *in time* T_0.

The plan for studying system (1)–(4) controllability is as follows. First, we exploit the standard methods (Naimark 1969). With operator A we relate an entire matrix function G whose zeros coincide with the set $\{0\} \cup \{\pm\sqrt{\lambda_n}\}$. Then, after some manipulation of function G, we produce a GF of an exponential family close to one arising from the problem of moments. With the help of the GF, we demonstrate the $\mathcal{L}$-basis property of the family in space $L^2(0, T_0; \mathbb{C}^N)$ and thereby prove system (1)–(4) controllability as well.

To construct the GF, let us introduce functions $\varphi_s(x, k)$, $s = 1, 2, \ldots, N$, as solutions to the Cauchy problem

$$\varphi_s''(x, k) + k^2 \rho_s(x)\varphi_s(x, k) = 0, \qquad (9)$$

$$\varphi_s(0, k) = 0, \qquad \varphi_s'(0, k) = k.$$

Set

$$\psi_s(k) := \varphi_s(l_s, k), \qquad \zeta_s(k) := k^{-1}\varphi_s'(l_s, k), \qquad s = 1, 2, \ldots, N.$$

Since functions $\varphi_s(x, k)$ are entire in k and $\varphi_s(x, k)|_{k=0} = 0$, ψ_s and ζ_s are also entire functions. Moreover, equality $\varphi_s(-x, k) = -\varphi_s(x, k)$ implies

$$\psi_s(-k) = -\psi_s(k), \qquad \zeta_s(-k) = \zeta_s(k). \qquad (10)$$

We need some information on the asymptotics of solutions φ_s.

Proposition VII.2.3 (Fedoryuk 1983: chap. 2, sec. 3). *Uniform in* $x \in [0, l_s]$ *asymptotics at* $|k| \to \infty$, $k \in \mathbb{C}$, *is valid:*

$$\varphi_s(x, k) = [\rho_s(x)/\rho_s(0)]^{-1/4} \sin(kq_s(x)) + \mathcal{O}(k^{-1}\exp(q_s(x)|\mathrm{Im}\, k|)),$$

$$\varphi_s'(x, k) = [\rho_s(x)\rho_s(0)]^{-1/4} k \cos(kq_s(x)) + \mathcal{O}(\exp(q_s(x)|\mathrm{Im}\, k|))$$

with $q_s(x) := \int_0^x \sqrt{\rho_s(\xi)}\, d\xi$.

Let us look for the eigenfunctions Φ_n of operator A in the form

$$\Phi_n(x) = \{\varphi_s(x, \omega_n)\eta_n^{(s)}\}_{s=1}^N,$$

where $\omega_n := \sqrt{\lambda_n} > 0$, while $\eta_n^{(s)}$ are complex numbers. Conventionally, we write such a set of functions φ_s defined for every s on the appropriate segment $[0, l_s]$ as a vector function

$$\Phi_n = \text{diag}[\varphi_s(x, \omega_n)]\eta_n, \qquad \eta_n = \{\eta_n^{(s)}\}_{s=1}^N \in \mathbb{C}^N. \tag{11}$$

From the definition of functions φ_s, it follows that Φ_n satisfy the required differential equations. Substituting Φ_n into the boundary conditions, we arrive at the equations

$$\psi_1(\omega_n)\eta_n^{(1)} - \psi_2(\omega_n)\eta_n^{(2)} = 0,$$
$$\psi_1(\omega_n)\eta_n^{(1)} - \psi_3(\omega_n)\eta_n^{(3)} = 0,$$
$$\cdots\cdots\cdots\cdots\cdots\cdots\cdots\cdots$$
$$\psi_1(\omega_n)\eta_n^{(1)} - \psi_N(\omega_n)\eta_n^{(N)} = 0,$$
$$\zeta_1(\omega_n)\eta_n^{(1)} + \zeta_2(\omega_n)\eta^{(2)} + \cdots + \zeta_N(\omega_n)\eta^{(N)} = 0.$$

It is natural now to introduce an entire matrix function

$$G(k) = \begin{pmatrix} \psi_1(k) & -\psi_2(k) & 0 & \cdot & 0 \\ \psi_1(k) & 0 & -\psi_3(k) & \cdot & 0 \\ \psi_1(k) & 0 & & \cdot & -\psi_N(k) \\ \cdot & \cdot & \cdot & \cdot & \cdot \\ \zeta_1(k) & \zeta_2(k) & \zeta_3(k) & \cdot & \zeta_N(k) \end{pmatrix}.$$

Lemma VII.2.4.

(a) *Any eigenfunction of operator A is of the form* (11) *with $\eta_n \in \text{Ker } G(\omega_n)$.*

(b) *Conversely, any function of the form* $\text{diag}[\varphi_s(x, k)]\eta$ *with $\eta \in \text{Ker } G(k_0)$ and $\eta \neq 0$, $k_0 \neq 0$, is an eigenfunction of operator A corresponding to the eigenvalue k_0^2.*

PROOF OF THE LEMMA.

(a) Let $\varphi_{ns}(x)$ denote the components of eigenfunction $\Phi_n(x)$. Obviously, $\varphi_{ns}(x)$ satisfies equation (9) for $k = \omega_n$ and condition $\varphi_{ns}(0) = 0$. So $\varphi_{ns}(x)$ differs from $\varphi_s(x, \omega_n)$ by a constant factor $\varphi_{ns}'(0)/\omega_n$. Setting

$$\eta_n = \omega_n^{-1}\{\varphi_{ns}'(0)\} \in \mathbb{C}^N,$$

we write Φ_n in the form (11). Boundary conditions for φ_{ns} at the linking point imply equality $G(\omega_n)\eta_n = 0$.

(b) For any complex number k_0 and any $\eta = \{\eta^{(s)}\}_{s=1}^N$, vector function

$$\Phi(x, k_0) := \mathrm{diag}[\varphi_s(x, k_0)]\eta = \{\varphi_s(x, k_0)\eta^{(s)}\}_{s=1}^N$$

satisfies the equation

$$-\Phi'' = k_0^2 \, \mathrm{diag}[\rho_s]\Phi$$

and boundary condition $\Phi(0, k_0) = 0$. If, along with this, $G(k_0)\eta = 0$, then the components of vector function $\Phi(x, k)$ also meet conditions (3), (4). If $k_0 \neq 0$ and $\eta \neq 0$, Φ is an eigenfunction of operator A corresponding to eigenvalue k_0^2. The lemma is proved.

Let us elucidate conditions on vectors η_n pending the requirement that the family of eigenfunctions $\{\Phi_n\}$ should be orthonormal in H. Let μ be an eigenfrequency of system (1)–(4) (i.e., μ^2 is an eigenvalue of operator A), let multiplicity of μ be equal to χ, and $\Phi_{\mu1}, \Phi_{\mu2}, \ldots, \Phi_{\mu\chi}$ be corresponding eigenfunctions, $\Phi_{\mu j} = \mathrm{diag}[\varphi_s(x, \mu)]\eta_{\mu j}$, $\eta_{\mu j} \in \mathrm{Ker}\, G(\mu)$. If we set

$$d_{\mu s} := \int_0^{l_s} \varphi_s^2(x, \mu)\rho_s \, dx,$$

we have

$$(\varphi_{\mu p}, \varphi_{\mu q})_H = \sum_{s=1}^N d_{\mu s}\eta_{\mu p}^{(s)}\bar{\eta}_{\mu q}^{(s)} = \delta_p^q. \tag{12}$$

By Proposition 3, numbers $d_{\mu s}$ are bounded and bounded away from zero uniformly in s and μ:

$$0 < c \leq |d_{\mu s}| \leq C. \tag{13}$$

Now if D denotes the operator generated by bilinear form

$$\sum_{s=1}^N d_{\mu s}\eta^{(s)}\bar{\zeta}^{(s)}$$

on the subspace $\mathfrak{N}^\mu := \mathrm{Ker}\, G(\mu)$, then, in view of (12), operator $\sqrt{D}$ transforms the set $\{\eta_{\mu j}\}_{j=1}^\chi$ to an orthonormal basis in $\{\hat{\eta}_{\mu j}\}_{j=1}^\chi$ of subspace $\mathfrak{N}^\mu$.

From (13) we conclude that family $\{\eta_{\mu j}\}_{j=1}^\chi$ is a basis in subspace $\mathfrak{N}^\mu$ uniformly in μ:

$$c \sum_{j=1}^\chi |c_j|^2 \leq \left\langle\!\left\langle \sum_{j=1}^\chi c_j\eta_{\mu j} \right\rangle\!\right\rangle^2 \leq C \sum_{j=1}^\chi |c_j|^2. \tag{14}$$

That is, the basis constants (Definition II.5.4) are bounded and bounded away from zero uniformly in μ.

On the basis of this study of the structure of eigenfunctions Φ_n, let us introduce, in keeping with Section III.3 and (8), the exponential family

$$\tilde{\mathscr{E}}_T = \{e_n\}_{n \in \mathbb{K}} \subset L^2(0, T; \mathbb{C}^N),$$

$$e_n = \eta_n \exp(i\omega_n t), \qquad \eta_n = |\omega_n^{-1}| \Phi'_{|n|}(0).$$

Here, we have set $\omega_{-n} = -\omega_n$ for $n \in \mathbb{N}$. As in Section 1, factor $|\omega_n^{-1}|$ shows that we are investigating the set of reachability $\mathscr{R}(T)$ in space $\mathscr{W}_0$. As a space $\mathscr{H}_0$, we also take the space $\mathscr{W}_0$. From Theorem III.3.10 we find that B controllability of system (1)–(4) in space $\mathscr{W}_0$ follows from the $\mathscr{L}$-basis property of family $\tilde{\mathscr{E}}_T$ (in $L^2(0, T; \mathbb{C}^N)$). It is more convenient for us to deal with a somewhat modified family built by the zeros of matrix function $\tilde{G}(k) := G(k - i)$. Let us write σ for the set of zeros of $\det \tilde{G}(k)$ (it is obvious that $\sigma = \{\omega_n + i\} \cup \{i\}$, since $\det G(0) = 0$). For any $\mu \in \sigma$, choose in $\mathfrak{N}^\mu = \operatorname{Ker} \tilde{G}(\mu)$ orthonormal basis $\{\hat{\eta}_{\mu j}\}_{j=1}^{\chi_\mu}$, $\chi_\mu := \dim \mathfrak{N}^\mu$, and set

$$\hat{\mathscr{E}}_T := \{e^{i\mu t} \hat{\eta}_{\mu j}\}_{\mu \in \sigma}^{j=1, 2, \ldots, \chi_\mu} \subset L^2(0, T; \mathbb{C}^N).$$

Lemma VII.2.5. If $\hat{\mathscr{E}}_T \in (LB)$, then $\tilde{\mathscr{E}}_T \in (LB)$.

PROOF. Family $\hat{\mathscr{E}}_T$ differs from the family $\tilde{\mathscr{E}}_T$ by

  (i) the shift of the spectrum to the upper half-plane,
 (ii) the presence of additional elements corresponding to the point $k = 0$,
(iii) the different choice of vectors in subspaces $\operatorname{Ker} G(\omega_n) = \operatorname{Ker} \tilde{G}(\omega_n + i)$.

Two former differences obviously cannot damage the implication we are proving. As for (iii), we have already compared the expansions over various bases in subspace $\operatorname{Ker} G(\omega_n) = \operatorname{Ker} \tilde{G}(\omega_n + i)$ (see (12)–(14)). The lemma is proved.

Next we examine the controllability of system (1)–(5) in the following way. As already mentioned, B controllability of the system is implied by the force of the results from Chapter III, by the $\mathscr{L}$-basis property of exponential family $\tilde{\mathscr{E}}_T$. As we have just proved, the $\mathscr{L}$-basis property of $\tilde{\mathscr{E}}_T$ is in turn provided by the $\mathscr{L}$-basis property of family $\hat{\mathscr{E}}_T$ constructed by zeros of the $\tilde{G}$ function. The proof of the latter (in space $L^2(0, T; \mathbb{C}^N)$)

falls naturally into two parts:

(1) the investigation of the $\mathscr{L}$-basis property of family $\hat{\mathscr{E}}_\infty$ in $L^2(0, \infty; \mathbb{C}^N)$ and

(2) the proof of the projection from $\bigvee \hat{\mathscr{E}}_\infty$ to $L^2(0, \infty; \mathbb{C}^N)$ to be isomorphic.

We first study the second problem, and to do this we construct a GF of family $\hat{\mathscr{E}}_T$ on the basis of the further factorization of matrix function $\tilde{G}$. Then, using the spectral meaning of the ω_n and vectors η_n (see (11)), we prove that $\hat{\mathscr{E}}_\infty$ forms an $\mathscr{L}$-basis.

In the following theorem the inverse means the inverse matrix function: $\tilde{G}^{-1}(-k)\tilde{G}(k) = I$ for any k.

Theorem VII.2.6.

(a) *Function $\tilde{G}$ is bounded on $\mathbb{R}$ together with its inverse.*

(b) *There may be found an entire inner function Θ_0 in $\mathbb{C}_+$ such that function $F := \Theta_0 \tilde{G}$ can be factorized in the form*

$$F = F_e^+ \Pi = F_e^- \Theta, \tag{15}$$

where $F_e^\pm$ are outer functions in $\mathbb{C}^\pm$, respectively, Π is a BPP, and ESF Θ satisfies the estimate

$$\langle\langle \exp(ikT_0)\Theta^{-1}(k)\rangle\rangle \prec 1, \qquad k \in \mathbb{C}_+. \tag{16}$$

Let us discuss assertion (b) of the theorem. Function $\tilde{G}$, as one may see from the asymptotics of $\varphi_s(x, k)$, increases exponentially in both upper and lower half-planes. It so happens that the growth of $\tilde{G}$ in $\mathbb{C}_+$ can be removed by means of the ESF factor Θ_0 so accurately as to produce the operator function F bounded in $\mathbb{C}_+$, which has no exponentially decreasing factor in the upper half-plane (the inverse matrix function is bounded). In the lower half-plane, F is represented by a product of an outer function F_e^- and ESF Θ. Out of function F we then make a GF of family $P_\Theta \tilde{\mathscr{E}}_T := \{(k - \bar{\mu})^{-1}\eta_{\mu j}\}_{\mu \in \sigma}^{j=1, 2, \dots, \chi_\mu}$ and in this context estimate (16) will mean that the Fourier transform of space K_Θ contains $L^2(0, T_0; \mathbb{C}^N)$.

Denote matrix $\exp(ik \operatorname{diag}[L_s])$ by Θ_1.

Lemma VII.2.7.

(a) *For $\operatorname{Im} k \geq 1$, operator-function $G(k)\Theta_1(k)$ is bounded along with its inverse.*

(b) *For $\operatorname{Im} k \leq 1$, operator-function $G(k)\Theta_1^{-1}(k)$ and its inverse are bounded.*

Sublemma VII.2.8. Let matrix J be of the form

$$
J = \begin{pmatrix}
a_1 & -a_2 & 0 & \cdot & 0 \\
a_1 & 0 & -a_3 & \cdot & 0 \\
a_1 & 0 & & \cdot & -a_N \\
\cdot & \cdot & \cdot & \cdot & \cdot \\
b_1 & b_2 & b_3 & \cdot & b_N
\end{pmatrix}.
\tag{17}
$$

Then

$$
\det J = \sum_{s=1}^{N} b_s \prod_{j \neq s}^{N} a_j = a_1 a_2 \cdots a_N \sum_{s=1}^{N} b_s / a_s.
$$

The proof of this statement is carried out by means of the standard methods.

PROOF OF LEMMA 7. (a) Operator $G(k)\Theta_1(k)$ may be written in the form (17) with

$$
a_s(k) = \psi_s(k) e^{ikL_s}, \qquad b_s(k) = \zeta_s(k) e^{ikL_s}.
$$

Proposition 3 implies

$$
a_s(k) = \beta_s e^{ikL_s} \sin(kL_s) + o(1),
$$

$$
b_s(k) = \gamma_s e^{ikL_s} \cos(kL_s) + o(1), \qquad k \in \mathbb{C}_+;\ \beta_s, \gamma_s > 0
$$

and this ensures that $G(k)\Theta_1(k)$ will be bounded in $\mathbb{C}_+$. Using Sublemma 8, we arrive at the formula

$$
\det[G(k)\Theta_1(k)] = \sum_{s=1}^{N} b_s(k) \prod_{j \neq s,\, j=1}^{N} a_j(k)
$$

$$
= \left(\prod_{s=1}^{N} \beta_s e^{ikL_s} \sin(kL_s) \right) \sum_{s=1}^{N} \frac{\gamma_s}{\beta_s} \cotan(kL_s) + o(1).
\tag{18}
$$

Let us show that

$$
|\det[G(k)\Theta_1(k)]| \succ 1, \qquad \text{Im } k \geq 1.
\tag{19}
$$

Since the zeros of $\det G(k)$ lie on the real axis, for $\text{Im } k \geq 1$, estimate (19) follows from the similar estimate of the main term in (18):

$$
\left| \prod_{s=1}^{N} \beta_s e^{ikL_s} \sin(kL_s) \sum_{s=1}^{N} \frac{\gamma_s}{\beta_s} \cotan(kL_s) \right| \succ 1, \qquad \text{Im } k \geq 1.
\tag{20}
$$

Each of the factors $e^{ikL_s} \sin(kL_s)$ satisfies the equality

$$|e^{ikL_s} \sin(kL_s)| = \tfrac{1}{2}|1 - e^{2ikL_s}| \geq \tfrac{1}{2}(1 - e^{-2L_s})$$

(for $\operatorname{Im} k \geq 1$).

Sublemma VII.2.9. *For* $\operatorname{Im} z \geq \delta > 0$, *we have*

$$|\operatorname{Im} \cotan(z)| \geq c(\delta) > 0.$$

PROOF OF THE SUBLEMMA. For $\operatorname{Im} z \to +\infty$, we have $\cotan(z) \to -i$. Therefore, taking into account that $\cotan(z)$ is a periodic function, it is enough to establish that $\operatorname{Im} \cotan(z) \neq 0$ for $z \notin \mathbb{R}$. Let

$$\cotan(z) = i(e^{iz} + e^{-iz})/(e^{iz} - e^{-iz}) =: p \in \mathbb{R}.$$

Since $\exp(2iz) = (p + i)/(p - i)$, $|\exp(2iz)| = 1$ and therefore $z \in \mathbb{R}$. The sublemma is proved.

Applying Sublemma 9 to the sum over s in (20), we obtain

$$\left| \operatorname{Im} \sum_{s=1}^{N} \frac{\gamma_s}{\beta_s} \cotan(kL_s) \right| \geq \sum_{s=1}^{N} \frac{\gamma_s}{\beta_s} c(L_s) > 0$$

for $\operatorname{Im} k \geq 1$. Hence, estimate (20), and therefore (19), is valid. The latter and the boundedness of function $G\Theta_1$ in $\mathbb{C}_+$ imply assertion (a) of the lemma.

To check assertion (b), let us exploit the implication

$$G(-k) = -\operatorname{diag}[1, 1, \ldots, 1, -1]G(k)$$

of equality (10). Since $\Theta_1^{-1}(-k) = \Theta_1(k)$, function $G(k)\Theta_1^{-1}(k)$ is obviously bounded for $\operatorname{Im} k \leq -1$ if $G\Theta_1$ is bounded for $\operatorname{Im} k \geq 1$, as guaranteed by assertion (a). The lemma is proved.

PROOF OF THEOREM 6. Assertion (a) of the theorem follows directly from assertion (6b) of the lemma, since the explicit form of function $\Theta_1(k) = \exp\{ik \operatorname{diag}[L_s]\}$ reveals that it is bounded together with its inverse for $\operatorname{Im} k = -1$.

We start to prove assertion (b) by obtaining a factorization of function $\tilde{G}$ in $\mathbb{C}_+$. Proposition 3 gives that functions $\exp(ikT_0/2)\psi_s(k)$ and $\exp(ikT_0/2)\zeta_s(k)$ are bounded in $\mathbb{C}_+$. Therefore function $\exp(ikT_0/2)\tilde{G}(k)$ is bounded in $\mathbb{C}_+$ and may be factorized as

$$e^{ikT_0/2} \tilde{G} = \Theta_2 \tilde{F}_e^+ \Pi \tag{21}$$

with Θ_2 being an ESF, and $\tilde{F}_e^+$ being an outer operator function bounded in $\mathbb{C}_+$ according to Subsection II.1.2.9. Set

$$\Theta_0 := e^{ikT_0/2}\,\Theta_2^{-1}(k) \tag{22}$$

and verify that function Θ_0 is bounded in $\mathbb{C}_+$ (then it is an ESF). This function is bounded on the real axis since it is unitary-valued on it. By Lemma 7(a), function $\tilde{G}^{-1}(k)$ is bounded for $\operatorname{Im} k \geq 2$. Formula (21) provides $\Theta_0 = \tilde{F}_e^+ \Pi \tilde{G}^{-1}$, so Θ_0 is also bounded for $\operatorname{Im} k \geq 2$. All that remains is to check that it is bounded in the strip $0 \leq \operatorname{Im} k \leq 2$. Let us take advantage of the fact that multiplication of Θ_0 by $e^{ik\alpha}$ with large enough α leads to an ESF and therefore $\langle\langle \Theta_0\, e^{ik\alpha} \rangle\rangle \leq 1$. Hence, Θ_0 is bounded in the strip.

We just proved the factorization of $\tilde{G}$ in $\mathbb{C}_+$.

To prove the required factorization in $\mathbb{C}_-$, we need some statement about the change of the factorization order.

Lemma VII.2.10. Let $\mathscr{F}_e^-$ be an outer operator function in $\mathbb{C}_-$, let $\mathscr{F}_e^-$ and $[\mathscr{F}_e^-]^{-1}$ be bounded in $\mathbb{C}_-$, and let Θ be an ESF. Then outer operator function $\tilde{\mathscr{F}}_e^-$ in $\mathbb{C}_-$ and ESF $\tilde{\Theta}$ may be found such that

$$\Theta \mathscr{F}_e^- = \tilde{\mathscr{F}}_e^- \tilde{\Theta}. \tag{23}$$

PROOF OF THE LEMMA. Function $\exp(-ika)\Theta(k)$ is bounded in $\mathbb{C}_-$ and so for some $a > 0$ factorization

$$e^{-ika}\,\Theta \mathscr{F}_e^- = \tilde{\mathscr{F}}_e^-\Theta_3^{-1} \tag{24}$$

is possible (Θ_3 is an ESF). By setting $\tilde{\Theta}(k) = \exp(ika)\Theta_3^{-1}(k)$, one may see that functions $\tilde{\mathscr{F}}_e^-$ and $\tilde{\Theta}$ satisfy equation (23). Let us demonstrate that $\tilde{\Theta}^{-1}$ is bounded in $\mathbb{C}_-$. From (23) we derive the equality

$$[\tilde{\Theta}]^{-1} = [\mathscr{F}_e^-]^{-1}\Theta^{-1}\tilde{\mathscr{F}}_e^-.$$

Function $[\mathscr{F}_e^-]^{-1}$ is bounded in $\mathbb{C}_-$ by the condition of the lemma, while $\tilde{\mathscr{F}}_e^-$ – by relation (24) and the condition. That is why function $\tilde{\Theta}^{-1}$ is bounded in $\mathbb{C}_-$. If we now apply formula (3) of Section II.1, we end up with the boundedness of function $\tilde{\Theta}$ in $\mathbb{C}_+$. The lemma is proved.

By Lemma 7(b), function $\tilde{G}(k)\Theta_1^{-1}(k)$ is bounded in $\mathbb{C}_-$ along with its inverse. Therefore it is an outer function $\tilde{F}_e^-$; that is, $F = \Theta_0 \tilde{F}_e^- \Theta_1$. With the help of Lemma 10, one is able to change the order of the first two cofactors:

$$\Theta_0 \tilde{F}_e^- = F_e^- \Theta_4 \qquad (25)$$

and to get $F = F_e^- \Theta_4 \Theta_1$. Thus factorization (15) is established for $\Theta = \Theta_4 \Theta_1$.

It only remains to check the validity of estimate (16). Since function $\Theta_1^{-1}(k) \exp(ikT_0/2)$ is bounded in $\mathbb{C}_+$, it suffices to prove the estimate

$$\langle\langle \Theta_4^{-1}(k) \, e^{ikT_0/2} \rangle\rangle \prec 1, \qquad k \in \mathbb{C}_+. \qquad (26)$$

Function Θ_4 appears following the change in the order of factorization in (25). Therefore $\Theta_4^{-1} = [\tilde{F}_e^-]^{-1} \Theta_0^{-1} F_e^-$. Outer functions $[\tilde{F}_e^-]^{-1}$ and F_e^- are bounded in $\mathbb{C}_-$, so that instead of (26), we can verify that function $\Theta_0^{-1}(k) \, e^{ikT_0/2}$ is bounded. By the force of (22), this function equals ESF $\Theta_2(k)$. Theorem 6 is proved completely.

Now it is possible to demonstrate Lemma 1. Function $\det G(k)$ is an STF by Theorem 6. Then (see Proposition II.1.28) set $\{\omega_n\}_{n \in \mathbb{K}}$ of its zeros is a finite unification of separable sets. The assertion of Lemma 1 now follows from Lemma II.1.21 and Lemma III.2.4.

In contrast to a GF, function F has inner functions Π and Θ as right cofactors. To use the results from Section II.3, let us proceed with the so-called associated (adjoint) functions, which we will mark by the sign $\check{}$: $\check{\Phi}(k) = \Phi^*(-k)$. Function Φ is an outer function (a BPP or an ESF) if and only if the associated function $\check{\Phi}$ is an outer function (a BPP or an ESF, respectively), since these properties are governed by the properties of scalar function $\det \Phi$.

Lemma VII.2.11. A factorization

$$\check{F} = \check{\Pi} \check{F}_e^+ = \check{\Theta} \check{F}_e^-$$

holds with functions $\check{F}$ and $\check{F}^{-1}$ uniformly bounded on $\mathbb{R}$ and

$$\check{\Pi}^*(\mu_n)\eta_n = 0, \qquad \langle\langle \check{\Theta}^{-1}(k) \, e^{ikT_0} \rangle\rangle \prec 1, \qquad k \in \mathbb{C}_+ \qquad (27)$$

(remember, $\mu_n = \omega_n + i$).

PROOF OF THE LEMMA. Functions $\check{F}$ and $\check{F}^{-1}$ are uniformly bounded on $\mathbb{R}$, since it is true for functions F and F^{-1} in the light of Theorem 6(a). Let us check relations (27). Since $-\mu_n = \bar{\mu}_{-n}$, $\check{\Pi}^*(\mu_n) = \Pi(-\bar{\mu}_n) = \Pi(\mu_{-n})$. By definition, $-\mu_n = \eta_{-n}$, so Ker $\check{\Pi}^*(\mu_n) = $ Ker $\Pi(\mu_n)$.

Function $\exp(ikT_0)\Theta^{-1}(k)$ is bounded in $\mathbb{C}_+$ by Theorem 6(b). Since the norm of an operator equals that of the adjoint one, associated function

$\exp(ikT_0)\check{\Theta}^{-1}(k)$ is bounded in the upper half-plane as well. The lemma is proved.

Corollary VII.2.12. *Orthogonal projector from $\bigvee \hat{\mathscr{E}}_\infty$ to $L^2(0, T_0; \mathbb{C}^N)$ is an isomorphism on its image.*

PROOF. Using the fact that the nullsubspace of self-adjoint operator A contains only eigenfunctions, it is possible to show that zeros of $G(k)$ and, hence, zeros of $\check{\Pi}(k)$, are semisimple. In view of Remark II.3.15, we do not need this fact and do not prove it. Now Lemma 11 and Corollary II.3.22 imply that operator $P_{\hat{\Theta}}|_{K_{\check{\Pi}}}$ is an isomorphism of $K_{\check{\Pi}}$ and $K_{\hat{\Theta}}$. Space $K_{\check{\Pi}}$ contains simple fractions $(k - \bar{\mu}_n)^{-1}\mu_n$. Space $K_{\hat{\Theta}}$ is contained in space K_{T_0} by Lemma 12. Therefore the restriction of operator P_{T_0} on $\bigvee\{(k - \bar{\mu}_n)^{-1}\eta_n\}$ is an isomorphism on its image. By the transition from simple fractions to exponentials, we obtain the assertion of our corollary.

Thus, we have just finished the first stage of the study of the basis property of family $\tilde{\mathscr{E}}_T$. Lemma 5 and Corollary 12 yield an implication

$$\hat{\mathscr{E}}_\infty \in (LB) \Rightarrow \tilde{\mathscr{E}}_T \in (LB).$$

To check the $\mathscr{L}$-basis property for family $\hat{\mathscr{E}}_\infty$, we use the following idea. If $\hat{\mathscr{E}}_\infty \notin (LB)$, then, roughly speaking, there are some exponentials with close frequencies $\omega_n + i$ and close vectors η_n. But then, eigenfunctions Φ_n are also close, which contradicts their orthogonality in space W_0.

To implement this program, we need some estimates of solutions to systems of linear equations. Since we are also interested in their application in the next section, we present them with more generality than is required by the problem under consideration.

Lemma VII.2.13. *Let $\mathscr{A}(x)$ be positive definite and continuously differentiable on the segment $[0, a]$ matrix function, and let $\mathscr{Y}(x, k)$ be a solution to the Cauchy problem:*

$$\mathscr{Y}''(x, k) = -k^2 \mathscr{A}(x)\mathscr{Y}(x, k), \qquad \mathscr{Y}(0, k) = 0, \qquad \mathscr{Y}'(0, k) = kI. \quad (28)$$

Then function $\mathscr{Y}(x, k)$ is uniformly bounded in $k \in \mathbb{R}$:

$$\langle\langle \mathscr{Y}(x, k) \rangle\rangle \prec 1, \qquad x \in [0, a], k \in \mathbb{R},$$

and uniformly continuous in $k \in \mathbb{R}$: for any $\varepsilon > 0$, constant C may be found such that an inequality holds

$$\langle\langle \mathscr{Y}(x, k) - \mathscr{Y}(x, \tilde{k}) \rangle\rangle < C|k - \tilde{k}|, \qquad x \in [0, a], k \in \mathbb{R}, \tilde{k} \in \mathbb{R}.$$

PROOF OF THE LEMMA. Let us move from the second-order system (28) to the first-order system

$$\mathscr{L}'(x, k) = k\mathscr{B}(x)\mathscr{L}(x, k), \qquad \mathscr{L}(0, k) = I_{2N}, \tag{29}$$

$$\mathscr{B} := \begin{pmatrix} 0 & I_N \\ -\mathscr{A} & 0 \end{pmatrix}.$$

Here, I_N and I_{2N} are unit operators in $\mathbb{C}^N$ and $\mathbb{C}^{2N}$, respectively.

One should check whether the matrix function $\mathscr{L}(x, k)$ is bounded and uniformly continuous, since $\mathscr{Y}(x, k)$ is the upper right $N \times N$ block of matrix $\mathscr{L}$.

Set

$$\delta k := k - \tilde{k}, \qquad \delta\mathscr{L}(x, k, \tilde{k}) := \mathscr{L}(x, k) - \mathscr{L}(x, \tilde{k}).$$

Then

$$(\delta\mathscr{L}(x, k, \tilde{k}))' = k\mathscr{B}(x)\delta\mathscr{L}(x, k, \tilde{k}) + \delta k\mathscr{B}(x)\mathscr{L}(x, k), \qquad \delta\mathscr{L}(0, k) = 0$$

and from here

$$\delta\mathscr{L}(x, k, \tilde{k}) = \int_0^x \mathscr{L}(x, k)\mathscr{L}^{-1}(\xi, k)[\delta k\mathscr{B}(\xi)\mathscr{L}(\xi, \tilde{k})] \, d\xi. \tag{30}$$

Matrix function $\mathscr{L}(x, k)\mathscr{L}^{-1}(\xi, k)$ is the solution to equation (29), which turns to I_{2N} at the point $x = \xi$. To estimate this function, let us make a substitution,

$$\mathscr{L} = \begin{pmatrix} I & -I \\ i\sqrt{\mathscr{A}} & i\sqrt{\mathscr{A}} \end{pmatrix} \mathscr{L}_1.$$

Then we have an equation

$$\mathscr{L}_1' = \left[ik\begin{pmatrix} \sqrt{\mathscr{A}} & 0 \\ 0 & -\sqrt{\mathscr{A}} \end{pmatrix} - \mathscr{B}_1 \right] \mathscr{L}_1 \tag{31}$$

in which all the $N \times N$ blocks of the matrix $\mathscr{B}_1$ equal $\sqrt{\mathscr{A}^{-1}}(d/dx)\sqrt{\mathscr{A}}$.

Proposition VII.2.14 (Hartman 1964: chap. IV, lemma 4.2). *A solution to the vector equation*

$$\varkappa'(x) = \tilde{\mathscr{B}}(x)\varkappa(x), \qquad \varkappa(\xi) = \varkappa_0$$

is estimated by the largest, $\kappa_+(x)$, and the smallest, $\kappa_-(x)$, eigenvalues of the Hermitian part $(\tilde{\mathscr{B}}(x) + \tilde{\mathscr{B}}^*(x))/2$ of $\mathscr{B}$, as follows: for $x > \xi$

$$\langle\langle\varkappa(x)\rangle\rangle \leq \langle\langle\varkappa_0\rangle\rangle \exp\left(\int_\xi^x \kappa_+(s)\,ds\right),$$

$$\langle\langle\varkappa(x)\rangle\rangle \geq \langle\langle\varkappa_0\rangle\rangle \exp\left(\int_\xi^x \kappa_-(s)\,ds\right),$$

$$(32)$$

and for $x < \xi$

$$\langle\langle\varkappa(x)\rangle\rangle \leq \langle\langle\varkappa_0\rangle\rangle \exp\left(-\int_\xi^x \kappa_+(s)\,ds\right),$$

$$\langle\langle\varkappa(x)\rangle\rangle \geq \langle\langle\varkappa_0\rangle\rangle \exp\left(-\int_\xi^x \kappa_-(s)\,ds\right).$$

Now we can exploit Proposition 14 to estimate $\mathscr{L}_1$. Apply estimate (32) to equation (31). If we observe that the Hermitian part of the matrix of the equation does not depend on k, we obtain matrix function $\mathscr{L}_1$ to be bounded uniformly in $k \in \mathbb{R}$. Hence, matrix function $\mathscr{L}(x, k)\mathscr{L}^{-1}(\xi, k)$ is also uniformly bounded, and the assertion of the lemma follows from (30).

Let us now state the main result on the controllability of the system of connected strings.

Let us prolong the proof of Theorem 2. As already mentioned, B controllability of system (1)–(4) is equivalent to the $\mathscr{L}$-basis property of family $\tilde{\mathscr{E}}_{T_0}$, which in turn follows from the $\mathscr{L}$-basis property of family $\hat{\mathscr{E}}_\infty$.

Suppose that $\hat{\mathscr{E}}_\infty \notin (LB)$. From the $\mathscr{L}$-basis property criterion on the semiaxis (Theorem II.2.12), it follows that for any $\varepsilon > 0$ there may be found set Ω_ε of points v_j, $j \in \mathcal{N} \subset \mathbb{N}$, with the total number of points not greater than $N + 1$ such that

(i) hyperbolic distance between any two points of Ω_ε is not larger than ε, and

(ii) vector family $\{\eta_j\}_{j \in \mathcal{N}}$ is almost linearly dependent in a sense that

$$\min_{n \in \mathcal{N}} \varphi_{\mathbb{C}^N}\left(\eta_n, \bigvee_{m \in \mathcal{N}, m \neq n} \eta_n\right) \leq \varepsilon.$$

Since points v_j lie on the line Im $k = 1$, in (i), instead of the hyperbolic metrics, one may take the Euclidean one. Moreover, (ii) implies that Gram matrix $\Gamma_{\mathcal{N}}$ is "almost degenerate." So (i) and (ii) can be replaced by

(i') $|\omega_m - \omega_n| \leq \varepsilon$, $m, n \in \mathcal{N}$, and (ii') for some $c = \{c_n\}_{n \in \mathcal{N}}, \sum_{n \in \mathcal{N}} |c_n|^2 = 1$, the inequality

$$\langle \Gamma_{\mathcal{N}} c, c \rangle_{\mathcal{N}} := \left\langle \left\langle \sum_{n \in \mathcal{N}} c_n \eta_n \right\rangle \right\rangle^2 \leq \varepsilon \tag{33}$$

is valid.

Let Γ_Φ denote the Gram matrix for family $\{\Phi_n\}_{n \in \mathcal{N}}$ of eigenvectors of operator A. Applying Lemma II.2.3 to this family and using representation (11), we arrive at the inequality

$$\langle \Gamma_\Phi c, c \rangle_{\mathcal{N}} \leq 2 \max_{s, n, s \leq N, n \in \mathcal{N}} \|\varphi_s(x, \omega_n)\|^2_{L^2_{\rho_s}(0, l_s)} < \Gamma_{\mathcal{N}} c, c \rangle_{\mathcal{N}}$$

$$+ 2\langle\langle \Gamma_{\mathcal{N}} \rangle\rangle_{\mathcal{N}} \max_{s \leq N, m, n \in \mathcal{N}} \|\varphi_s(x, \omega_m) - \varphi_s(x, \omega_n)\|^2_{L^2_{\rho_s}(0, l_s)}. \tag{34}$$

Using Lemma 13 (for a scalar situation) with functions $\varphi_s(x, k)$, which satisfy equation (9), we derive an inequality

$$\|\varphi_s(x, \omega_m) - \varphi_s(x, \omega_n)\|^2_{L^2_{\rho_s}(0, l_s)} \leq C\varepsilon, \qquad m, n \in \mathcal{N}, s = 1, 2, \ldots, N. \tag{35}$$

Then (33)–(35) lead to the conclusion that Gram matrix Γ_Φ of a subfamily $\{\Phi_n\}_{n \in \mathcal{N}}$ of family $\{\Phi_n\}_{n \in \mathbb{N}}$ is "almost degenerate": for some vector $c = \{c_n\}_{n \in \mathcal{N}}, \sum_{n \in \mathcal{N}} |c_n|^2 = 1$,

$$\langle \Gamma_\Phi c, c \rangle_{\mathcal{N}} \leq C_1 \varepsilon. \tag{36}$$

On the other hand, owing to the orthogonality of eigenfunctions, we have

$$\langle \Gamma_\Phi c, c \rangle_{\mathcal{N}} = \left\| \sum_{n \in \mathcal{N}} c_n \Phi_n \right\|^2 = \sum_{n \in \mathcal{N}} |c_n|^2 = 1.$$

Constant C_1 in (36) does not depend on ε, so we have a contradiction and so the theorem is proved.

Remark VII.2.15. If we change variable: $x \mapsto x/L_s$ for each of the strings, we obtain a problem for $x \in [0, 1]$ (for any s). Condition (3), (4) turns into

$$y_1(1, \cdot) = y_2(1, \cdot) = \cdots = y_N(1, \cdot),$$

$$l_1^{-1} \frac{\partial}{\partial x} y_1(1, \cdot) + l_2^{-1} \frac{\partial}{\partial x} y_2(1, \cdot) + \cdots + l_N^{-1} \frac{\partial}{\partial x} y_N(1, \cdot) = 0. \tag{37}$$

In the obtained system, energy does not conserve, since the boundary condition (37) is not self-adjoint. This is connected with the fact that the operator of the change of variable is an isomorphism but is not an isometry. The differential operator that arises is similar to the self-adjoint one; its eigenfunctions form a Riesz basis. The approach in Chapter III is also applicable in this case.

Remark VII.2.16. Formal application of the microlocal approach developed by Bardos, Lebeau, and Rauch (1992) for multidimensional hyperbolic equations to the string system (1)–(4) corroborates our result on B controllability in the time T_0. We suppose that the result can be refined as follows. There are sets $\Delta_s \subset \mathbb{R}$ such that $\sum \mathrm{mes}\ \Delta_s = 2 \sum L_s$ and the system is B-controllable with supp $u_s \subset \Delta_s$. This corresponds to the fact that the indicator diagram width of det $G(k)$ is equal to $2 \sum L_s$. N. Burq gave some arguments corroborating this hypothesis (private communication).

3. Control of multichannel acoustic system

3.1. Consider the following system of equations for vector function $y(x, t)$:

$$\mathscr{A}(x) \frac{\partial^2 y(x, t)}{\partial t^2} = \frac{\partial^2 y(x, t)}{\partial x^2}, \qquad 0 < x < l, 0 < t < T. \tag{1}$$

$$y(0, t) = u(t), \qquad u \in L^2(0, T; \mathbb{C}^N), \qquad y(l, t) = 0, \tag{2}$$

$$y(x, 0) = \frac{\partial}{\partial t} y(x, 0) = 0. \tag{3}$$

Here, $\mathscr{A}(x)$ is a C^2-diagonable positive-definite matrix function. That is,

$$\mathscr{A}(x) = U(x) \operatorname{diag}[\rho_n(x)] U^{-1}(x), \tag{4}$$

with $U \in C^2[0, l]$; $\rho_n \in C^2[0, l]$, $n = 1, 2, \ldots, N$, and $\rho_n(x) > 0$, for $x \in [0, l]$, U being a unitary-valued matrix function.

Let us state without proof conditions which are sufficient for a smooth diagonalization. In particular, if $\mathscr{A}(x)$ is analytical, it is C^∞-diagonable (Kato 1966: chap. 2).

Proposition VII.3.1 (Ivanov 1989). Let

(a) $\mathscr{A} \in C^p[0, l]$,

(b) *the eigenvalues $\rho_n(x)$ of matrix $\mathscr{A}(x)$ belong to $C^p[0, l]$,*

(c) *the multiplicity of any zero of any function R_j*

$$R_j := \sum_{i, i \neq j} (\rho_i(x) - \rho_j(x)), \qquad j = 1, 2, \ldots, N,$$

is not greater than r, $r \leq p$.

Then matrix $\mathscr{A}(x)$ is C^q-diagonable for $q = p - r$; that is, formula (4) holds with $U \in C^q[0, l]$.

Let us also demand that the following condition be valid: if at $x = x_0$ eigenvalues $\rho_i(x)$ and $\rho_j(x)$ coincide (the so-called turning point), then in some neighborhood of the x_0 representation,

$$\rho_i(x) - \rho_j(x) = (x - x_0)^\kappa \varphi(x) \qquad (5)$$

is true, with $\kappa = \kappa_{ij}(x_0) \in \mathbb{N}$ being the multiplicity of zero of the function $\rho_i(x) - \rho_j(x)$; $\varphi(x) = \varphi_{ij}(x, x_0)$ is continuously differentiable about x_0, and $\varphi_{ij}(x_0) \neq 0$.

System (1), (2) describes N wave channels interconnected by means of the transmission coefficients $\mathscr{A}_{ij}(x)$. At the boundary point $x = 0$, a control, u, is specified while a zero regime is prescribed at $x = 1$. If matrix function U is constant, then, after the change of variables $y(x, t) = Uv(x, t)$, the system splits into N noncoupled channels

$$\text{diag}[\rho_n(x)] \frac{\partial^2 v(x, t)}{\partial t^2} = \frac{\partial^2 v(x, t)}{\partial x^2}, \qquad 0 < x < l, 0 < t < T. \qquad (6)$$

To solve a control problem for system (1), (2), we construct a GF of an exponential family, just as in Section 2, and we study its behavior in $\mathbb{C}$ and demonstrate B controllability of the system for the time T_0

$$T_0 := 2 \int_0^l \langle\langle \mathscr{A}(x) \rangle\rangle^{1/2} \, dx = 2 \int_0^l \max_{n=1, 2, \ldots, N} \{\sqrt{\rho_n(x)}\} \, dx.$$

If we consider system (1), (2) according to the plan of Chapter III, we set

$$H := L^2_{\mathscr{A}(x)}(0, l; \mathbb{C}^N), \qquad \|v\|^2_H = \int_0^l \langle \mathscr{A}(x)v(x), v(x) \rangle \, dx,$$

$$V := H^1_0(0, l; \mathbb{C}^N).$$

Determine on space $V \times V$ a bilinear form

$$a[\varphi, \psi] := \int_0^l \langle \varphi'(x), \bar{\psi}'(x) \rangle \, dx,$$

which generates (see Chapter III, Section 1) operator A

$$(A\varphi, \psi)_H = a[\varphi, \psi]; \qquad \varphi \in \mathscr{D}(A), \psi \in V$$

$$(A\varphi)(x) = -\mathscr{A}(x)^{-1} \frac{d^2}{dx^2} \varphi(x)$$

with the domain $H^2(0, l; \mathbb{C}^N) \cap H_0^1(0, l; \mathbb{C}^N)$. We denote the eigenvalues and the normed eigenfunctions of this operator by λ_n and Φ_n, $n \in \mathbb{N}$. Operator B has the form

$$\langle Bv, \varphi \rangle_* = \langle v(0), \bar{\varphi}'(0) \rangle, \qquad \varphi \in W_2 D(A),$$

(see Section 2 and Subsection III.2.2).

As in Sections 1, 2, the solution of system (1), (2), (3) satisfies the relation

$$\left\{ y, \frac{\partial}{\partial t} y \right\} \in C(0, T; \mathcal{W}_0), \tag{7}$$

$$\mathcal{W}_0 = W_0 \oplus W_{-1} = H \oplus H^{-1}(0, l; \mathbb{C}^N).$$

This fact will be verified after the construction of the STF with zeros $\{\pm\sqrt{\lambda_n}\}$.

3.2. Consider matrix equation

$$-Y''(x, k) = k^2 \mathcal{A}(x) Y(x, k) \tag{8}$$

with conditions

$$Y(0, k) = 0, \qquad Y'(0, k) = kI. \tag{9}$$

Set $G(k) := Y(l, k)$, which is obviously an entire function of parameter k of the exponential type (see Proposition 2.14). In the scalar case, $G(k)$ coincides with function $\varphi_n(x, k)$ (see formula (9) of Section 2) for $\rho_n(x) = \mathcal{A}(x)$. It is clear that $G(-k) = -G(k)$ and $G(0) = 0$. One easily sees $G(k) = kG_0(k)$, where $G_0(0)$ is a nondegenerate matrix, so $k = 0$ is a semisimple zero. The following serves as an analog of Lemma 2.5.

Lemma VII.3.2.

(a) *Any eigenfunction of operator A is of the form*

$$\Phi_n = Y(x, \sqrt{\lambda_n}) \eta_n$$

with $\eta_n \in \text{Ker } G(\sqrt{\lambda_n})$.

(b) *Conversely, any function of the form $Y(x, k_0)\eta$ is the eigenfunction of operator A corresponding to eigenvalue k_0^2 if $\eta \in \text{Ker } G(k_0)$, $\eta \neq 0$, and $k_0 \neq 0$.*

If μ is an eigenfrequency of system (1), (2) of multiplicity κ (μ^2 is an eigenvalue of operator A), and $\Phi_{\mu 1}, \Phi_{\mu 2}, \ldots, \Phi_{\mu\kappa}$ are corresponding eigenfunctions, then vectors $\eta_{\mu 1}, \eta_{\mu 2}, \ldots, \eta_{\mu\kappa}$ in the representation

$$\Phi_{\mu n} = Y(x, \mu) \eta_{\mu n}$$

satisfy following orthogonality conditions

$$\int_0^l \langle \mathcal{A}(x)\Phi_{\mu n}, \Phi_{\mu m} \rangle \, dx = \delta_n^m. \tag{10}$$

Let us introduce a vector exponential family

$$\tilde{\mathscr{E}}_T = \{e_n\}_{n \in \mathbb{K}} \subset L^2(0, T; \mathbb{C}^N), \qquad e_n(t) = \eta_n \exp(i\omega_n t),$$

$$\eta_n = |\omega_n^{-1}|\Phi'_{|n|}(0), \qquad \omega_n := \text{sgn}(n)\sqrt{\lambda_{|n|}}.$$

In light of the results of Chapter III, the properties of this family determine the controllability type of system (1), (2) in space $\mathcal{W}_0$.

We also introduce family $\hat{\mathscr{E}}_T$ constructed by zeros of function $\tilde{G}(k) := G(k - i)$:

$$\hat{\mathscr{E}}_T := \{e^{i\mu t} \hat{\eta}_{\mu j}\}_{\mu \in \sigma}^{j=1,2,\ldots,\kappa_\mu} \subset L^2(0, l; \mathbb{C}^N).$$

Here, $\sigma = \{\omega_n + i) \cup \{i\}$ is the set of zeros of $\tilde{G}(k)$, $\hat{\eta}_{\mu j}$ are the orthonormal basis in Ker $G(\mu)$, and κ_μ is multiplicity of zero μ.

We write $\hat{\mathcal{X}}$ for the family of simple fractions obtained out of exponential family $\{e^{i\mu t} \hat{\eta}_{\mu j}\}_{\mu \in \sigma}^{j=1,2,\ldots,\kappa_\mu} \subset L^2(0, \infty; \mathbb{C}^N)$ by the $\mathscr{F}$ transform (see Subsection II.1.1.5).

We need results concerning the asymptotics of $G(k)$ for real k and the behavior of $G(k)$ for Im $k \to \infty$.

Theorem VII.3.3.

(a) *For* $|k| \to \infty$ *a uniform in* Im k, $|\text{Im } k| \le$ const, *asymptotics*

$$G(k) = U(l) \, \text{diag}\left[a_n \sin\left(k \int_0^l \sqrt{\rho_n(s)} \, ds \right) \right] U^{-1}(0) + o(1) \tag{11}$$

is valid where a_n *are some constants and* $a_n \ne 0$, $n = 1, 2, \ldots, N$.

(b) *There may be found an ESF* Θ *and an entire matrix-function* $X(k)$ *nondegenerate in* $\mathbb{C}$ *such that*

   (i) *matrix-functions* $X(k)$ *and* $X^{-1}(k)$ *are bounded on the real axis,*

   (ii) *matrix function* $F := X\tilde{G}$ *can be factorized in the form*

$$F = F_e^+ \Pi = F_e^- \Theta, \tag{12}$$

where $F_e^\pm$ *are outer functions in* $\mathbb{C}^\pm$, *respectively,* Π *is a BPP constructed by zeros of* $\tilde{G}$, *and*

   (iii) *ESF* Θ *satisfies the estimate*

$$\langle\langle e^{ikT_0} \Theta^{-1}(k) \rangle\rangle \prec 1, \qquad k \in \mathbb{C}_+. \tag{13}$$

The proof of Theorem 3 is found below, since it requires the detailed study of the equation (8) solution for $|k| \to \infty$.

If one considers this theorem to be valid, the further investigation of B controllability almost completely repeats the one performed in Section 2 for a system of strings connected at a single point. For instance, relation (7) (and hence the embedding $\mathcal{R}(T) \subset \mathcal{W}_0$) is implied by the fact that set $\{\omega_n\}$ is a unification of a finite number of separable sets, since $\{\omega_n\} \cup \{0\}$ is the set of zeros of an STF det G.

Moreover, the implication

$$\text{If } \hat{\mathscr{E}}_T \in (LB), \text{ then } \tilde{\mathscr{E}}_T \in (LB)$$

is demonstrated with the help of the same arguments as in Lemma 2.5 with the exploitation of the orthogonality condition (10) and the asymptotics (11).

Theorem VII.3.4. *System* (1), (2) *is B-controllable in space* $\mathcal{W}_0$ *in time* T_0.

Remark VII.3.5. *If* $U(x)$ *is a constant matrix, then system* (1), (2) *is* B-*controllable in the time* T_1

$$T_1 = 2 \max_n \int_0^l \sqrt{\rho_n(x)} \, dx \leq T_0.$$

Indeed, the change $y(x, t) = Uv(x, t)$ yields the system (6), that is, the system of N disconnected strings with the largest optical length $T_1/2$. Since the single nth string of optical length L_n is controllable in time $2L_n$ (see Russell 1967, or Chapter V, Section 2), system (1), (2) is B-controllable in time T_1. In addition, we can demand that the nth component of control $v(x, t)|_{x=0} = U^{-1}u(t)$ be zero at

$$t > T^{(n)}, \qquad T^{(n)} := 2 \int_0^l \sqrt{\rho_n(x)} \, dx \leq T_1.$$

Remark VII.3.6. Avdonin, Belishev, and Ivanov (1991a) studied an inverse problem for the equation for vector function $y^f(x, t)$

$$\frac{\partial^2 y^f(x, t)}{\partial t^2} = \frac{\partial^2 y^f(x, t)}{\partial x^2} + Q(x) y^f(x, t), \qquad 0 < x < \infty, 0 < t < T,$$

$$y^f(x, 0) = y_t^f(x, 0) = 0, \qquad y^f(0, t) = f(t).$$

By the inverse problem data, namely, the response operator (Dirichlet to Neumann map)

$$f \mapsto y_x^f(0, t), \qquad t \in [0, 2T],$$

one should have to find potential $Q(x)$ for $x \in [0, T]$. The way to solve the problem (see Belishev 1989; Avdonin, Belishev, and Ivanov 1991a, 1991b) is to use the so-called controllability on a filled domain: the property

$$\{y^f(x, t) \mid f \in L^2(0, T; \mathbb{C}^N)\} = L^2(0, T; \mathbb{C}^N).$$

(By means of the source $f(t)$ of class L^2, we can obtain in time T any profile of the wave of class L^2 on the part of the axis where the wave has already arrived.) To solve this we proved that a system with Dirichlet boundary condition at $x = T$

$$\frac{\partial^2 y(x, t)}{\partial t^2} = \frac{\partial^2 y(x, t)}{\partial x^2} + Q(x) y(x, t), \qquad 0 < x < T, \, 0 < t < T,$$

$$y(x, t)|_{t<0} = 0, \qquad y(x, t)|_{x=0} = u \in L^2(0, T; \mathbb{C}^N), \qquad y(x, t)|_{x=T} = 0,$$

is B-controllable in space $L^2(0, T; \mathbb{C}^N) \oplus H^{-1}(0, T; \mathbb{C}^N)$ in time $2T$. From this it is not difficult to extract controllability on a filled domain.

3.3. We now enter upon the proof of Theorem 3. At this point, we study in detail the behavior of solutions to equation (8). We are interested in two questions:

(1) the asymptotics of the solution with k tending to infinity in the strip $|\text{Im } k| \leq \text{const}$ (this is required to verify that $G(k)$ and $G^{-1}(k)$ are bounded on the line $\text{Im } k = \text{const} \neq 0$); and
(2) the exponential growth of the solution at $|\text{Im } k| \to \infty$, which is needed to extract singular factors in G.

Remark VII.3.7. Ivanov and Pavlov (1978) and Ivanov (1978, 1983a, 1983b) investigated the vector Regge problem (the problem of resonance scattering) for the equation

$$\mathscr{A}(x) u_{tt}(x, t) = u_{xx}(x, t), \qquad x \in (0, \infty),$$

with $\mathscr{A}(x) \equiv I$ for $x > l$. In the study of the completeness and the basis property of the family of the so-called resonance state, the vector exponential family constructed by zeros of Jost function $M(k)$ plays the principal role ($M(k)$ is the value at $x = 0$ of the solution of equation (8) obeying conditions $Y(l, k) = I$, $Y'(l, k) = -ikI$). The tools designed to investigate equation (8) in those studies may be applied to $G(k)$ as well.

Our treatment of solutions to equation (8) is as follows.

We first transform (8) into the first-order system with the diagonal principal part (in k) of the form

$$ik \operatorname{diag}[\sqrt{\rho_1}, \sqrt{\rho_2}, \ldots, \sqrt{\rho_N}, -\sqrt{\rho_1}, -\sqrt{\rho_2}, \ldots, -\sqrt{\rho_N}].$$

The next step consists of "splitting apart" the equation. That is, it is reduced to two equations of lower dimension whose principal parts are

$$\pm ik \operatorname{diag}[\sqrt{\rho_1}, \sqrt{\rho_2}, \ldots, \sqrt{\rho_N}].$$

The splitting is accomplished by choosing a special linear transform of the unknown function (depending on k).

The first step in the transformation of (8) is to replace it by an equation of the first order in the standard way:

$$Z' = \begin{pmatrix} 0 & I \\ -\mathscr{A} & 0 \end{pmatrix} Z, \tag{14}$$

where $Z(x, k)$ is a matrix-valued function of dimension $2N$. A solution Z of this equation is related to linearly independent solutions Y_1 and Y_2 of (8) by the obvious equality

$$Z = \begin{pmatrix} Z_{11} & Z_{12} \\ Z_{21} & Z_{22} \end{pmatrix} = \begin{pmatrix} Y_1 & Y_2 \\ k^{-1}Y_1' & k^{-1}Y_2' \end{pmatrix}.$$

In particular, solution $Y(x, k)$ of (8), (9) coincides with the block Z_{12} of the solution of (14) satisfying initial condition $Z(0, k) = I_{2N}$ (I_{2N} the unit matrix of $2N$ dimension).

Lemma VII.3.8. The transformation

$$Z = \begin{pmatrix} I & -I \\ i\sqrt{\mathscr{A}} & i\sqrt{\mathscr{A}} \end{pmatrix} \begin{pmatrix} U & 0 \\ 0 & U \end{pmatrix} Z_1 \tag{15}$$

converts (14) into

$$Z_1' = \left[ik \begin{pmatrix} \mathscr{R} & 0 \\ 0 & -\mathscr{R} \end{pmatrix} + \begin{pmatrix} \mathscr{F} & \mathscr{G} \\ \mathscr{G} & \mathscr{F} \end{pmatrix} \right] Z_1, \tag{16}$$

where

$$\mathscr{R} := \operatorname{diag}[\sqrt{\rho_1}, \sqrt{\rho_2}, \ldots, \sqrt{\rho_N}],$$

$$\left. \begin{aligned} \mathscr{G} &:= -\tfrac{1}{2} U^{-1} \sqrt{\mathscr{A}^{-1}} (\sqrt{\mathscr{A}})' U \\ &= -\tfrac{1}{2} \mathscr{R} U^{-1} U' \mathscr{R}^{-1} + \tfrac{1}{2} (\log \mathscr{R})' + U^{-1} U', \\ \mathscr{F} &:= \mathscr{G} - U^{-1} U', \end{aligned} \right\} \tag{17}$$

Lemma 8 is proved by direct substitution.

Many studies have dealt with the analysis of the asymptotics at $k \to \infty$ of solutions $\mathcal{B}(x, k)$ to the equation

$$\mathcal{Y}' = k\mathcal{B}(x, k)\mathcal{Y}, \qquad \mathcal{B}(x, y) = \mathcal{B}(x) + \mathcal{O}(k^{-1})$$

(see, e.g., Wasov 1965, 1985). Such analyses run into serious difficulties when $\mathcal{B}(x)$ has multiple eigenvalues. In our situation, the system matrix possesses a characteristic structure; namely, its principal term is a diagonal matrix with real entries, and the entries naturally fall into two isolated groups, $\{\sqrt{\rho_n}\}$ and $\{-\sqrt{\rho_n}\}$. Inside a group, coincidences of entries may occur. These peculiarities enable us to reveal a substitution

$$Z_1(x, k) = W(x, k)Z_2(x, k), \qquad x \in [0, l], \qquad \text{Im } k \geq \text{const},$$

such that the matrix of the equation for Z_2 becomes block diagonal. We speak about the splitting of the system just in this sense. Note that in our case we manage to perform this splitting not locally, but on the whole segment $[0, l]$ at once.

The standard procedure for finding W leads to the solution of a nonlinear equation with a principal (in k) linear part and a quadratic nonlinearity. We shall seek W in the form

$$W(x, k) = \begin{pmatrix} I & V_+(x, k) \\ V_-(x, k) & I \end{pmatrix} \tag{18}$$

proceeding with the requirement that equation (16) for function Z_2 should become

$$Z_2' = \begin{pmatrix} ik\mathcal{R}(x) + \mathcal{D}_+(x, k) & 0 \\ 0 & -ik\mathcal{R}(x) + \mathcal{D}_-(x, k) \end{pmatrix} Z_2. \tag{19}$$

Substitution of $Z_1 = WZ_2$ into (16) together with (19) yields four matrix equations

$$\mathcal{D}_\pm(x, k) = \mathcal{F}(x) + \mathcal{G}(x)V_\pm(x, k), \tag{20}$$

$$\left. \begin{aligned} V_\pm'(x, k) = &\pm ik[V_\pm(x, k)\mathcal{R}(x) + \mathcal{R}(x)V_\pm(x, k)] + \mathcal{G}_+ \\ &+ [V_\pm(x, k)\mathcal{F}(x) + \mathcal{F}(x)V_\pm(x, k)] - V_\pm(x, k)\mathcal{G}(x)V_\pm(x, k) \end{aligned} \right\}. \tag{21}$$

The matter thus reduces to the investigation of equations (21). For the entries $v_{mn}^\pm$ of $V_\pm$, the equations take the form of a system

$$(v_{mn}^\pm)' = \pm ik(\sqrt{\rho_m} + \sqrt{\rho_n})v_{mn}^\pm - g_{mn}(v^\pm), \qquad 1 \leq m, n \leq N, \tag{22}$$

where $v^{\pm}$ are vector-functions $\{v^{\pm}_{mn}\}^{N}_{m,n=1}$, $g_{mn}(v)$ is a polynomial of the second degree in variables $\{v_{mn}\}$ with the coefficients independent of k. Equation (22) can be written as a system for vector functions $\{v^{\pm}_{mn}\}$

$$(v^{\pm})'(x, k) = \pm ikR(x)v^{\pm}(x, k) + D(x) + Q(x)v^{\pm}(x, k) + \Gamma(x, k^{\pm}) \quad (23)$$

in which $R = \mathrm{diag}[\sqrt{\rho_m} + \sqrt{\rho_n}]$ and the components of quadratic vector function $\Gamma(x, v)$ are

$$\langle \Gamma_n(x)v(x, k), v(x, k) \rangle,$$

with matrix-valued functions $\Gamma_n(x)$.

Lemma VII.3.9. There exist matrix solutions $V_+(x, k)$ and $V_-(x, k)$ of equation (21), which are entire functions of k for any given x, for which the conditions

$$V_+(0, k) = 0,$$
$$V_-(l, k) = 0, \quad (24)$$

are valid, and for $\mathrm{Im}\, k \geq \mathrm{const}$ *and* $x \in [0, l]$ *we have*

$$\langle\langle V_+(x, k) \rangle\rangle \prec 1/(1 + |k|), \qquad \langle\langle V_-(x, k) \rangle\rangle \prec 1/(1 + |k|). \quad (25)$$

We confine ourselves to the solution $V_+(x, k)$. The study of V_- may be carried out in similar fashion. The local existence and analyticity of V_+ follow from well-known theorems (Hartman 1964; Kamke 1959). Let us prove global existence of V_+ and check the estimates (25).

In what follows we repeatedly use the formula connecting a fundamental solution of a homogeneous matrix equation and a solution of an inhomogeneous one. If

$$\mathcal{Y}'_0(x) = \mathcal{B}(x)\mathcal{Y}_0(x), \qquad \det \mathcal{Y}_0(0) \neq 0, \quad (26)$$

$$\mathcal{Y}'(x) = \mathcal{B}(x)\mathcal{Y}(x) + \mathcal{S}(x), \qquad \mathcal{Y}(0) = \mathcal{Y}_1, \quad (27)$$

then

$$\mathcal{Y}(x) = \mathcal{Y}_1 + \int_0^x \mathcal{Y}_0(x)\mathcal{Y}_0^{-1}(s)\mathcal{S}(s)\, ds. \quad (28)$$

The formula will be used to obtain an integral equation for solving a homogeneous equation with parameter k, if we are to extract the principal part of the solution.

Sublemma VII.3.10. For the solution of the problem

$$w' = (ikR(x) + Q(x))w + D(x), \qquad w|_0 = 0, \qquad (29)$$

the estimate

$$\langle\langle w(x, k)\rangle\rangle \prec 1/(1 + |k|)$$

is valid.

PROOF OF THE SUBLEMMA. Let us consider an auxiliary problem

$$w_0' = (ikR(x) + \text{diag}[Q_{nn}(x)])w_0, \qquad w_0|_0 = 0,$$

having the explicit solution

$$w_0(x, k) = \exp\left(\int_0^x \text{diag}[ikr_{nn} + Q_{nn}]\, ds\right). \qquad (30)$$

Using (26)–(28) we obtain the integral equation

$$w(x, k) = \int_0^x w_0(x, k)w_0^{-1}(s, k)(Q - \text{diag}[Q_{nn}])(D(s) + w(s, k))\, ds$$

$$=: w^0(x, k) + (Kw)(x, k),$$

where K is the integral operator

$$(Kf)(x, k) = \int_0^x K(x, s, k)f(s)\, ds,$$

$$K(x, s, k) := w_0(x, k)w_0^{-1}(s, k)(Q(s) - \text{diag}[Q_{nn}(s)]).$$

We arrive at a Volterra integral equation with an operator acting in space of continuous matrix-functions, which is bounded uniformly in k, Im $k \geq$ const. To prove the sublemma we check that w_0 is "small"

$$\langle\langle w_o(x, k)\rangle\rangle \prec 1/(1 + |k|) \qquad (31)$$

(uniformly in $x \in [0, l]$ and k, Im $k \geq$ const).

From (30) it follows that ij-entry of w_o is presented as a phase integral

$$w_{ij}^0(x, k) = \int_0^x \tilde{w}_{ij}^0(s)\, e^{ik\varphi_i(x, s)}\, ds$$

with phase

$$\varphi_i(x, s) := \int_s^x r_{ii}(t)\, dt = \int_s^x \sqrt{\rho_i(t)}\, dt.$$

Since ρ_i is positive, the phase has no stationary points and therefore (31) is valid. The sublemma is proved.

Let us return to equation (23). From a comparison of problems (23), (24), and

$$u' = (ikR(x) + Q(x))u, \qquad u|_0 = I, \tag{32}$$

we obtain an integral equation for v_+

$$v_+(x, k) = (\Omega v_+)(x, k)$$

(see (26)–(28)) with (nonlinear) operator Ω acting in space of continuous vector functions:

$$(\Omega v_+)(x, k) := \int_0^x u(x, k)u^{-1}(s, k)D(s, k)\, ds$$

$$+ \int_0^x u(x, k)u^{-1}(s, k)\Gamma(s, v_+)\, ds. \tag{33}$$

Here, v is a vector function in $C(0, l; \mathbb{R}^{N^2})$.
We now prove two estimates:

$$\langle\langle(\Omega v)(x, k)\rangle\rangle \le C_1|k^{-1}| + C_2\|v\|^2, \tag{34}$$

$$\langle\langle(\Omega v_1)(x, k) - (\Omega v_2)(x, k)\rangle\rangle \le C_3\|v_1 - v_2\|(\|v_1\| + \|v_2\|). \tag{35}$$

Here, $v(x)$, $v_1(x)$, and $v_2(x)$ belong to $\mathbb{R}^{N^2}$, $\langle\langle\cdot\rangle\rangle$ is the norm in $\mathbb{R}^{N^2}$, and $\|\cdot\|$ is the norm in $C(0, l; \mathbb{R}^{N^2})$.

Indeed, the first summand in (33) satisfies equation (29) and therefore coincides with w. So it has already been estimated in Sublemma 10 as $\mathcal{O}(|k^{-1}|)$. To estimate the second summand, we notice that matrix-valued function $u(x, k)u^{-1}(s, k)$ is the solution to (32) on the segment $[s, l]$ with the condition $u(x, k)u^{-1}(s, k) = I$. So taking into account Proposition 2.14, we see that the estimate (34) follows from boundedness of Γ

$$\langle\langle\Gamma(x, v)\rangle\rangle \prec \langle\langle v\rangle\rangle^2$$

uniformly in x, k.

Estimate (35) is the direct consequence of the definition of quadratic vector function Γ, since

$$\langle\Gamma_j v_1, v_1\rangle - \langle\Gamma_j v_2, v_2\rangle = \langle\Gamma_j v_1, v_1 - v_2\rangle + \langle\Gamma_j(v_1 - v_2), v_2\rangle.$$

Suppose that functions v_1 and v_2 are such that

$$\langle\langle v_{1,2}(x)\rangle\rangle \le C_4|k^{-1}|$$

and we choose C_4, proceeding with the requirement that

$$\langle\!\langle\!\langle (\Omega v_1)(x, k) - (\Omega v_2)(x, k)\rangle\!\rangle\!\rangle \le \rho(\|v_1 - v_2\|), \qquad 0 < \rho < 1,$$

$$\langle\!\langle\!\langle (\Omega v_{1,2})(x, k)\rangle\!\rangle\!\rangle \le C_4 |k^{-1}|.$$

Using estimates (34), (35), it is easy to conclude that such a constant C_4 exists if $|k|$ is large enough.

So, constriction of operator Ω (acting in $C(0, l; \mathbb{R}^{N^2})$) on the set $\{v \mid \langle\!\langle v(x)\rangle\!\rangle \le C_4 |k^{-1}|\}$ is a contraction. Therefore the solution of (22) (and consequently the solution of (21)) exists on the set and satisfies the requirements of the lemma. The proof of Lemma 9 is completed.

We now turn to the fundamental solutions of equations (14) and (19). The solutions are related by

$$Z = \begin{pmatrix} I & -I \\ i\sqrt{\mathscr{A}} & i\sqrt{\mathscr{A}} \end{pmatrix} \begin{pmatrix} U & 0 \\ 0 & U \end{pmatrix} \begin{pmatrix} I & V_+(x, k) \\ V_-(x, k) & I \end{pmatrix} Z_2$$

(see (15) and (18)).

It is also evident that the solution of (19) can be sought in the form

$$Z_2(x, k) = \begin{pmatrix} \Psi_+(x, k) & 0 \\ 0 & \Psi_-(x, k) \end{pmatrix},$$

where $\Psi_\pm$ are solution of

$$\left.\begin{array}{c} \Psi'_\pm(x, k) = (\pm ik\mathscr{R}(x) + \mathscr{D}_\pm(x, k))\Psi_\pm(x, k) \\ \Psi_\pm(0, k) = I \end{array}\right\}. \tag{36}$$

This makes solution Z of (14) with $Z(0, k) = I_{2N}$ take the form

$$Z(x, k) = \begin{pmatrix} I & -I \\ i\sqrt{\mathscr{A}(x)} & i\sqrt{\mathscr{A}(x)} \end{pmatrix} \begin{pmatrix} U(x) & 0 \\ 0 & U(x) \end{pmatrix}$$

$$\times \begin{pmatrix} I & V_+(x, k) \\ V_-(x, k) & I \end{pmatrix} \begin{pmatrix} \Psi_+(x, k) & 0 \\ 0 & \Psi_-(x, k) \end{pmatrix}$$

$$\times \left[\begin{pmatrix} I & -I \\ i\sqrt{\mathscr{A}(0)} & i\sqrt{\mathscr{A}(0)} \end{pmatrix} \begin{pmatrix} U(0) & 0 \\ 0 & U(0) \end{pmatrix} \begin{pmatrix} I & V_+(0, k) \\ V_-(0, k) & I \end{pmatrix}\right]^{-1}$$

Corollary VII.3.11. Solution $Y(x, k)$ of (8) with $Y(0, k) = 0$, $Y'(0, k) = ikI_N$, *is expressed in terms of $V_\pm$, $\Psi_\pm$ by the formula*

$$Y(x, k) = \frac{U(x)}{2i}\left[(I - V_-(x, k)\Psi_+(x, k) + (V_+(x, k) - I)\Psi_-(x, k)\right.$$

$$\left. \times (I - V_-(0, k))\right]\mathscr{R}^{-1}(0)U^{-1}(0), \tag{37}$$

and for Im $k \le$ const solution $Y(x, k)$ has asymptotics

$$Y(x, k) = \frac{U(x)}{2i}\left[\Psi_+(x, k) - \Psi_-(x, k)\right]\mathscr{R}^{-1}(0)(U^{-1}(0) + \mathcal{O}(|k|^{-1})).$$

PROOF. By multiplying matrices and carrying out elementary transformations, we obtain (37). The asymptotic expression is provided by Lemma 9. The corollary is proved.

Note that for $N = 1$, systems (36) have explicit solutions

$$\Psi_\pm(x, k) = \exp\left[\pm ik \int_0^x \sqrt{\rho(s)}\, ds + \int_0^x \mathscr{D}_\pm(s, k)\, ds\right].$$

Solution Ψ_+ (Ψ_-) is of the exponential growth in $\mathbb{C}_-$ ($\mathbb{C}_+$) and decreases exponentially in $\mathbb{C}_+$ ($\mathbb{C}_-$), respectively. This fact remains valid for the vector case as well.

Now we study the asymptotics of solutions $\Psi_\pm$ in a strip parallel to the real axis.

Lemma VII.3.12. For $|\text{Im } k| \le$ const, functions $\Psi_\pm(x, k)$ have the asymptotics

$$\Psi_\pm(x, k) = \Psi_\pm^0(x, k) + \mathcal{O}(|k^{-\alpha}|) \tag{38}$$

and

$$\Psi_\pm^0(x, k) = \text{diag}\left[\sqrt[4]{\rho_n(x)/\rho_n(0)}\, d_n(x) \exp\left(\pm ik \int_0^x \sqrt{\rho_n(\xi)}\, d\xi\right)\right] + \mathcal{O}(|k^{-1}|) \tag{39}$$

where

$$d_n(x) := \exp\left(-\frac{1}{2}\int_0^x (U^{-1}(\xi)U'(\xi))_{nn}\, d\xi\right)$$

and $1/\alpha$ is the maximal order of turning points of matrix functions $\mathscr{A}$,

$$\frac{1}{\alpha} := \left(\max_{x_0;\, i,j=1,2,\ldots,N} \kappa_{ij}(x_0)\right),$$

(*and* $\kappa_{ij}(x_0)$ *is the exponent in* (5) – *the order of turning point* x_0 *or the multiplicity of zero* x_0 *of function* $\rho_i(x) - \rho_j(x)$).

PROOF OF LEMMA 12. To investigate solutions $\Psi_\pm(x, k)$ we construct an integral equation starting from equations

$$(\Psi_\pm^0(x, k))' = (\pm ik \, \mathrm{diag}[\sqrt{\rho_n(x)}] + \mathrm{diag}[\mathscr{D}_{nn}^\pm(x, k)])\Psi_\pm^0(x, k),$$

$$\Psi_\pm^0(0, k) = I_N,$$

$$\mathscr{D}_{nn}^\pm := (\mathscr{D}_\pm)_{nn}.$$

Using (26)–(28) and the explicit formula

$$\Psi_\pm^0(x, k) = \exp \mathrm{diag}\left(ik \int_0^x \sqrt{\rho_n(s)} \, ds + \int_0^x \mathscr{D}_{nn}^\pm(s, k) \, ds \right) \quad (40)$$

we arrive at

$$\Psi_\pm(x, k) = \Psi_\pm^0(x, k) + \int_0^x \Psi_\pm^0(x, k)[\Psi_\pm^0(s, k)]^{-1}$$

$$\times [\mathscr{D}^\pm(s, k) - \mathrm{diag}[\mathscr{D}_{nn}^\pm(s, k)]] \, ds = \Psi_\pm^0(x, k)$$

$$+ \int_0^x \exp\left[\mathrm{diag}\left(ik \int_s^x \sqrt{\rho_n(\xi)} \, d\xi + \int_s^x \mathscr{D}_{nn}^\pm(\xi, k) \, d\xi \right) \right]$$

$$\times [\mathscr{D}^\pm(s, k) - \mathrm{diag}[\mathscr{D}_{nn}^\pm(s, k)]]\Psi_\pm(s, k) \, ds, \quad (41)$$

or, in the operator form,

$$\Psi_\pm = \Psi_\pm^0 + \mathscr{K}^\pm \Psi_\pm. \quad (42)$$

Here $\mathscr{K}^\pm$ is the matrix integral operator acting in space of continuous matrix functions

$$(\mathscr{K}^\pm \Psi)(x, k) = \int_0^x \exp[\mathrm{diag}(ik\psi_n(x, s))]\mathscr{K}_0^\pm(x, s)\Psi(s) \, ds,$$

where

$$\mathscr{K}_0^\pm(x, s) := \exp\left[\mathrm{diag}\left(\int_s^x \mathscr{D}_{nn}^\pm(\xi, k) \, d\xi \right) \right][\mathscr{D}^\pm(s, k) - \mathrm{diag}[\mathscr{D}_{nn}^\pm(s, k)]],$$

and phase $\psi_n(x, s)$ has the form

$$\psi_n(x, s) = \int_s^x \sqrt{\rho_n(\xi)} \, d\xi.$$

It follows from (42) that the "error" $\Psi_\pm - \Psi_\pm^0$ satisfies

$$\Psi_\pm - \Psi_\pm^0 = \mathscr{K}^\pm \Psi_\pm^0 + \mathscr{K}^\pm (\Psi_\pm - \Psi_\pm^0). \tag{43}$$

Since Im k is bounded, operators $\mathscr{K}^\pm$ prove to be uniformly bounded. Now, using an estimate for Volterra equation (43), we find

$$\|\Psi_\pm - \Psi_\pm^0\| \leq C\|\mathscr{K}^\pm \Psi_\pm^0\|,$$

where constant C does not depend on x and k.

We now write out ij entry of matrix $\mathscr{K}^\pm \Psi_\pm^0$. It is enough to take a nondiagonal entry, since $(\mathscr{K}^\pm \Psi_\pm^0)_{jj} = 0$ (see (40) and (41)).

$$(\mathscr{K}^\pm \Psi_\pm^0)(x, k) = \int_0^x \mathscr{D}_{ij}^\pm (s, k) \exp\left(\int_0^s \mathscr{D}_{jj}^\pm(\xi, k)\, d\xi + \int_s^x \mathscr{D}_{ii}^\pm(\xi, k)\, d\xi \right)$$

$$\times \exp\left(ik \int_0^s \sqrt{\rho_j(\xi)}\, d\xi + ik \int_s^x \sqrt{\rho_i(\xi)}\, d\xi \right) ds$$

$$=: \int_0^x \tilde{\mathscr{D}}_{ij}^\pm (s, k)\, e^{ik f_{ij}(x, s)}\, ds.$$

The phase of this integral is

$$f_{ij}(x, s) = \int_0^s \sqrt{\rho_j(\xi)}\, d\xi + \int_s^x \sqrt{\rho_i(\xi)}\, d\xi.$$

Let us show that

$$(\mathscr{K}^\pm \Psi_\pm^0)_{ij} = \mathcal{O}(|k^{-\alpha}|) \tag{44}$$

uniformly in x, $x \in [0, l]$. The decrease of $(\mathscr{K}^\pm \Psi_\pm^0)_{ij}$ depends on multiplicity of zeros of

$$\frac{\partial}{\partial s} f_{ij}(x, s) = \sqrt{\rho_j(s)} - \sqrt{\rho_i(s)}.$$

The multiplicity of a zero x_0 of the function $\sqrt{\rho_j(s)} - \sqrt{\rho_i(s)}$ is equal to the multiplicity of the zero of the function $\rho_j(s) - \rho_i(s)$ and does not exceed α. By the Watson lemma (see, e.g., Fedoryuk 1977: p. 31) we obtain (44) and, in view of (43), also (38).

Let us check the equality (39). Equations (17) and (20) and Lemma 9 imply

$$\mathscr{D}_{jj}^\pm (x, k) = \mathscr{F}_{jj}(x) + \mathcal{O}(|k|^{-1}) = \tfrac{1}{4}(\log \rho_j(x))' - \tfrac{1}{2}(U^{-1}(x)U'(x))_{jj} + \mathcal{O}(|k|^{-1}).$$

Together with (40) they give us (39). Lemma 12 is proved.

By taking (39) into account, we obtain the following corollary from Lemmas 12 and 10.

Corollary VII.3.13. For $|\text{Im } k| \le$ const *an asymptotic expression is valid*

$$Y(x, k) = U(x) \, \text{diag}\left[(\rho_n(x)\rho_n(0))^{1/4} \exp\left(-\frac{1}{2} \int_0^x (U^{-1}(\xi)U'(\xi))_{nn} \, d\xi \right) \right]$$

$$\times \, \text{diag}\left[\sin\left(k \int_0^x \sqrt{\rho_n(\xi)} \, d\xi \right) \right] U^{-1}(0) + \mathcal{O}(|k^{-\alpha}|). \tag{45}$$

Remark VII.3.14. Formula (45) may be obtained from equation (16) by the same method as the asymptotics of $\Psi_\pm(x, k)$. Thus, no splitting is needed to investigate the behavior in the strip $|\text{Im } k| \le$ const.

Remark VII.3.15. Kucherenko (1974) has presented a particular case of the fact that under rather general assumptions the main asymptotic term of the equation

$$\mathcal{Y}' = ik\mathcal{B}(x)\mathcal{Y}$$

at $|k| \to \infty$, $k \in \mathbb{R}$, $\mathcal{B}^*(x) = \mathcal{B}(x)$ is independent of the change in multiplicites of the eigenvalues of matrix $\mathcal{B}$. Ivanov and Pavlov (1978) proved this for the general situation, and it has since been reestablished at least twice (Rosenblum 1979; Gingold and Heieh 1985).

Having studied the asymptotics (formula (45)) and representations of solution $Y(x, k)$ of equation (8) (formulas (37) and (36) and the estimates (25)) we are finally able to prove Theorem 3.

3.4. Proof of Theorem 3.

(a) The asymptotics (11) of G follows directly from (45). Note also that this asymptotics, as well as the absence of zeros of $\det \tilde{G}(k)$ on $\mathbb{R}$, implies that $\tilde{G}(k)$ and $\tilde{G}^{-1}(k)$ are bounded for real k.

(b) Let us write (37) for $x = l$ in the form

$$G(k) = L_+(k)\Psi_+(l, k)R_+(k) + L_-(k)\Psi_-(l, k)R_-(k). \tag{46}$$

Here $L_\pm$ and $R_\pm$ are entire matrix functions bounded at $\text{Im } k \ge$ const such that for $|k| \to \infty$, $\text{Im } k \ge$ const, there exist the limits

$$L_\pm(k) \to L_\pm^0, \qquad R_\pm(k) \to R_\pm^0, \tag{47}$$

and constant matrices $L_\pm^0$ and $R_\pm^0$ entering these relations are nondegenerate.

Let us first prove the assertion about the separation of a singular factor in the solution $\Psi_-(l, k)$.

Lemma VII.3.16. There exists an ESF Θ_+ such that operator function $\tilde{F}_+^e := \Theta_+(k)\Psi_-(l, k)$ is an outer function in $\mathbb{C}_+$ and

$$\langle\langle\Theta_+^{-1}(k)\, e^{ikT_0/2}\rangle\rangle \le 1, \qquad k \in \mathbb{C}_+. \tag{48}$$

PROOF. $\Psi_-^{-1}(l, k)$ is the value at $x = 0$ of the solution to equation (36) with the unit data at $x = l$. Indeed, function $\tilde{\Psi}(x, k) := \Psi_-(x, k)\Psi_-^{-1}(l, k)$ is the solution of (36) obeying conditions $\tilde{\Psi}(l, k) = I$ and $\tilde{\Psi}(0, k) = \Psi_-^{-1}(l, k)$. Solution $\tilde{\Psi}(x, k)$ is bounded in $\mathbb{C}_+$ by Proposition 2.14, and so $\Psi_-^{-1}(l, k)$ is bounded in $\mathbb{C}_+$. Liouville's formula for solution provides

$$\det \Psi_-^{-1}(l, k) = \exp\left(ik \int_0^1 \operatorname{Tr} \mathscr{R}(s)\, ds + \int_0^1 \operatorname{Tr} \mathscr{D}_-(s, k)\, ds\right). \tag{49}$$

Since $\mathscr{D}_-(s, k)$ is bounded in $\mathbb{C}_+$, a BPP is absent in the factorization of the bounded matrix function $\Psi_-^{-1}(l, k)$:

$$\Psi_-^{-1}(l, k) = F_e^+(k)\Theta_+(k),$$

and F_e^+ is bounded in $\mathbb{C}_+$. Moreover, (49) implies that $|\det F_e^+(k)|$ is bounded away from zero in $\mathbb{C}_+$ and hence $\tilde{F}_e^+ := [F_e^+]^{-1}$ is a bounded outer operator function.

From Proposition 2.14, the estimate in $\mathbb{C}_+$ for the equation (36) also follows:

$$\langle\langle\Psi_-^{-1}(l, k)\rangle\rangle \prec \exp\left(\operatorname{Im} k \int_0^l \max \sqrt{\rho_j(s)}\, ds\right) = \exp(\operatorname{Im} kT_0/2).$$

Then,

$$\langle\langle\Theta_+^{-1}(k)\, e^{ikT_0/2}\rangle\rangle = e^{-\operatorname{Im} kT_0/2}\langle\langle\Psi_-^{-1}[F_e^+]^{-1}\rangle\rangle$$

$$\le \langle\langle\Psi_-^{-1}(l, k)\langle\langle F_e^+\rangle\rangle^{-1}\, e^{-\operatorname{Im} kT_0/2}\rangle\rangle \prec 1.$$

The lemma is proved.

Our next step is to find an ESF Θ_1 such that $\Theta_1 G$ is an operator function bounded in $\mathbb{C}_+$ without a singular factor (it is more convenient for us to work with G; a transition to $\tilde{G}$ will be accomplished later).

We look for Θ_1 starting from the condition

$$\Theta_+ L_-^{-1} = \tilde{L}_-^{-1} \Theta_1, \tag{50}$$

where $\tilde{L}_-^{-1}$ is a matrix function bounded in $\mathbb{C}_+$ without a singular factor. If L_- has no zeros at $\operatorname{Im} k \geq 0$, then Θ_1 and $\tilde{L}_-$ are determined from the problem of the reverse-order factorization (see Subsection II.1.2.9) and we have

$$\tilde{L}_- \Theta_1 = \Theta_+ L_-. \tag{51}$$

Notice that $|\det \tilde{L}_-| = |\det L_-|$, and hence function $\tilde{L}_-$, as well as L_-, is bounded in $\mathbb{C}_+$ along with its inverse.

Now let L_- have zeros. In view of (47), their number is finite, and one is able, by multiplying (50) by an appropriately chosen scalar rational function, to arrive again at the problem of the factorization of bounded analytical functions.

The function Θ_1 determined by (50) satisfies the same inequality (48) as function Θ_+. Indeed, from (51) and (48) we have

$$\langle\langle \Theta_1^{-1}(k)\, e^{ikT_0/2} \rangle\rangle = \langle\langle L_- \Theta_+^{-1}\, e^{ikT_0/2} \tilde{L}_-^{-1} \rangle\rangle$$

$$\leq \langle\langle \tilde{L}_-^{-1}(l, k) \rangle\rangle \langle\langle L_- \rangle\rangle \langle\langle \Theta_+^{-1}\, e^{ikT_0/2} \rangle\rangle \prec 1;$$

that is,

$$\langle\langle \Theta^{-1}(k)\, e^{ikT_0/2} \rangle\rangle \leq 1, \qquad k \in \mathbb{C}_+. \tag{52}$$

Let us now demonstrate that matrix function $\Upsilon(k) := \Theta_1(k) G(k)$ is bounded in $\mathbb{C}_+$ and has no singular factor there. Indeed, by (51) and Lemma 16,

$$\Upsilon = \Theta_1 L_+ \Psi_+ R_+ + L_1 \tilde{F}_e^+ R_-. \tag{53}$$

The first summand here exponentially decreases at $\operatorname{Im} k \to \infty$ according to Proposition 2.14:

$$\langle\langle \Psi_+(l, k) \rangle\rangle \prec \exp(-\varepsilon \operatorname{Im} k), \qquad \varepsilon := \int_0^l \min \sqrt{\rho_j}(s)\, ds.$$

As already noted, the matrix functions $L_-^{-1}, [\tilde{F}_e^+]^{-1}$ are bounded for large enough k. Therefore, from relation (52) we have

$$\det \Upsilon(k) = \det[L_1 \tilde{F}_e^+ R_-] \det[I + [L_1 \tilde{F}_e^+ R_-]^{-1} \mathcal{O}(e^{-\varepsilon \operatorname{Im} k})]$$

$$= \det[L_1 \tilde{F}_e^+ R_-](1 + \mathcal{O}(e^{-\varepsilon \operatorname{Im} k}))$$

and we conclude that Υ has no singular factor in $\mathbb{C}_+$.

We now examine $\Theta_1(-k)G(k)$ in the lower half-plane, where we exploit relation $G(k) = -G(-k)$. Then from (46) we have

$$\Theta_1(-k)G(k) = -\Theta_1(-k)G(-k) = -\Upsilon(-k). \qquad (54)$$

The above arguments about Υ show that $\Theta_1(-k)G(k)$ is bounded in $\mathbb{C}_-$ and does not contain any singular factor there.

Now we move from $G(k)$ to $\tilde{G}(k) = G(k - i)$. Set $X(k) := \Theta_1(k - i)$ and check that $F(k) := X(k)\tilde{G}(k)$ satisfies factorization condition (12) and estimate (13). First, we prove the lemma, which is needed for the factorization of X.

Lemma VII.3.17. Let $\tilde{\Theta}$ be an ESF in $\mathbb{C}_+$. Then for any $z_0 \in \mathbb{C}$ function $\Theta_0(k) := \tilde{\Theta}(k - z_0)$ is an operator function bounded in $\mathbb{C}_+$ such that Θ_0 and Θ_0^{-1} are bounded on $\mathbb{R}$. Furthermore, if $\tilde{\Theta}_0^{-1} \exp(ika)$ is bounded in $\mathbb{C}_+$, then $\Theta_0^{-1} \exp(ika)$ is also bounded in $\mathbb{C}_+$.

PROOF. Like any ESF, function $\tilde{\Theta}(k)$ is the value at $x = 1$ of the solution to the problem (see Subsection II.1.2.4)

$$\tilde{\mathscr{Y}}'(x, k) = ik\mathscr{B}(x)\tilde{\mathscr{Y}}(x, k), \qquad \tilde{\mathscr{Y}}(0, k) = I,$$

with a summable Hermitian matrix function $\mathscr{B}(x)$. Then $\Theta_0(k)$ coincides with the value at $x = 1$ of the solution to the problem

$$\mathscr{Y}'(x, k) = [ik\mathscr{B}(x) - iz_0\mathscr{B}(x)]\mathscr{Y}(x, k), \qquad \mathscr{Y}(0, k) = I.$$

Proposition 2.14 implies the boundedness of Θ_0 in $\mathbb{C}_+$; that is, Θ_0 may be factorized in the form $\Theta_0 = \hat{F}_e^+\hat{\Theta}$. From the Liouville formula it follows that $|\det \hat{F}_e^+|$ is separated from zero and therefore $[\hat{F}_e^+]^{-1}$ is a bounded outer operator function in $\mathbb{C}_+$. By means of similar arguments concerning ESF $\tilde{\Theta}_0^{-1} \exp(ika)$, we find that $\Theta_0^{-1} \exp(ika)$ is bounded and complete the proof of Lemma 17.

From the lemma it follows that both $X(k)$ and $X(k)^{-1}$ are bounded on $\mathbb{R}$; that is, assertion (i) of Theorem 3(b) is proved.

Let us now use $F(k) = \Upsilon(k - i)$. By virtue of Lemma 17, F is bounded in the half-plane $\operatorname{Im} k > 1$. In the strip $0 \le \operatorname{Im} k \le 1$, the boundedness of F is provided by the asymptotics of $G(k)$, and Lemma 17 applied to Θ_1. Since $\det[G(k)\Theta_1(k)]$ does not contain a singular factor in $\mathbb{C}_+$, the same goes for $\det[\Theta_1(k - i)G(k - i)]$. Thus we have

$$F(k) = F_e^+(k)\Pi(k)$$

and the factorization of F in $\mathbb{C}_+$ is delivered.

Let us study the factorization of F in $\mathbb{C}_-$. Set $\Theta_0(k) = \Theta_1(k)\Theta_1^{-1}(-k)$. Now we proved that

$$\langle\langle\Theta_0^{-1}(k) e^{ikT_0}\rangle\rangle \prec 1, \qquad k \in \mathbb{C}_+. \tag{55}$$

Indeed,

$$\langle\langle\Theta_0^{-1}(k) e^{ikT_0}\rangle\rangle \leq \langle\langle\Theta_1^{-1}(k) e^{ikT_0/2}\rangle\rangle\langle\langle\Theta_1(-k) e^{ikT_0/2}\rangle\rangle.$$

The first cofactor is bounded by (52). For the second one, we have, from unitary property and analyticity of Θ_1, the relation

$$\Theta_1(-k) = (\Theta_1^{-1}(-\bar{k}))^*$$

(see formula (3) in Section II.1). Since the norm of matrix is equal to the norm of its conjugate, estimate (55) is true.

Function $\Theta_0^{-1}(k - i)F(k)$ is equal to

$$\Theta_0^{-1}(k - i)\Theta_1(k - i)G(k - i) = \Theta_1(-k + i)\Theta_1^{-1}(k - i)\Theta_1(k - i)G(k - i)$$

$$= \Theta_1(-k + i)G(k - i)$$

and, according to what is proved (see (54)), is an outer operator function F_e^- in $\mathbb{C}_-$. Referring to Θ_0 Lemma 17, we end up with the factorization

$$\Theta_0(k - i) = \Theta(k)\hat{F}_e^-(k),$$

where Θ is an ESF such, by (52), that

$$\langle\langle\Theta^{-1}(k) e^{ikT_0}\rangle\rangle \prec 1, \qquad k \in \mathbb{C}_+.$$

Now, $F = \Theta\hat{F}_e^- F_e^-$ and Theorem 3 is proved completely.

4. Controllability of a nonhomogeneous string controlled at the ends

We consider an initial boundary-value problem describing vibrations of a string:

$$\rho(x)\frac{\partial^2 y(x, t)}{\partial t^2} = \frac{\partial^2 y(x, t)}{\partial x^2}, \qquad 0 < x < l, 0 < t < T. \tag{1}$$

$$y(0, t) = u_1(t) \in L^2(0, T), \qquad y(l, t) = u_2(t) \in L^2(0, T), \tag{2}$$

$$y(x, 0) = \frac{\partial}{\partial t} y(x, 0) = 0. \tag{3}$$

Here, u_1, u_2 are the controls, and we assume ρ to be a positive function of the class $C^2[0, l]$. This problem of control has been formulated by D. L. Russell. In this section we study reachability set $\mathcal{R}(T)$ of system (1), (2).

To bring the system into the scope of the scheme in Chapter III, we set

$$H := L^2_\rho(0, l), \qquad V := H^1_0(0, l),$$

$$a[\varphi, \psi] := \int_0^l \varphi'(x)\bar{\psi}'(x)\, dx; \qquad \varphi, \psi \in V.$$

Operator A, corresponding to bilinear form $a[\varphi, \psi]$, is

$$(A\varphi)(x) := -\frac{1}{\rho(x)}\frac{d^2}{dx^2}\varphi(x)$$

with the domain $D(A) = H^2(0, l) \cap H^1_0(0, l)$. We write λ_n and φ_n, $n \in \mathbb{N}$, for the eigenvalues of A and its normed in $L^2_\rho(0, l)$ eigenfunctions, respectively.

Let us consider controls u_1 and u_2 to be components of a vector function belonging to the space $U_T = L^2(0, T; \mathbb{C}^2)$ and introduce operator $B: \mathbb{C}^2 \mapsto W_{-2}$ by means of the formula

$$\langle Bv, \psi \rangle_* = v_1\bar{\psi}'(0) - v_2\bar{\psi}'(l), \qquad \psi \in W_2 = D(A) \qquad (4)$$

(compare it with operator B from Section 1).

The eigenvalues λ_n are separable, as is well known from the asymptotics presented below. Therefore, by the force of Lemmas II.1.21 and III.2.4, the solution $y(x, t)$ of the problem (1)–(3) enjoys the property

$$\left\{y, \frac{\partial}{\partial t}y\right\} \in C(0, T, \mathcal{W}_0), \qquad (5)$$

$$\mathcal{W}_0 = W_0 \oplus W_{-1} = L^2_\rho(0, l) \oplus H^{-1}(0, l).$$

Theorem VII.4.1. System (1), (2) is B-controllable in time

$$T_0 = \int_0^l \sqrt{\rho(x)}\, dx$$

in space $\mathcal{W}_0$.

Avdonin and Ivanov (1983) examined a similar problem in a system with the Neumann boundary control

$$\left(\frac{\partial}{\partial x}y(0, t) = u_1(t), \frac{\partial}{\partial x}y(l, t) = u_2(t)\right).$$

Their approach was based on the asymptotics of the exponential family, and it allowed them to obtain controllability in time $T > T_0$. For $T = T_0$, the system with the Neumann control is not controllable since the corresponding family of exponentials is excessive in $L^2(0, T; \mathbb{C}^2)$ (for a homogeneous string, this fact is established immediately). In the situation we are analyzing now, the controllability of system (1), (2) for $T = T_0$ takes place, but to prove it, the knowledge of the asymptotics of λ_n and φ_n is no longer sufficient. We obtain the controllability of system (1), (2) by representing the string as two elastically connected strings, which enables us to use the results of Section 2 regarding the controllability of N strings connected at one point.

PROOF OF THEOREM 1. Take the point $x_0 \in (0, l)$ separating the parts $[0, x_0]$ and $[x_0, l]$ of the string with the same optical length $T_0/2$.

We introduce a unitary mapping

$$Q: L_\rho^2(0, l) \ni y \mapsto (y_1, y_2) \in L_{\rho_1}^2(0, x_0) \oplus L_{\rho_2}^2(0, l - x_0),$$

where

$$\rho_1(x) = \rho(x), \qquad y_1(x) = y(x), \qquad x \in (0, x_0),$$

$$\rho_2(x) = \rho(l - x), \qquad y_2(x) = y(l - x), \qquad x \in (0, l - x_0).$$

The image (φ_1, φ_2) of differentiable function φ under the action of Q satisfies conditions $\varphi_1(x_0) = \varphi_2(l - x_0)$ and $\varphi_1'(x_0) + \varphi_2'(l - x_0) = 0$ following from continuity and differentiability $c \cdot \varphi$ at x_0. In a sense, one can view conditions (3) and (4) in Section VII.2 as the smoothness conditions.

We proceed following the scheme of Chapter III, that is, in terms of spaces H, V, bilinear form $a[\cdot, \cdot]$ and operator B. We use the subscript I to mark these and other items for the system (1), (2); the subscript II stands for the values involved in relations (1)–(4) of Section VII.2 for $N = 2$. The equalities

$$QH_I = H_{II}, \qquad QV_I = V_{II}, \qquad a_I[\varphi, \psi] = a_{II}[Q\varphi, Q\psi], \qquad \varphi, \psi \in V_I \quad (6)$$

follow directly from the definitions. Relation (6) implies that operators A_I and A_{II} generated by the forms a_I and a_{II}, respectively, are unitary equivalent: $A_I = Q^*A_{II}Q$. In particular, eigenfunctions $\varphi_{n,I}$ and $\varphi_{n,II}$ of these operators are also related by the Q mapping: $\varphi_{n,II} = Q\varphi_{n,I}$. Hence the spaces $W_{r,I}$ and $W_{r,II}$ generated by operators A_I and A_{II} are related similarly:

$$W_{r,II} = QW_{r,I}. \tag{7}$$

Here, Q means an operator from $W_{r,I}$ into $W_{r,II}$ acting as

$$Q(\textstyle\sum c_n \varphi_{n,I}) = \sum c_n \varphi_{n,II}.$$

For $r = 0$, this definition coincides with the primary one. Finally, operators B_I and B_{II} acting from $\mathbb{C}^2$ to $W_{-2,I}$ or $W_{-2,II}$, respectively, are related in the following way:

$$\langle B_I v, \varphi \rangle_* = \langle B_{II} v, Q\varphi \rangle_*, \qquad v \in \mathbb{C}^2, \qquad \varphi \in W_{2,I} = D(A_I).$$

The equality is the immediate implication of the definitions.

The previous consideration makes it clear that if one writes systems (1), (2), and (1)–(4) as system (12) in Section III.3 their trajectories starting from zero are interrelated by $y_{II}(\cdot, t) = Q y_I(\cdot, t)$. Therefore, for the reachability sets of the systems we have

$$\mathcal{R}_{II}(T) = Q\mathcal{R}_I(T) \qquad \text{and so} \quad \mathcal{R}_I(T) = Q^{-1}\mathcal{R}_{II}(T).$$

In view of Theorem 2.14, the reachability set $\mathcal{R}_{II}(T)$ coincides with the phase space $\mathcal{W}_{0,II}$. Using (7), we establish the assertion of our theorem.

The controllability of system (1), (2) is proved without investigating the exponential family arising in this problem (such an investigation was carried out in Section 2). We now examine this family in order to answer two questions:

(i) What is the set of controls transferring the system in time $T \geq T_0$ from the zero state to itself?
(ii) What is the form of the reachability set $\mathcal{R}(T)$ for $T < T_0$?

In what immediately follows, the subscript T in the notations of various families indicates the space $\mathcal{U}_T = L^2(0, T; \mathbb{C}^N)$ in which the families are considered. So, $\tilde{\mathcal{E}}_T$ denotes the family of exponentials corresponding to the control problem for the system (1), (2) in space $\mathcal{W}_0$:

$$\tilde{\mathcal{E}}_T = \{e_n\}_{n \in \mathbb{K}} \subset \mathcal{U}_T, \qquad e_n = \eta_n \exp(i\omega_n t),$$

$$\omega_n = \operatorname{sgn}(n)\sqrt{\lambda_{|n|}}, \qquad \eta_n = |\omega_n|^{-1}\begin{pmatrix} \varphi'_{|n|}(0) \\ -\varphi'_{|n|}(l) \end{pmatrix}.$$

We are writing out the known asymptotic expressions for frequencies ω_n and vectors η_n (Fedoryuk 1983: chap. 2, sec. 10), which are not difficult to derive from Proposition 2.3 as well:

$$\omega_n = 2\pi n/T_0 + \mathcal{O}(1/n), \tag{8}$$

$$\left.\begin{array}{cc} \eta_{2n} = \eta^{(1)} + \mathcal{O}(1/n), & \eta_{2n+1} = \eta^{(2)} + \mathcal{O}(1/n), \\[2mm] \eta^{(1)} = \begin{pmatrix} \alpha \\ \beta \end{pmatrix}, & \eta^{(2)} = \begin{pmatrix} \alpha \\ -\beta \end{pmatrix}, \end{array}\right\} \tag{9}$$

where

$$\alpha := \frac{2\rho^{1/4}(0)}{\sqrt{T_0}}, \qquad \beta := \frac{2\rho^{1/4}(l)}{\sqrt{T_0}}.$$

Let us now introduce the exponential family corresponding to the main term of asymptotics (8) and (9). It is convenient to denote this family by some other letter, namely,

$$\Xi_T := \{\xi_n\}_{n \in \mathbb{K}}, \qquad \xi_n = \eta_n^0 \exp(2int/T_0),$$

$$\eta_{2n}^0 := \zeta^{(1)} := \begin{pmatrix} 1 \\ 1 \end{pmatrix}, \qquad \eta_{2n+1}^0 := \zeta^{(2)} := \begin{pmatrix} 1 \\ -1 \end{pmatrix}.$$

We first study the properties of family Ξ_T.

Theorem VII.4.2.

(a) *Family* $\Xi_{T_0} \cup \{\xi_0\}$, $\xi_0(t) := \zeta^{(1)}$, *forms a Riesz basis in* $\mathcal{U}_{T_0}$.
(b) *For* $T < T_0$, *there exists a subfamily* $\Xi_T^{(0)} \subset \Xi_T$ *that forms a Riesz basis in* $\mathcal{U}_T$, *and family* $\Xi_T \backslash \Xi_T^{(0)}$ *is an infinite one.*

PROOF. According to the decomposition of family Ξ_T into two series

$$\Xi_T^{(1)} = \{\xi_{2n}\}_{n \in \mathbb{K}}, \qquad \Xi_T^{(2)} = \{\xi_{2n+1}\}_{n \in \mathbb{Z}},$$

the space $\mathcal{U}_T$ may also be considered split into two subspaces $L_T^{(1)} \oplus L_T^{(2)}$, with $L_T^{(j)}$ consisting of the elements of the form $f(t)\zeta^{(j)}$, $j = 1, 2$, and $f \in L^2(0, T)$. Actually, we have two scalar problems for families

$$\Xi_{T_0, sc}^{(1)} := \{\exp(i4nt/T_0)\}_{n \in \mathbb{K}} \quad \text{and} \quad \Xi_{T_0, sc}^{(2)} := \{\exp[i(4n + 2)t/T_0]\}_{n \in \mathbb{Z}}$$

in $L^2(0, T)$ instead of a vector one. The first family complemented by $1(t) := 1$ constitutes an orthogonal basis in $L^2(0, T_0)$, and the second one is also an orthogonal basis in $L^2(0, T_0)$. The families are almost normed and therefore both $\Xi_{T_0}^{(1)} \cup \{\zeta_0\}$ and $\Xi_{T_0}^{(2)}$ form Riesz bases in $L_T^{(j)}$, $j = 1, 2$, respectively. Thus, assertion (a) is proved.
(b) Apply Theorem II.4.18 to scalar families $\Xi_{T, sc}^{(1)} \cup 1(t)$ and $\Xi_{T, sc}^{(2)}$.

According to this theorem, there exists a subset $\mathcal{N}^{(1)}$ of even integers and a subset $\mathcal{N}^{(2)}$ of odd integers such that families $\{\xi_n\}_{n \in \mathcal{N}^{(j)}}$ constitute Riesz bases in space $L_T^{(j)}$, $j = 1, 2$, respectively. In addition, the sets $2\mathbb{Z}\backslash\mathcal{N}^{(1)}$ and $(2\mathbb{Z} + 1)\backslash\mathcal{N}^{(2)}$ are both infinite. We may assume $0 \notin \mathcal{N}^{(1)}$, which means that the additional element ζ_0 does not belong to the basis family $\{\xi_n\}_{n \in \mathcal{N}^{(1)}}$. Indeed, in the scalar case one can replace, without loss of the basis property, the "exponential" $1(t) = \exp(0 \cdot t)$ by any other one from $\{\xi_n\}_{n \notin \mathcal{N}^{(1)}}$. Now all the requirements of assertion (b) are fulfilled. The theorem is proved.

Recall that the dimension of the space orthogonal to all the elements of family Ξ is called the codimension of Ξ.

Theorem VII.4.3.

(a) For $T \geq T_0$, family $\tilde{\mathscr{E}}_{T_0}$ is an $\mathscr{L}$-basis in $\mathscr{U}_{T_0}$ of a unit codimension for $T = T_0$ and of an infinite codimension for $T > T_0$.

(b) For $T < T_0$, there exists a subfamily $\tilde{\mathscr{E}}_T^0 \subset \tilde{\mathscr{E}}_T$ such that $\tilde{\mathscr{E}}_T^0$ constitutes a Riesz basis in $\mathscr{U}_T$ and family $\tilde{\mathscr{E}}_T\backslash\tilde{\mathscr{E}}_T^0$ is infinite.

PROOF.

(a) For $T \geq T_0$ the $\mathscr{L}$-basis property follows from B controllability of system (1)–(3) (see Theorems 1 and III.3.10).

Let Z be an operator acting in $\mathbb{C}^2$ and mapping the basis $\{\eta^{(1)}, \eta^{(2)}\}$ of space $\mathbb{C}^2$ into the basis $\{\zeta^{(1)}, \zeta^{(2)}\}$. Families Ξ_T and $Z\tilde{\mathscr{E}}_T$ are asymptotically close in the sense of Theorem II.5.9:

$$\omega_n - i2\pi n/T_0 \xrightarrow[|n| \to \infty]{} 0, \qquad \langle\langle \eta_n - \eta_n^0 \rangle\rangle \xrightarrow[|n| \to \infty]{} 0.$$

Let $T = T_0$. By Theorem II.5.5, family $\{\xi_n\}_{|n| \leq M} \cup \{Ze_n\}_{|n| > M}$ forms a Riesz basis for large enough M, inasmuch as $\{\xi_0\} \cup \Xi_{T_0}$ forms a basis. Therefore, we are able to conclude that

$$\text{codim}\{Ze_n\}_{|n| > M} = 2M + 1.$$

Since operator $f(\cdot) \mapsto Zf(\cdot)$ is evidently an isomorphism of $\mathscr{U}_{T_0}$, we have

$$\text{codim}\{e_n\}_{|n| > M} = \text{codim}\{Ze_n\}_{|n| > M} = 2M + 1.$$

Family $\tilde{\mathscr{E}}_{T_0}$ is an $\mathscr{L}$-basis; hence elements $e_{\pm 1}, e_{\pm 2}, \ldots, e_{\pm M}$ do not belong to $\bigvee \{e_n\}_{|n|>M}$. Therefore

$$\text{codim } \tilde{\mathscr{E}}_{T_0} = \text{codim} \bigvee \{e_n\}_{|n|>M} - 2M = 1.$$

The fact that the codimension of $\tilde{\mathscr{E}}_T$ is infinite for $T > T_0$ follows from Theorem II.5.9(a).

(b) Let $\Xi_T^{(0)} = \{\xi_n\}_{n\in\mathscr{N}}$ be a basis subfamily whose existence is guaranteed by Theorem 2(b). Along with it, the set $\mathbb{K}\backslash\mathscr{N}$ is infinite. Family $\{Ze_n\}_{n\in\mathscr{N}}$ is asymptotically close to $\Xi^{(0)}$, and so, by Theorem II.5.5, family $\{\xi_n\}_{|n|\le M} \cup \{Ze_n\}_{|n|>M}$ constitutes a Riesz basis for large enough M. Thus family $\{Ze_n\}_{n\in\mathscr{N};|n|>M}$ is an $\mathscr{L}$-basis of a finite codimension. According to Theorem II.5.9(b), family $Z\tilde{\mathscr{E}}_{T_0}$ is complete in $\mathscr{U}_T$, and we can supplement family $\{Ze_n\}_{n\in\mathscr{N};|n|>M}$ with $Ze_{n_1}, Ze_{n_2}, \ldots, Ze_{N_r}$ to complete it while preserving the $\mathscr{L}$-basis property. Then

$$\{e_n\}_{n\in\mathscr{N};|n|>M} \cup \{e_{n_j}\}_{j=1}^r$$

forms a Riesz basis in $\mathscr{U}_T$. The theorem is proved.

Now we are able to answer the questions about system (1), (2) posed earlier.

Theorem VII.4.4.

(a) *For $T \ge T_0$, there exist nonzero controls transferring system (1), (2) from the zero state into itself. The set of such controls for $T = T_0$ constitutes a one-dimensional subspace (in $\mathscr{U}_{T_0}$), and for $T > T_0$ its dimension is infinite.*

(b) *For $T < T_0$, the reachability set is a proper subspace of $\mathscr{W}_0$ with an infinite codimension.*

PROOF.

(a) Control u transfers the system from the zero state to the zero one if and only if $u^T = u(T - t)$ is orthogonal to $\tilde{\mathscr{E}}_T$ (see moment equalities (23) in Section III.2). By Theorem 3, the dimension of the space orthogonal to $\tilde{\mathscr{E}}_T$ is 1 for $T = T_0$ and turns into infinity for $T > T_0$.

Assertion (b) follows immediately from Theorems VII.4.3(b), III.2.3, and I.2.1(e).

References

Agmon, S. 1965. *Lectures on Elliptic Boundary Value Problems*. Princeton, N.J.: Van Nostrand.

Avdonin, S. A. 1974a. On Riesz bases of exponentials in L^2. *Vestnik Leningradskogo Universiteta, Ser. Mat., Mekh., Astron.* 13:5–12 (Russian); English transl. in *Vestnik Leningrad Univ. Math.* 7:(1979) 203–11.

Avdonin, S. A. 1974b. On the question of Riesz bases consisting of exponential functions in L^2. *Zapiski Nauchnykh Seminarov Leningradskogo Otdeleniya Mat. Inst. Steklova (LOMI)* 39:176–7 (Russian); English transl. in *J. Soviet Math.* 8 (1977):130–1.

Avdonin, S. A. 1975. *Controllability of Systems Described by Hyperbolic Type Equations*. Dep. VINITI no. 2430, 14 p. Leningrad (Russian).

Avdonin, S. A. 1977a. On exponential problem of moments. *Zapiski Nauchnykh Seminarov Leningradskogo Otdeleniya Mat. Inst. Steklova (LOMI)* 77:193–4 (Russian).

Avdonin, S. A. 1977b. *Riesz Bases of Exponentials and Control Problems of Distributed Parameter Systems*. Ph.D. diss. Leningrad Univ. (Russian).

Avdonin, S. A. 1980. On controllability of distributed parameter systems. *Vestnik Leningradskogo Universiteta, Ser. Mat., Mekh., Astron.* 19:5–8 (Russian).

Avdonin, S. A. 1982. Controllability of a singular string. In *Problemy Mekhaniki Upravlyaemogo Dvizheniya* (Problems of control mechanics), pp. 3–8. Perm': Perm' Univ. (Russian).

Avdonin, S. A. 1991. Existence of basis subfamilies of a Riesz basis from exponentials. *Vestnik Leningradskogo Universiteta, Ser. Mat., Mekh., Astron.* 15:101–5 (Russian); English transl. in *Vestnik Leningrad Univ. Math.*

Avdonin, S. A., M. I. Belishev, and S. A. Ivanov. 1991a. Boundary control and a matrix inverse problem for the equation $u_{tt} - u_{xx} + V(x)u = 0$. *Matematicheskii Sbornik.* 182(3):307–31 (Russian); English transl. in *Math. USSR Sbornik.* 72:(1992) 287–310.

Avdonin, S. A., M. I. Belishev, and S. A. Ivanov. 1991b. Dirichlet boundary control in filled domain for multidimensional wave equation. *Avtomatika* 2:86–90 (Russian); English transl. in *Journal of Automation and Information Sciences* 24:(1992) 76–80.

Avdonin, S. A., and O. Ya. Gorshkova. 1986. On controllability of parabolic systems with time delays. In *Uravneniya v Chastnykh Proizvodnykh* (Partial Differential Equations), pp. 53–5. Leningrad: Leningrad Pedagog. Inst. (Russian).

Avdonin, S. A., and O. Ya. Gorshkova. 1987. Controllability of multi-dimensional parabolic systems with time delays. In *Matematicheskaya Fizika* (Mathematical Physics), pp. 95–9. Leningrad: Leningrad Pedagog. Inst. (Russian).

Avdonin, S. A., and O. Ya. Gorshkova. 1992. Controllability and quasi-controllability of parabolic systems with delay. *Differentsial'nye Uravneniya*, no. 3 (Russian); English transl. in *Diff. Equations* 28:(1992) 374–83.

Avdonin, S. A., M. Horvath, and I. Joó. 1989. Riesz bases from elements of the form $x_k\, e^{i\lambda x_k}$. *Vestnik Leningradskogo Universiteta, Ser. Mat., Mekh., Astron.* 22:3–7 (Russian); English transl. in *Vestnik Leningrad Univ. Math.*

Avdonin, S. A., and S. A. Ivanov. 1983. Riesz bases of exponentials in a space of vector-valued functions and controllability of a nonhomogeneous string. *Teoriya Operatorov i Teoriya Funktsii* (Operator Theory and Function Theory), vol. 1, pp. 62–8. Leningrad: Leningrad Univ. (Russian).

Avdonin, S. A., and S. A. Ivanov. 1984. Series bases of exponentials and the problem of the complete damping of a system of strings. *Doklady Akad. Nauk SSSR* 275:355–8 (Russian); English transl. in *Soviet Phys. Dokl.* 29:(1984) 182–4.

Avdonin, S. A., and S. A. Ivanov. 1988. Families of exponentials and wave equation in a parallelepiped. In *5th Conf. on Complex Anal. Halle. Dec. 12–17 1988: Abstracts*, p. 3. Halle: Halle-Wittenberg Martin-Luther-Univ.

Avdonin, S. A., and S. A. Ivanov. 1989a. A generating matrix-valued function in problem of controlling the vibration of connected strings. *Doklady Akad. Nauk SSSR* 307(5):1033–7 (Russian); English transl. in *Soviet Math. Dokl.* 40:(1990) 179–83.

Avdonin, S. A., and S. A. Ivanov. 1989b. *Controllability of Distributed Parameter Systems and Families of Exponentials.* Kiev: UMK VO (Russian).

Avdonin, S. A., and S. A. Ivanov. 1995. Boundary controllability problems for the wave equation in a parallelepiped. *Applied Math. Letters* 8:97–102.

Avdonin, S. A., S. A. Ivanov, and A. Z. Ishmukhametov. 1991. A quadratic problem of optimal control of the vibrations of a string. *Doklady Akad. Nauk SSSR.* 316(4):781–5 (Russian); English transl. in *Soviet Math. Dokl.* 43:(1991) 154–8.

Avdonin, S. A., S. A. Ivanov, and I. Joó. 1989. On Riesz bases from vector exponentials: I, II. *Annales Univ. Sic. Budapest.* 32:101–26.

Avdonin, S. A., S. A. Ivanov, and I. Joó. 1990. Initial and pointwise control of the vibrations of a rectangular membrane. *Avtomatika* 6:86–90 (Russian); English transl. in *Journ. of Automation and Information Sciences.*

Avdonin, S. A., and I. Joó. 1988. Riesz bases of exponentials and sine type functions. *Acta Math. Hung.* 51:3–14.

Bardos, C., G. Lebeau, and J. Rauch. 1988a. *Contrôle et Stabilization dan les Problémes Hyperboliques*, Appendix II in J.-L. Lions, ed., *Contrôlabilité Exacte Perturbations et Stabilization de Systemes Distribués*, vol. 1, pp. 492–537. Paris: Masson.

Bardos, C., G. Lebeau, and J. Rauch. 1988b. Un example d'utilisation des notions de propagation pour le contrôle et la stabilisation des problèmes hiperboliques. *Rend. Sem. Mat. Univ. Pol. Torino, Fasc. Spec.* 11–32.

Bardos, C., G. Lebeau, and J. Rauch. 1992. Sharp sufficient condition for the observation, control and stabilization of waves from the boundary. *SIAM J. Control Optim.* 30:1024–65.

Bari, N. K. 1951. Biorthogonal systems and bases in Hilbert space. *Uchenye Zapiski Moskovskogo Gosudarstvennogo Univ.* 4(148):68–107 (Russian).

Belishev, M. I. 1989. Wave bases in multidimensional inverse problems. *Matematicheskiĭ Sbornik* 180(5):584–602 (Russian); English transl. in *Math. USSR Sbornik* 67:(1990)

Berezanskiĭ, Yu. M. 1965. *Decomposition in Eigenfunctions of Selfadjoint Operators.* Kiev: Naukova dumka (Russian).

Birman, M. Sh., and M. Z. Solomyak. 1977. Asymptotics of spectrum of differential equations. *Itogi Nauki i Tekhniki* 14:5–59 (Russian).

Birman, M. Sh., and M. Z. Solomyak. 1980. *Spectral Theory of Selfadjoint Operators in Hilbert Space.* Leningrad: Leningrad Univ. (Russian); English transl. Reidel, 1987.

Boas, R. M. 1954. *Entire Functions.* New York: Academic Press.

Buslaeva, M. V. 1974. The root subspaces of a contraction that can be generated by an inner function. *Zapiski Nauchnykh Seminarov Leningradskogo Otdeleniya Mat. Inst. Steklova (LOMI)* 42:78–84 (Russian); English transl. in *J. Soviet Math.* 9:(1978).

Butkovskiĭ, A. G. 1965. *Theory of Optimal Control of Distributed Parameter Systems.* Moscow: Nauka (Russian); English transl. 1969.

Butkovskiĭ, A. G. 1975. *Methods of Optimal Control of Distributed Parameter Systems.* Moscow: Nauka (Russian).

Butkovskiĭ, A. G. 1976. Application of certain results from number theory to the finite control and controllability problem in distributed systems. *Doklady Akad. Nauk SSSR* 227(2):309–11 (Russian); English transl. in *Soviet Phys. Dokl.* 21:(1976) 124–6.

Carleson, L. 1958. An interpolation problem for bounded analytic functions. *Amer. J. Math.* 80:921–30.

Cassels, J. W. S. 1957. *An Introduction to Diophantine Approximation.* Cambridge Tracts in Mathematics and Mathematical Physics, no. 45. Cambridge: Cambridge University Press.

Chen, G., M. C. Delfour, A. M. Krall, and G. Payre. 1987. Modelling, stabilization and control of serially connected beams. *SIAM J. Control Optim.* 25:526–46.

Curtain, R. F., and A. Pritchard. 1978. *Infinite Dimensional Linear System Theory.* Berlin: Springer.

Duffin, R. J., and J. J. Eachus. 1942. Some notes on an expansion theorem of Paley and Wiener. *Bull. Amer. Math. Soc.* 48:850–5.

Duffin, R. J., and A. C. Schaffer. 1952. A class of of nonharmonic Fourier series. *Trans. Amer. Math. Soc.* 72:341–66.

Duren, P. 1970. *Theory of H² Spaces.* New York: Academic Press.

Egorov, A. I. 1978. *Optimal Control of Heat and Diffusion Processes.* Moscow: Nauka (Russian).

Egorov, A. I., and V. N. Shakirov. 1983. Optimal control problem of quasilinear hyperbolic system. In *Optimal'noe Upravlenie Mekhanicheskimi Sistemami* (Optimal Control of Mechanical Systems), pp. 22–32. Leningrad: Leningrad Univ. (Russian).

Egorov, Yu. V. 1963a. Some problems of optimal control theory. *Zhurnal Vycheslitel'noi Matematiki and Matematicheskoi Fiziki* 3(5):883–904 (Russian).

Egorov, Yu. V. 1963b. Optimal Control in Banach Space. *Doklady Akad Nauk SSSR* 150:241–4 (Russian).

Emanuilov, O. Yu. 1990. Exact controllability of hyperbolic equations. I, II.

Avtomatika 3:10–13; 4:9–16 (Russian); English transl. in *Journ. of Automation and Information Sciences.*

Fattorini, H. O. 1966. Control in finite time of differential equations in Banach space. *Comm. Pure. Appl. Math.* 19:17–35.

Fattorini, H. O. 1967. On complete controllability of linear systems. *J. Diff. Equations* 3:391–402.

Fattorini, H. O. 1968. Boundary control systems. *SIAM J. Control* 6:349–85.

Fattorini, H. O. 1975. Boundary control of temperature distributes in a parallelipepidon. *SIAM J. Control.* 13:1–13.

Fattorini, H. O. 1979. Estimates for sequences biorthogonal to certain complex exponentials and boundary control of the wave equation. In *Lecture Notes in Control and Inform. Sci.* 2:111–24.

Fattorini, H. O., and D. L. Russell. 1971. Exact controllability theorems for linear parabolic equations in one space dimension. *Arch. Rat. Mech. Anal.* 43:272–92.

Fattorini, H. O., and D. L. Russell. 1974. Uniform bounds on biorthogonal functions for real exponentials and application to the control theory of parabolic equations. *Quart. Appl. Math.* 32:45–69.

Fedoryuk, M. V. 1977. *Saddle-point Method.* Moscow: Nauka (Russian).

Fedoryuk, M. V. 1983. *Asymptotic Methods for Ordinary Linear Differential Equations.* Moscow: Nauka (Russian).

Fleming, W. H. 1988. Report of the Panel of Future Directions in Control Theory: A Mathematical Perspective. In *Soc. Ind. Appl. Math.* (Philadelphia) 1–98.

Fuhrmann, P. A. 1972. On weak and strong reachability and controllability of infinite-dimensional linear systems. *J. Optim. Theory Appl.* 9:77–89.

Gal'chuk, L. I. 1968. Optimal control of systems described by parabolic equations. *Vestnik Moskovskogo Universiteta,* 3:21–28 (Russian).

Garnett, J. B. 1981. *Bounded Analytic Functions.* New York: Academic Press.

Gingold, H., and P.-F. Heieh. 1985. A global study of a Hamilton system with multiturning points. In *Lecture Notes in Math.* 1151:164–71.

Gokhberg, I. Ts., and M. G. Krein. 1965. *Introduction to the Theory of Linear Nonselfadjoint Operators in Hilbert Space.* Moscow: Fiztekhizdat (Russian); English transl., Providence, R.I.: American Math. Soc. (1969).

Golovin, V. D. 1964. The biorthogonal decompositions in linear combination of exponential functions for L^2 space. *Zapiski Mekhanicheskogo Fakul'teta Khar'kovskogo Gosudarstvennogo Universiteta i Khar'kovskogo Matematicheskogo Obshchestva* 30:18–24 (Russian).

Graham, K. D., and D. L. Russell, 1975. Boundary value control of the wave equation in a spherical region. *SIAM J. Contr. Optim.* 13:174–96.

Grisvard, P. 1987. Contrôlabilité exacte avec conditions mêlés. *C. R. Acad. Sci. Sér. I* 305:363–6.

Gubreev, G. M. 1987. A basis property of families of the Mittag-Leffler functions. Dzhrbashyan's transform and the Muckenhoupt condition. *Funktsional. Anal. i Prilozhen.* 20(3):71–2.

Hansen, S. W. 1991. Bounds on functions biorthogonal to sets of complex exponentials: Control of damped elastic systems. *J. Math. Anal. Appl.* 158:487–508.

Haraux, A. 1988. Contrôlabilité exacte d'une membrane rectangulaire au moyen d'une fonctionnelle analytique localisee. *C. R. Acad. Sci. Paris, Sér. I,* 306:125–8.

Haraux, A., and J. Jaffard. 1991. Pointwise and spectral control of plate vibrations. *Revista Mat. Iberoamericana.* 7(1):1–24.

Haraux, A., and V. Komornik. 1985. Oscillations of anharmonic Fourier series and the wave equation. *Revista Mat. Iberoamericana.* 1:55–77.

Haraux, A., and V. Komornik. 1991. On the vibrations of rectangular plates. *Proceedings of the Royal Society of Edinburgh.* 119A:47–62.

Hartman, F. 1964. *Ordinary Differential Equations.* New York: John Wiley & Sons.

Helson, H. 1964. *Lectures on Invariant Subspaces.* New York: Academic Press.

Helson, H., and G. Szego. 1960. A problem in prediction theory. *Annal. Math. Pure Appl.* 51:107–38.

Ho, L. F. 1986. Observabilité frontière de l'equation des ondes. *C. R. Acad. Sci. Paris,* Sér. I, 302:443–6.

Hoffman, K. 1962. *Banach Spaces of Analytic Functions.* Englewood Cliffs, N.J.: Prentice-Hall.

Hörmander, L. 1968. The spectral function of an elliptic operator. *Acta Math.* 121:193–218.

Hörmander, L. 1976. *Partial Differential Equations.* (Grundlehren der Mathematishen Wissenschaften. Bd. 116). Berlin: Springer.

Hrushchev, S. V. 1987. Unconditional bases in $L^2(0, a)$. *Proceedings of the American Mathematical Society.* 99(4):651–6.

Hrushchev, S. V., N. K. Nikol'skiĭ, and B. S. Pavlov. 1981. Unconditional bases of reproducing kernels. *Lecture Notes in Math.* 864:214–335. Berlin: Springer.

Hunt, R. A., B. Muckenhoupt, and R. L. Wheeden. 1973. Weighted norm inequalities for the conjugate function and Hilbert transform. *Trans. Amer. Math. Soc.* 176:227–51.

Ingham, A. E. 1934. A note on Fourier transforms. *J. London Math. Soc.* 9:29–32.

Ivanov, S. A. 1978. Completeness of resonance state system of matrix polar operator. *Vestnik Leningradskogo Universiteta, Ser. Mat., Mekh., Astron.* 19:43–8 (Russian); English transl. in *Vestnik Leningrad Univ. Math.* 11:(1983).

Ivanov, S. A. 1983a. Completeness and basis property of resonance state system of operator $-A^{-1}(x)(d^2/dx^2)$. *Vestnik Leningradskogo Universiteta, Ser. Mat., Mekh., Astron.* 13:89–90 (Russian).

Ivanov, S. A. 1983b. The Regge problem for vector-functions. *Teoriya Operatorov i Teoriya Funktsii* (Theory of Operators and Theory of Functions), vyp. 1, pp. 68–86. Leningrad: Leningrad Univ. (Russian).

Ivanov, S. A. 1983c. *Vector Exponential Systems and Generating Function.* Dep. in VINITI 26.08.83, no. 4692 (Russian).

Ivanov, S. A. 1985. Bases of rational vector-valued functions and Carleson sets. *Doklady Akad. Nauk Armyanskoĭ SSR.* 80(1):20–25 (Russian).

Ivanov, S. A. 1987. The best rational approximations by vector-valued functions in Hardy spaces. *Matematicheskiĭ Sbornik.* 133(1):134–42 (Russian); English transl. in *Math. USSR Sbornik.* 61:(1988) 137–45.

Ivanov, S. A. 1989. Smooth diagonalization of Hermite matrix-valued functions. *Ukrainskiĭ Matematicheskiĭ Zhurnal.* 41(11):1569–72 (Russian).

Ivanov, S. A., and B. S. Pavlov. 1978. Carleson series of resonances in the Regge problem. *Izvestiya Akad. Nauk SSSR, Ser. Mat.* 42:26–55; English transl. in *Math. USSR Izv.* 12 (1978).

Joó, I. 1987a. On the vibration of a string. *Studia Sci. Math. Hung.* 22:1–9.

Joó, I. 1987b. On the reachability set of a string. *Acta Math. Hung.* 49:203–11.

Kadets, M. I. 1964. The exact value of Paley–Wiener constant. *Doklady Akad. Nauk. SSSR.* 155(6):1253–4 (Russian).

Kamke, E. 1959. *Differentialgleichungen. Lösungsmethoden und Lösungen. I. (Gewöhnliche Differentialgleichungen)*. (Mathematik und ihre Anwendungen in Physik und Technik. Reihe A. Bd. 18, 6th ed. Leipzig: Akad. Verlag-Ges. Geest & Portig.

Kato, T. 1966. *Perturbation Theory for Linear Operators* (Die Grundlehren der Mathematishen Wissenshaften. Bd. 132. Berlin: Springer.

Katsnelson, B. E. 1971. On bases in exponential functions for L^2 spaces. *Funktsional'nyi Analiz i ego Prilozheniya*. 5(1):37–74 (Russian).

Komornik, V. 1989a. On the vibrations of square membrane. *Proc. Royal Soc. Edinburgh* 111A:13–20.

Komornik, V. 1989b. Exact controllability in short time for the wave equation. *Ann. Inst. Henry Poincaré*. 6:153–64.

Koosis, P. 1980. *Introductions to H^p Space*. London Math. Soc. Lecture Note Series, no. 40. Cambridge University Press.

Korobov, V. I., and G. M. Sklyar. 1987. Optimal time problem and power problem of moments. *Matematicheskii Sbornik*. 134:186–206 (Russian).

Krabs, W. 1992. On moment theory and controllability of one-dimensional vibrating systems and heating processes. *Lecture Notes in Control and Information Sciences*, 173. Berlin: Springer.

Krabs, W., G. Leugering, and T. I. Seidman. 1985. On boundary controllability of vibrating plate. *Appl. Math. Optim.* 13:205–29.

Krasovskii, N. N. 1968. *The Theory of Motion Control*. Moscow: Nauka (Russian).

Kucherenko, V. V. 1974. Asymptotics of solution of equation $A(x, -i (\partial/\partial x))u = 0$ with $h \to 0$ for characteristics of varying multiplicity. *Izvestiya Akad. Nauk SSSR Ser. Mat.* 38(3):625–62 (Russian).

Kuzenkov, O. A., and V. I. Plotnikov. 1979. *On Fourier Method for a Vector Parabolic Equation*. Dep. VINITI No. 3393, Kazan'.

Lagnese, J. E. 1983. Control of wave processes with distributed controls supported on a subregion. *SIAM J. Contr. Optim.* 21(1):68–85.

Lagnese, J. E., G. Leugering, and E. J. P. G. Schmidt. 1993. Controllability of planar network of Timoshenko beams. *SIAM J. Control Optim.* 31:780–811.

Lagnese, J. E., and J.-L. Lions. 1989. *Modeling Analysis and Control of Thin Plates*. New York: Springer.

Landis, E. M. 1956. On some properties of solutions of elliptic equations. *Doklady Akad. Nauk. SSSR.* 107:640–3 (Russian).

Lasiecka, I., J.-L. Lions, and R. Triggiani. 1986. Nonhomogeneous boundary value problems for second order hyperbolic operators. *J. Math. Pures et Appl.* 65(2):35–93.

Lasiecka, I., and R. Triggiani, 1981. A cosine operator approach to modeling $L^2(0, T; L^2(0, T))$ boundary input hyperbolic equations. *Appl. Math. Optim.* 7:35–93.

Lasiecka, I., and R. Triggiani. 1989a. Trace regularity of the solutions of the wave equation with homogeneous Neumann boundary conditions and supported away from the boundary. *J. Math. Anal. Appl.* 141:49–71.

Lasiecka, I., and R. Triggiani. 1989b. Exact controllability of the wave equation with Neumann boundary control. *Appl. Math. Optim.* 19:243–90.

Lax, P. D., and N. Milgram. 1954. Parabolic equations. *Ann. Math. Studies*. 33:167–90.

Lebeau, G. 1992. Control for hyperbolic equations. *Lecture Notes in Control and Information Sciences*. 185:160–83. Berlin: Springer.

Lebeau, G., and L. Robbiano. 1994. Controle exact de l'equation de la chaleur. Preprint 94-37. Univ. de Paris-Sud Math.

Leont'ev, A. F. 1976. *Series of Exponentials.* Moscow: Nauka (Russian).

Leugering, G., and E. J. P. G. Schmidt. 1989. On the control of networks of vibrating strings and beams. In *Proc. 28th. IEEE Conf. on Decision and Control,* vol. 3, 2287–90.

Levin, B. Ya. 1956. *Distribution of Zeros of Entire Functions.* Moscow: Gostekhizdat (Russian).

Levin, B. Ya. 1961. On Riesz bases of exponentials in L^2. *Zapiski Matematicheskogo Otdeleniya Fiziko-matematicheskogo Fakul'teta Kharkovskogo Universiteta* 27 (Ser. 4): 39–48 (Russian).

Levin, B. Ya., and I. V. Osrovskiĭ. 1979. On small perturbations of zero set of sine-type functions. *Izvestiya Akad. Nauk SSSR Ser. Mat.* 43:87–110 (Russian).

Levinson, N. 1940. Gap and density theorems. *Amer. Math. Soc. Coll. Publ.* 26.

Lions, J.-L. 1968. *Contrôle Optimal des Systemes Governes par des Equations aux Derivees Partielles.* Paris: Dunod and Gauthier-Villars; English transl. in Berlin: Springer (1971).

Lions, J.-L. 1983. *Contrôle des Systémes Distribués Singuliers.* Paris: Gauthier-Villars.

Lions, J.-L. 1986. Contrôlabilité exact de systèmes distribuès. *C. R. Acad. Sci. Paris,* Sér. I, 302:471–5.

Lions, J.-L. 1988a. *Contrôlabilité Exacte Perturbations et Stabilization de Systémes Distribués,* vols. 1 and 2. Paris: Masson.

Lions, J.-L. 1988b. Exact controllability, stabilization and perturbations for distributed systems. *SIAM Review.* 30(1):1–68.

Lions, J.-L., and E. Magenes. 1968. *Problemes aux Limites Nonhomogenes et Applications,* vols. 1 and 2. Paris: Dunod.

Littman, W. 1986. Near optimal time boundary controllability for a class of hyperbolic equations. In *Proceedings of Conference on Distributed Parameter Control, Gainesville, GA.* New York: Springer.

Litvinov, V. G. 1987. *Optimization in Elliptic Boundary Problems with Applications to Mechanics.* Moscow: Nauka (Russian).

Lurie, K. A. 1975. *Optimal Control of Mathematical Physics Systems.* Moscow: Nauka (Russian).

Meyer, Y. 1985. Etude d'un modèle mathematique issu du contrôle des structure déformable. *Nonlinear Partial Dif. Equat. Appl. College de France Semin.* 7:234–42.

Mikhailov, V. P. 1979. *Partial Differential Equations.* Moscow: Nauka (Russian).

Minkin, A. M. 1991. Reflection on exponents, and unconditional bases of exponentials. *Algebra i Anal.* 3(5) (Russian); English transl. in *St. Petersburg Math. J.* 3(1992):1043–68.

Mizel, V. J., and T. I. Seidman. 1969. Observation and prediction for the heat equation. *J. Math. Anal. Appl.* 28:303–12.

Mizel, V. I., and T. I. Seidman. 1972. Observation and prediction for the heat equation, II. *J. Math. Anal. Appl.* 38:149–66.

Mizohata, S. 1958. Unicite du prolongement des solution pour quelque operateur differentiels paraboliques. *Mem. Colleg. Sci. Univ. Kyoto,* Ser. A, 1:219–39.

Naimark, M. A. 1969. *Linear Differential Operators.* 2d ed. Moscow: Nauka (Russian); English transl. of 1st ed., 1 and 2. New York: Ungar, 1967, 1968.

Narukawa, K., and T. Suzuki. 1986. Nonharmonic Fourier series and its applications. *Appl. Math. Optim.* 14:249–64.

Nefedov, S. A., and F. A. Sholokhovich. 1985. On a semi-group approach to boundary control problems. *Izvestiya Vuzov, Mat.* 12:37–42 (Russian).

Nikol'skiĭ, N. K. 1980. *A Treatise on the Shift Operator.* Moscow: Nauka (Russian); Engl. transl., Berlin: Springer, 1986.

Nikol'skiĭ, N. K., and B. S. Pavlov. 1970. Eigenvector bases on completely nonunitary contractions and the characteristic functions. *Izvestiya Akad. Nauk SSSR Ser. Mat.* 34:90–133 (Russian); English transl. in *Math. USSR Izv.* 4 (1970).

Nikol'skiĭ, S. M. 1969. *Approximations of Functions of Several Variables and Embedding Theorems.* Moscow: Nauka.

Paley, R. E. A. C., and N. Wiener. 1934. *Fourier Transform in the Complex Domain.* AMS Coll. Publ., vol. 19. New York.

Pavlov, B. S. 1971. On the joint completeness of eigenfunctions of contraction and its conjugation. *Problemy Matematicheskoi Fiziki* (Problems of Mathematical Physics). 5:101–12. Leningrad: Leningrad Univ. (Russian).

Pavlov, B. S. 1973. Spectral analysis of a differential operator with a "blurred" boundary condition. *Problemy Matematicheskoi Fiziki* (Problems of Mathematical Physics). 6:101–19. Leningrad: Leningrad Univ. (Russian).

Pavlov, B. S. 1979. Basicity of exponential system and Muckenhoupt condition. *Doklady Akad. Nauk. SSSR.* 247(1):37–40 (Russian); English transl. in *Soviet Math. Dokl.* 20:655–9.

Plotnikov, V. I. 1968. Energetic inequality and the property of eigenfunction system to be overdetermined. *Izvestiya Akad. Nauk SSSR, Ser. Mat.* 32(4):743–55 (Russian).

Polya, G., and G. Szegö. 1964. *Aufgaben und Lehrsätze aus der Analysis.* Berlin: Springer.

Potapov, V. P. 1955. Multiplicative structure of *J*-contractive matrix functions. *Trudy Moskovskogo Matematicheskogo Obshchestva.* 4:125–236 (Russian); English transl. in *Amer. Math. Soc. transl.* 15 (1960).

Privalov, I. I. 1950. *Boundary Properties of Analytic Functions.* Moscow: Gostekhizdat (Russian).

Redheffer, R. M. 1968. Elementary remarks on completeness. *Duke Math. J.* 35(1):1–62.

Reid, R. M., and D. L. Russell. 1985. Boundary control and stability of linear water waves. *SIAM J. Contr. Optim.* 23(1):111–21.

Riekstin'sh, E. Ya. 1991. *Asymptotics and Estimates of Roots of Equations.* Riga: Zinatne (Russian).

Rolewicz, S. (1970). On controllability of system of strings. *Studia Math.* 36(2):105–110.

Rosenblum, G. V. 1979. Fundamental solution and spectral asymptotic of systems with turning points. *Problemy Matematicheskoi Fiziki* (Problems of Math. Physics). 9:122–28. Leningrad: Leningrad Univ. (Russian).

Russell, D. L. 1967. Nonharmonic Fourier series in the control theory of distributed parameter systems. *J. Math. Anal. Appl.* 18:(3):542–59.

Russell, D. L. 1971a. Boundary value control theory of the higher dimensional wave equation. *SIAM J. Control.* 9:29–42.

Russell, D. L. 1971b. Boundary value control theory of the higher dimensional wave equation. Part II. *SIAM J. Control.* 9:401–19.

Russell, D. L. 1972. Control theory of hyperbolic equations related to certain inharmonic analysis and spectral theory. *J. Math. Anal. Appl.* 40:336–68.

Russell, D. L. 1973. A unified boundary control theory for hyperbolic and parabolic partial differential equations. *Studies in Appl. Math.* 52(3):189–211.

Russell, D. L. 1978. Controllability and stabilizability theory for linear partial differential equations: recent progress and open questions. *SIAM Review.* 20(4):639–739.

Russell, D. L. 1982. On exponential bases for the Sobolev spaces over an interval. *J. Math. Anal. Appl.* 87:528–50.

Sadovnichiĭ, V. A. 1979. *Operator Theory.* Moscow: Moskovsk. Univ.

Sakawa, Y. 1974. Controllability for partial differential equations of parabolic type. *SIAM J. Contr. Optim.* 12:389–400.

Schmidt, E. J. P. G. 1992. On the modeling and exact controllability of networks of vibrating strings. *SIAM J. Contr. Optim.* 30:229–45.

Schmidt, E. J. P. G., and N. Weck. 1978. On the boundary behavior of solutions to elliptic and parabolic equations. *SIAM J. Contr. Optim.* 16(4):593–8.

Schwartz, L. 1959. *Etude des Sommes d'Exponentialles.* 2d ed. Paris: Hermann.

Sedletskiĭ, A. M. 1982. Biorthogonal decomposition of functions in exponential series on real intervals. *Uspekhi Matematicheskikh Nauk.* 37(5):51–95 (Russian).

Seidman, T. I. 1976. Observation and prediction for heat equation. III. *J. Differential Equations.* 20:18–27.

Seidman, T. I. 1977. Observation and prediction for the heat equation: patch observability and controllability. IV. *SIAM J. Contr. Optim.* 15:412–27.

Seip, K. 1995. On the connection between exponential bases and certain related sequences in $L^2(-\pi, \pi)$. *J. Functional Anal.* 130:131–60.

Sholokhovich, F. A. 1987. Epsilon-controllability of nonstationary linear dynamical systems in Banach space. *Differential' Uravneniya.* 23:475–80 (Russian).

Sz.-Nagy, B., and C. Foias. 1970. *Harmonic Analysis of Operators in Hilbert Space.* Rev. ed. Amsterdam: North-Holland.

Tataru, D. 1993. Unique continuation for solutions to pde's between Hörmander's and Holmgreen's theorems. Preprint. Department of Mathematics, Northwestern University.

Treĭl, S. R. 1986. Spatially compact eigenvector system forms a Riesz basis, if it is uniformly minimal. *Doklady Akad. Nauk. SSSR.* 288:308–12 (Russian).

Triggiani, R. 1978. On the relationship between first and second order controllable systems in Banach space. *Lecture Notes in Control and Inform. Sci.* 1:380–90.

Triggiani, R. 1991. Lack of exact controllability for the wave and plate equations with finitely many boundary controls. *Differential and Integral Equations.* 4(4):683–705.

Tsujioka, K. 1970. Remarks on controllability of second order evolution equations in Hilbert space. *SIAM J. Contr. Optim.* 8(1):90–9.

Vasil'ev, F. P., A. Z. Ishmukhametov, and M. M. Potapov. 1989. *The Generalized Moment Method in Optimal Control Problems.* Moscow: Moscow Univ. (Russian).

Vasyunin, V. I. 1976. On the number of Carleson series. *Zapiski Nauchnykh Seminarov Leningradskogo Otdeleniya Mat. Inst. Steklova (LOMI).* 65:178–82 (Russian).

Vasyunin, V. I. 1977. Unconditionally convergent spectral decomposition and interpolation problem. *Trudy Matematicheskogo Inst. Steklova.* 130:5–49 (Russian); English transl. in *Proc. Steklov Inst. Math.* 4(1979).

Wang, P. K. C. 1975. Optimal control of parabolic systems with boundary conditions involving time delay. *SIAM J. Contr. Optim.* 13(2):274–93.

Wasow, W. 1965. *Asymptotic Expansion for Ordinary Differential Equations*. Ser. Pure and Applied Mathematics, 15. New York: John Wiley & Sons.
Wasow, W. 1985. *Linear Turning Point Theory*. Ser. Appl. Math. Sciences, 54. Berlin: Springer.
Weiss, H. 1973. Use of semi-groups for derivation of controllability and observability in time-invariant linear systems. *Int. J. Control*. 18(3):475–9.
Young, R. M. 1980. *An Introduction to Nonharmonic Fourier Series*. New York: Academic Press.
Zuazua, Z. 1987. Contrôlabilité exact d'un modele de plaques vibrantes en un temps arbitrairement petit. *C. R. Acad. Sci. Paris*, Sér. I, 307:173–6.

Index

Pages with definitions are typed in **bold**.

angle between subspaces, **18**–19, 22–3, 25, 31, 62, 78

Bari theorem, **27**, 30
$\mathscr{L}$-basis, *see* Riesz basis in the closure of its linear span
basis constants, **120**
Blaschke condition (B), **43**, 55, 64–7
Blaschke–Potapov product (BPP), **44**, 47
Blaschke product (BP), **43**
BMO class, **104**

Carleson condition (C), **52**–4, 56, 67
Carleson constant, **52**–4, 56, 223
Carleson–Newman condition (CN), **52**
Carleson series, **73**
Cartwright class, **60**–1, 90
controllability
  B controllability, **162**–6, **169**–70
  E controllability, **162**–6, **169**–70
  M controllability, **162**–6, **169**–70, 172
  UM controllability, **162**–6, **169**–70
  W controllability, **162**–6, **169**–70

generating function (GF)
  operator-valued, **80**–1, 83–8, 90, 93–4, 238
  scalar, **101**–4
  scalar with multiply zeros, **113**
Gram matrix, **28**–9, 63, 262–3

Hardy space
  in the unit disc, **39**–41
  in the upper halfplane, **40**–1
Helson–Szego condition, **95**, 104

Hilbert operator, **42**, 93–8
  boundedness of, *see* Helson–Szego condition; Muchenhoupt condition
Hilbert transform, **94**–5
Hilbert Uniqueness Method, 166–7, 170–1

inner function, **42**–5, 47

W-linear independence (W), **24**, 25–6, 29–30, 32, 138–41, 170
  and controllability, 164, 170
ω-linear independence, **24**, 26, 28, 141

minimality (M) (*see also* uniform minimality; *-uniform minimality)
  and controllability, 164–5, 170
  elements of, **25**–6, 29, 31–3, 35–7
  exponentials of 55, 59, 63–5, 67, 78, 81, 83–4, 99–101
    under perturbations, 105, 124
  simple fractions of, *see* exponentials of
subspaces of, **30**–3, 59, 63–5, 67
Muckenhoupt condition (A_2), **95**–7, 104

Nagy–Foias model operator, **49**, 50–2

orthogonalizer, **26**–7, 30–1, 120
outer function
  operator-valued, **46**–7
  scalar, **45**–6

Paley–Wiener theorem, 42, 48
problem of moments, **34**, 34–8
projector
  general form, **20**–1
  onto K_S subspaces, **48**
  skew, **20**–3

reachability set, **152**–4, **157**–60, 168, 171
Riesz projector, **42**, 94
  sine-type function (STF), **61**, 102
Riesz basis (*see also* Riesz basis in the
    closure of its linear span
Riesz basis (*cont.*)
  elements of, **26**, 31, 35, 63
  exponentials of, 59, 78–80, 93–4, 101–10,
    113–14, 121, 124
  simple fractions of, *see* exponentials of
  subspaces of, **30**–1, 63
Riesz basis in the closure of its linear span
    (*LB*)
  and controllability, 164, 170
  elements of, **26**–9, 31, 63

exponentials of on an interval, 59,
  78–80, 121, 124
exponentials of on the semiaxis, 56, 59,
  67, 74, 121
simple fractions of, *see* exponentials of
subspaces of, **30**–1, 63

uniform minimality (*see also*
    *-uniform minimality), **26**
*-uniform minimality (*UM*), **26**, 32–3,
  35–7
  and controllability, 164, 170
  exponentials of, 56, 59, 63
  simple fractions of, *see* exponential of

For EU product safety concerns, contact us at Calle de José Abascal, 56–1°,
28003 Madrid, Spain or eugpsr@cambridge.org.

www.ingramcontent.com/pod-product-compliance
Ingram Content Group UK Ltd.
Pitfield, Milton Keynes, MK11 3LW, UK
UKHW010853090126
466816UK00011B/216